T0203570

Blockchain and Artificial Intelligence Technologies for Smart Energy Systems

Present energy systems are undergoing a radical transformation, driven by the urgent need to address the climate change crisis. At the same time, we are witnessing the sharp growth of energy data and a revolution of advanced technologies, with artificial intelligence (AI) and Blockchain emerging as two of the most transformative technologies of our time. The convergence of these two technologies has the potential to create a paradigm shift in the energy sector, enabling the development of smart energy systems that are more resilient, efficient, and sustainable.

This book situates itself at the forefront of this paradigm shift, providing a timely and comprehensive guide to AI and Blockchain technologies in the energy system. Moving from an introduction to the basic concepts of smart energy systems, this book proceeds to examine the key challenges facing the energy system, and how AI and Blockchain can be used to address these challenges. Research examples are presented to showcase the role and impact of these new technologies, while the latest developed testbeds are summarised and explained to help researchers accelerate their development of these technologies.

This book is an indispensable guide to the current changes in the energy system, being of particular use to industry professionals, from researchers to management, looking to stay ahead of technological developments.

Hongjian Sun, Professor at Durham University, UK. He is the Head of Durham Smart Grid Laboratory and leads Smart Grid research at Durham, with over 180 papers in refereed journals and international conferences. His research has been funded by EU H2020, EU ERDF, EPSRC, BEIS, Ofgem, Innovate UK, and industry. He is also the Editor-in-Chief for the *IET Smart Grid* journal.

Weiqi Hua, Assistant Professor at the University of Birmingham, UK. He took postdoctoral positions at the University of Oxford, UK, and Cardiff University, UK, after receiving his Ph.D. from the University of Durham, UK. He is an Editorial Board Member of *Applied Energy*, and Editorial Board Member of *Oxford Open Energy*.

Minglei You, Assistant Professor at the University of Nottingham, UK. He is also a member of the Power Electronics, Machines and Control (PEMC) group. He received his Ph.D. from Durham University in 2019, as a recipient of the Durham Doctoral Scholarship. From 2019 to 2021, he was a Postdoctoral Research Associate with Durham University and Loughborough University.

Blockchain and Artificial Intelligence Technologies for Smart Energy Systems

Hongjian Sun, Weiqi Hua and Minglei You

CRC Press
Taylor & Francis Group
Boca Raton London New York

CRC Press is an imprint of the
Taylor & Francis Group, an **informa** business

A CHAPMAN & HALL BOOK

First edition published 2024
by CRC Press
6000 Broken Sound Parkway NW, Suite 300, Boca Raton, FL 33487-2742

and by CRC Press
4 Park Square, Milton Park, Abingdon, Oxon, OX14 4RN

CRC Press is an imprint of Taylor & Francis Group, LLC

© 2024 Hongjian Sun, Weiqi Hua and Minglei You

ISBN: 978-0-367-77127-0 (hbk)
ISBN: 978-0-367-77250-5 (pbk)
ISBN: 978-1-003-17044-0 (ebk)

DOI: 10.1201/9781003170440

Typeset in Latin Modern font
by KnowledgeWorks Global Ltd.

Publisher's note: This book has been prepared from camera-ready copy provided by the authors.

This book is dedicated to my beloved father, Mr. Liqun Hua, whose unwavering love and support continue to inspire me; whose spirit inhabits my cells.
Dr Weiqi Hua
Oxford, United Kingdom
March 2023

Contents

Section III Testbeds for Smart Energy Systems

List of Figures

List of Tables

Foreword

Present energy systems are undergoing a radical transformation, driven by the urgent need to address the climate change crisis and rapid development of information and communication technologies. With the connection of more distributed energy resources, the electrification of transportation and heating, and the increasingly flexible demand enabled by smart meters and energy storage, the energy system is becoming more complex and interconnected than ever before.

At the same time, we are witnessing sharp growth of energy data and revolutions of advanced technologies, with artificial intelligence (AI) and Blockchain emerging as two of the most transformative technologies of our time. AI is empowering machines to learn rapidly from huge volumes of data and reason in ways that were once the exclusive domain of human intelligence; while Blockchain is revolutionising the way we store and share data, making it more secure, transparent, and efficient for processing data.

The convergence of these two technologies has the potential to create a paradigm shift in the energy sector, enabling the development of smart energy systems that are more resilient, efficient, and sustainable. By leveraging the power of AI and Blockchain technologies, we can create much more intelligent energy systems that can anticipate and respond to fast-changing energy demands (e.g., due to electric vehicle charging or extreme weather), optimise energy production and consumption (e.g., local production and consumption in the form of prosumers and their energy trading), and enable the integration of more renewable energy sources.

This book, titled *Blockchain and Artificial Intelligence Technologies for Smart Energy Systems*, is a timely and comprehensive guide to the applications of AI and Blockchain technologies to the energy system. It provides a detailed overview of the latest developments in these two fields and their potential impact on the energy system. The book begins by introducing the basic concepts of smart energy systems: AI and Blockchain. It then goes on to examine the key challenges faced by the energy system, and how AI and Blockchain can be used to address these challenges. Some research examples are presented to showcase the role and impact of AI and Blockchain technologies. The latest developed testbeds are summarised and explained to help researchers accelerate their research and development of relevant technologies.

One of the key themes of the book is the emerging role of prosumers and their peer-to-peer energy trading. The authors explore how Blockchain can be used to enable intelligent and decentralised energy management systems that can optimise energy generation and consumption, reduce energy waste and carbon emissions, and improve the overall efficiency of the energy system. Another important theme of the book is the role of AI for enabling smart multi-vector energy systems with high uncertainties.

The authors examine how AI can be used to develop multi-agent tools and energy hubs to manage the variability and intermittency of renewable energy generation, and uncertainties in multiple energy vectors including transport and heat.

Throughout the book, the authors provide practical examples and case studies to illustrate the potential applications of AI and Blockchain in the energy system. They also discuss the regulatory implications of these technologies, and the challenges that must be overcome to realise their full potential.

In conclusion, *Blockchain and Artificial Intelligence Technologies for Smart Energy Systems* is an important and timely contribution to the literature on clean energy, AI, and Blockchain technologies. It provides a comprehensive guide to the potential applications of AI and Blockchain in the energy sector, and explores how these technologies can be used to create more sustainable and efficient energy systems. The book is an essential read for anyone interested in the future of energy and the role of AI and Blockchain technologies in creating a more sustainable world.

Preface

Since the Industrial Revolution, the economic growth of economic powers (e.g., the UK and the USA) has been closely linked to their increasing energy demand. However, this increase in energy demand has also given rise to the climate crisis, which poses a threat to the safety of billions of people on the planet. The climate crisis is not limited to global warming; it includes various alarming occurrences such as sea-level rise resulting from melting glaciers, as well as more frequent extreme weather events like heatwaves, floods, and wildfires. Examples of such extreme weather events include: the 2020 wildfires in the western United States and Australia; the 2021 Henan floods in China; the 2021 European floods that affected several countries, including Austria, Germany, and Italy; and the 2022 storms in the UK.

Numerous countries have pledged to reduce greenhouse gas emissions to combat climate change. For instance, the UK aims to reach net-zero greenhouse gas emissions by 2050, while the EU has set the European Climate Law to achieve climate neutrality by the same year. These objectives represent one of the most urgent challenges for the coming decades, requiring significant reductions in greenhouse gas emissions across all energy sectors and systems, including power, heating, and transport sectors. The transport and heating sectors are especially challenging to decarbonise due to complex socio-technical problems, such as cultural patterns of mobility and the path-dependence of heat.

In the transport sector, millions of petrol and diesel vehicles are likely to be replaced with electric vehicles (EVs), powered by electricity from power systems. Many people believe that increasing low-carbon and renewable energy generation in power systems can reduce the majority of greenhouse gas emissions in this sector. However, the intermittency of renewable energy generation can compromise power system security, and uncoordinated EV charging could contribute to more power demand spikes in distribution networks and rising energy bills for consumers. Therefore, data-driven smart technologies are necessary to deliver precise and coordinated EV charging and synergise with intermittent renewable energy to scale up the use of both EVs and renewable energy to achieve the net-zero objective.

In the heating sector, heat production accounts for almost half of the total energy demand in many countries, contributing significantly to greenhouse gas emissions. Electrification of heat is one means of decarbonising heat demand, particularly when solar panels or other renewable energy sources are installed. However, daily and seasonal heat demand has the potential to disrupt electricity supplies if these loads are added to the power system. Smart energy systems could maximise the efficiency of such energy systems by coordinating heating demand in tandem with the available renewable energy.

Today's transformative technologies, such as artificial intelligence (AI) and Blockchain, have enormous potential to change businesses and transform entire industries. The energy system is probably the most important system that these smart technologies will change, and they are believed to play a vital role in stopping climate change and saving lives in the coming decades. This book aims to introduce key smart energy technologies, focusing on AI and Blockchain technologies.

Part I will present some fundamentals of smart energy systems, AI, and Blockchain. Part II will focus on the applications of these technologies in smart energy systems, such as peer-to-peer energy trading, Energy Internet, smart local energy market, multi-vector energy systems, and cyber-physical systems. Part III will introduce our latest efforts on building laboratory testbeds of integrating both software and hardware for testing smart energy technologies.

DISCLAIMER

Due to the time limit, there might be errors or typos in this book, but as authors we tried our best to correct them. If you found any errors or typos, please feel free to contact us.

Prof. Hongjian Sun, at Durham, UK
26 February 2023

Author Bios

Prof. Hongjian Sun received his Ph.D. in Electronic and Electrical Engineering at the University of Edinburgh (UK) and then took postdoctoral positions at King's College London (UK) and Princeton University (USA). Since April 2013, he has been with the Department of Engineering at the University of Durham. He is a Professor (Chair), a Chartered Engineer, a Fellow of the Durham Energy Institute, and a Fellow of the Higher Education Academy. Prof. Sun is the Head of the Durham Smart Grid Laboratory and leads Smart Grid research activities focused on: (i) demand-side management and demand response, and (ii) renewable energy sources integration and virtual power plants.

He has an established track record of publishing high-quality scientific articles, with more than 180 papers in refereed journals and international conferences. He has made contributions to and coauthored the IEEE 1900.6a-2014 Standard, secured two granted patents, and published five book chapters and two books. Prof. Sun has been the Principle Investigator (PI) or Co-Investigator (Co-Is) on a number of projects, with funding successes from the EU H2020, EU ERDF, UK EPSRC, UK BEIS, Ofgem, Innovate UK, and industry. Particularly relevant work includes the recently coordinated projects EU H2020 "TESTBED" (no. 734325, €882k) and "TESTBED2" (no. 872172, €1.4M) (both as the PI) which focus on developing decentralised optimisation methods and AI tools for ensuring the scalability of smart grid services. Moreover, he established extensive experience in modelling virtual power plants through the previously funded EPSRC project "TOPMOST" (EP/P005950/1), as the PI. He has been one of the Co-Is for the 2M GBP "CHEDDAR" (EP/X040518/1) project on cloud and distributed computing, the 1.3M GBP EPSRC "DecarboN8" (EP/S032002/1) project on decarbonising the transport, and the 880k GBP Ofgem project "Collaborative Visual Data Twin" on developing Digital Twins for asset management. In the past, he successfully delivered several other relevant research projects, e.g., the 1M GBP UK BEIS funded "USER" project for developing AI tools for demand response of hot water tanks, the 238k GBP Innovate UK project "Electrical and thermal storage optimisation in a virtual power plant," and the €3.7M H2020 project "SmarterEMC2" for developing ICT tools for smart grids. He is also the Editor-in-Chief for the *IET Smart Grid* journal, leading 90+ editors worldwide.

Dr Weiqi Hua is an Assistant Professor with the Department of Electronic, Electrical and Systems Engineering at the University of Birmingham (UK) since 2023. He took postdoctoral positions at the University of Oxford (UK) and Cardiff University (UK). He received his Ph.D. from the University of Durham (UK) in 2021, co-funded by Hyundai Motor Company and Durham Research Doctoral Scholarship, with the thesis title of 'Smart Grid Enabling Low Carbon Future Power Systems Towards

Prosumers Era'. He was the visiting researcher to the Hellenic Telecommunications Organisation (OTE), Greece in 2019, and visiting researcher to the Chinese Academy of Sciences, China in 2018 and 2019. His research interests include: (i) energy system modelling and optimisation, (ii) renewable energy integration, (iii) digitalisation, digital twin, and machine learning for energy system analytics, (iv) energy policy and economics, and (v) Blockchain for peer-to-peer energy trading. Dr Hua is an Editorial Board Member for *Applied Energy* (Impact Factor:11.446; CiteScore: 20.4). He served as the Lead Guest Editor for the Special Issue on the topic of Blockchain-based local energy market in *IET Smart Grid*, and Guest Editor for the Special Issue on the topic of energy-water nexus in *Water*. He was a technical program committee (TPC) member for IEEE SmartGridComm 2020, and TPC member for IEEE International Smart Cities Conference 2021. He has been appointed as the sessional chair for IEEE SmartGridComm 2018, and sessional chair for IEEE International Smart Cities Conference 2017. He has published a chapter for the IntechOpen book *Microgrids and Local Energy Systems*. Dr Hua has undertaken research projects collaborating with academia and industries, including: i) EPSRC Analytical Middleware for Informed Distribution Networks (AMIDiNe), ii) BEIS Heat Pump Ready Programme, iii) EPSRC Maximising flexibility through multi-scale integration of energy system (MISSION), iv) Integrated heating and cooling networks with heat-sharing-enabled smart prosumers, v) Virtual Power Plant for Interoperable and Smart isLANDS (VPP4ISLANDS), vi) EU ERDF Solid Wall Insulation innovation (SWIi), and vii) Horizon 2020 Testing and Evaluating Sophisticated information and communication Technologies for enaBling scalablE smart griD Deployment (TESTBED).

Dr Minglei You is an Assistant Professor with the Power Electronics, Machine and Control (PEMC) Research Group at the University of Nottingham, UK. He received his Ph.D. on Statistical Quality of Service for Smart Grids from Durham University in 2019, as a recipient of the Durham Doctoral Scholarship. From 2019 to 2021, he was a Postdoctoral Research Associate with Durham University and Loughborough University. He developed the statistical communication latency analysis and provisioning frameworks for power systems, and artificial intelligence (AI)-based solutions for information and communication technologies (ICT) and power/energy systems. The research outputs from these contributed to both national and international projects, including the "Smart Grid: Empowering SG Market Actors through Information and Communication Technologies (SmarterEMC2)" project funded by European (EU) Horizon 2020, the "Electrical and Thermal Storage Optimisation in a Virtual Power Plant", and "Artificial Intelligence-enabled Massive Multiple-input multiple-output (AIMM)" projects funded by Innovate UK, the "Ubiquitous Storage Empowering Response" project funded by UK BEIS, and the "Signal Processing for 5G and beyond networks" project funded by EPSRC. He has designed and implemented two testbeds in total, including "Software Defined Smart Grid Testbed" hosted at Durham University and "AI-enhanced MassiveMIMO Testbed" hosted at Loughborough University. He collaborated with industrial partners on the application of AI in real-world systems, including BT funded by Innovate UK and CELTIC-NEXT (EU), Hyundai funded by Hyundai, and OTE (Greece) funded by EU 2020 Marie Sklodowska-Curie Actions (MSCA). He has been serving as a Technical

Program Committee (TPC) member and helped organise international conferences, including ICC 2020, WCNC 2019, ICC 2018, and VTC 2017, and was recognised as an exemplary reviewer for IEEE Wireless Communications Letters in 2020.

Contributors

Hongjian Sun
Department of Engineering
University of Durham, UK

Weiqi Hua
Department of Engineering Science; School of Engineering
University of Oxford, UK; University of Birmingham, UK

Minglei You
Department of Electrical and Electronic Engineering
University of Nottingham, UK

I

Fundamental Theories

CHAPTER 1

Smart Energy Systems

I n this chapter, we will be introducing the definition and fundamental components of smart energy systems. Section 1.1 will define smart energy systems and outline their advantages. Section 1.2 will provide a brief overview of the role of renewable energy sources in facilitating clean energy supply and the transition to net-zero emissions. In Section 1.3, we will describe the role of energy storage systems in addressing the intermittency of renewable energy sources and enhancing the local energy balance. We will then discuss smart metering, demand-side management, and home energy management systems in Sections 1.4 through 1.6, respectively. Finally, Section 1.7 will summarise the key points of this chapter and provide a vision about the role of artificial intelligence (AI) and Blockchain technologies in smart energy systems.

1.1   INTRODUCTION

Smart energy system could be defined as a multi-vector and integrated energy system (including power, heat/thermal, hydrogen and gas vectors) combined with energy storage technologies and empowered by information and communication technologies (ICT) to realise flexible and efficient operations of the whole system [1]. An example of smart energy systems is shown in Figure 1.1.

We are currently experiencing a transformation in the way energy systems work. In a conventional energy system, energy flows in one direction, generated by large-scale centralised generation stations, transmitted by the transmission and distribution networks, and delivered to passive energy consumers. These consumers are only able to receive energy from the energy networks, and they pay their energy bills according to the retail price as price takers.

However, with the transition towards a smart energy system, numerous distributed energy generation sources, energy storage devices, and electric vehicles are being added to the energy network, particularly on the demand-side. This means that now energy flows in both directions, either from the supply side to the demand-side or within the demand-side [2]. Moreover, passive consumers are now becoming active prosumers who generate clean energy locally to meet their energy demand or share energy with others. This transformation has created an opportunity for energy consumers to actively participate in the energy market and contribute to a more

DOI: 10.1201/9781003170440-1

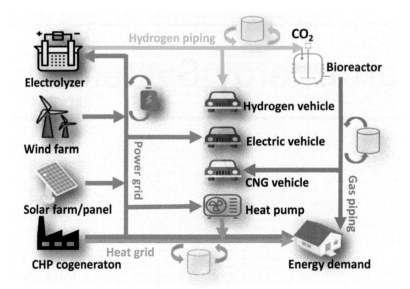

Figure 1.1 Smart energy system as a multi-vector integrated energy system. Multiple energy vectors include power, heat, hydrogen, and gas that aim to realise a low-carbon, flexible, and resilient energy system.

sustainable energy future [3]. Figure 1.2 presents a schematic illustration of the power vector in a smart energy system.

The main advantages of smart energy systems can be summarised as follows:

- Smart energy systems use ICT [4] to enable the real-time bidirectional information flows, which ensures timely monitoring of dynamic system states and

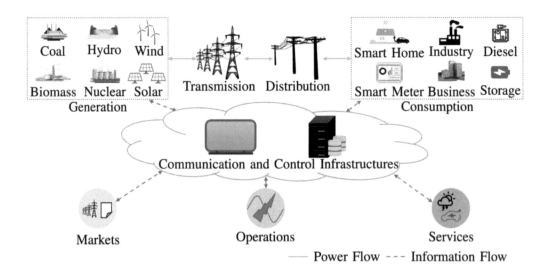

Figure 1.2 Schematic illustration of the power vector in a smart energy system. Both power and information flow bidirectionally, with active demand-side engagement.

supply–demand balance of energy. This guarantees system stability and security of supply.

- The accurate prediction and optimal dispatch of energy generation provide a solution for the intermittent issues of renewable energy sources. Hence, smart energy systems enhance the penetration of renewable energy so as to facilitate the net-zero energy transition [5].

- Smart energy systems enable the local energy supply–demand balance and reduce the transmission and distribution losses through integrating innovative technologies, such as the distributed generation, demand-side management, and peer-to-peer energy trading [6].

1.2 RENEWABLE ENERGY SOURCES

Renewable energy sources are derived from natural resources that are replenished within a reasonable time period [7]. These include wind, solar, hydro, geothermal, and biomass energy [8]. For power systems, as alternative energy sources to fossil fuels, the increasing penetration of renewable energy sources contributes to the enhanced energy security and improved power quality whilst reducing carbon emissions and operating costs [9]. For distribution networks, the deployment of distributed renewable energy sources can reduce carbon emissions and transmission losses, and improve the local energy supply–demand balance [10].

Nonetheless, the intermittency of renewable energy sources caused by weather conditions presents a challenge for their integration into power systems [11]. Smart energy systems are the key to overcoming this challenge, since the advanced communication and control infrastructures along with energy storage systems could enable the renewable energy sources to be optimally dispatched.

1.3 ENERGY STORAGE

The pressure of increasing intermittent renewable energy sources and growing energy demand drives the development of energy storage technologies [12]. These technologies include chemical energy storage, thermal energy storage, hydro pump energy storage, compressed air energy storage, flywheel energy storage, and superconducting magnetic energy storage [13]. They can flexibly "absorb" or "release" energy as and when required with the benefits of enhancing the system stability and the security of supply [14]. Typically energy storage technologies can reduce the need for investing in additional generation capacities, and contribute to financial savings and carbon emissions mitigation from conventional power generation [15]. The deployment of energy storage devices can also reduce the costs of updating transmission and distribution networks.

1.4 SMART METERING

Smart metering is an enabling technology for the demand-side management, optimal dispatch of distributed renewable energy sources, and peer-to-peer energy trading. With the smart meters installed, consumers can self-read the meter and control their energy usage, so as to adopt energy-efficient measures and save energy bills [30]. Smart metering (including smart meters and their communication and data storage infrastructures) facilitates distributed renewable energy sources to be integrated into power systems by providing an accurate forecast and optimal energy scheduling. In addition, smart metering is the key to the carbon emissions mitigation by changing consumption behaviours. From the short-term perspective, consumers can be incentivised by dynamic pricing to save their electricity bills, e.g., by using energy from energy storage devices when the electricity price is relatively high. From the long-term perspective, consumers can cost-effectively invest renewable energy sources and/or manage energy storage devices.

1.5 DEMAND-SIDE MANAGEMENT

As one of the key smart technologies, the demand-side management is defined as technologies of reshaping energy consumption behaviours of consumers through monetary incentives or educational programmes, in order to make power grid operations more reliable and cost-effective [16]. Traditionally power grids are designed for meeting the peak demand, thus leading to generation units often under-utilised with high operational costs. To make future power systems more affordable, the demand-side management is needed to mitigate and modify peak demand without the need of enhancing power networks, typically reducing the peak-to-average ratio of the demand. Recently the demand-side management is also used for dynamically balancing supply and demand to accelerate the adoption of renewable energy sources [17]. The strategies for delivering the demand-side management can be categorised as follows:

- *Energy efficiency*: Consumers can deliver the same tasks with lower energy demand by improving the energy efficiency of their loads [18].

- *Demand response*: The demand response refers to strategically reshaping consumption behaviours in response to the incentive of pricing signals [19]. The strategies of the demand response include curtailing energy demand during peak time periods and shifting energy demand from peak time periods to off-peak time periods, e.g., midnight or weekends [20].

- *Dynamic demand*: The operating cycles of loads can be optimally advanced or postponed by a few seconds or minutes without disturbing consumers, e.g., heat pumps, so as to increase the diversity factor and reduce the critical power mismatch for power systems [21].

Generally demand-side management can be divided into two different approaches: incentive-based demand-side management and price-based demand-side management.

The former one includes the direct load control, interruptible service, and demand-side bidding. They are designed for engaging with different types of consumers. The direct load control is typically offered to residential or small commercial customers who have some flexible loads that can be turned on/off given a short notice [22]. Examples of these flexible loads are air conditioner, water heater, and space heating which can tolerate inaccurate operation time to some degree. The interruptible services target large industrial or commercial energy consumers whose energy curtailment has huge impacts on the peak-to-average ratio and system contingency mitigation [23]. As a reward, they can be offered with bill/rate discounts for the retail tariff or free credits. However, penalties could be imposed if they fail to respond to the curtailment request in the predefine time period. Demand-side bidding (buy-back) is a mechanism to encourage the consumers' participation in energy markets by using their potentially curtailed energy consumption [24]. It can be offered to either large energy consumers or aggregators of many small energy consumers. These large energy consumers or aggregators can offer bids for the curtailment based on wholesale electricity market prices, generating additional incomes, thus reducing their total energy bills [25].

The price-based demand-side management focuses on modifying electricity prices to reflect the real system cost and operational status. It is believed that electricity prices change energy consumers' consumption patterns, i.e., there is more consumption if consumers are offered lower electricity prices and vice versa. The time-of-use pricing means that the electricity prices are different depending on when electricity is consumed [26], e.g., UK Economy-7 tariff offers cheap price for off-peak hours [27]. It is usually predefined on a day-ahead basis or even a month-ahead basis. The real-time pricing typically has different electricity prices for every hour, by linking prices with the dynamic balancing status of supply and demand in the wholesale market [28]. Consumers can be notified in a day-ahead manner, hour-ahead manner, or even 15-minutes-ahead manner. The critical time pricing is based on the time-of-use pricing, but added with a much higher price called as the critical peak price during a system contingency time period [29].

1.6   HOME ENERGY MANAGEMENT SYSTEM

Smart technologies can enhance the efficiency of home appliances by forming a smart home energy management system. It allows the consumers to monitor their energy generation and energy consumption, with devices such as roof-top solar panels and smart appliances, and to use energy in a cost-effective manner [31]. The home energy management systems consist of both the hardware and software elements. The hardware elements include sensors, actuators, controllers, and the communication network linking the smart appliances and consumers. The software elements can analyse historical consumption behaviours and generate optimal control functions based on users' preferences. For instance, receiving real-time pricing signals, the home energy management systems can help consumers strategically shift or curtail their loads as a kind of demand responses. Basic functions of home energy management systems include:

- Monitoring the home energy consumption at different granularities and time intervals, such as total home energy consumption vs. individual appliance usage within 1 minute, 1 hour, or 1 day.

- Monitoring environmental parameters, e.g., temperature and humidity, which can be used for setting home energy management strategies.

- Management of home appliances to meet consumers' needs through the automatic or manual control of appliances.

- Supporting the integration of renewable energy sources and energy storage devices.

- Interacting with external stakeholders of smart energy systems to realise advanced features, such as demand-side management or peer-to-peer energy trading.

1.7 CHAPTER SUMMARY

The smart energy system has become increasingly important in recent years due to the need to address climate change crisis and to achieve the Net Zero goal. This system can use a range of innovative technologies to optimise energy consumption and promote the use of renewable energy sources. One of the key benefits of the smart energy system is its ability to facilitate bidirectional communication between energy systems, distributed generators, and consumers. This communication allows for the optimisation of distributed generator operations and the active engagement of consumers. For example, smart meters can provide consumers with real-time information about their energy consumption, enabling them to adjust their behaviour and reduce their energy usage/bills and carbon emission.

From a controlling perspective, the smart energy system enables energy market participants to work together to achieve the overall benefits of whole systems. By cooperating, market participants can save energy bills for consumers, increase operating profits for generators, reduce carbon emissions, and enhance the security of supply. For instance, demand-side management programs can incentivise consumers to reduce their energy consumption during peak time, reducing the need for additional generation capacity and lowering costs. This cooperation between market participants can also help to optimise the use of renewable energy sources and energy storage devices, ensuring that energy is supplied when and where it is needed most.

Looking forward, the integration of AI and Blockchain technologies into the energy system is expected to further enhance its intelligence, efficiency, and effectiveness. AI can support the predictability, responsiveness, interoperability, and automation of the smart energy system, enabling it to respond more quickly and effectively to quick changes in energy demand and supply. Meanwhile, Blockchain can support secure data sharing and peer-to-peer energy trading, leading to reduced energy bills for consumers and increased integration of renewable energy resources. Ultimately, the integration of AI and Blockchain technologies into the energy system is expected to play a critical role in achieving the Net Zero goal and promoting sustainable development.

Bibliography

[1] S. H. R. Hosseini, A. Allahham, S. L. Walker, and P. Taylor, "Optimal planning and operation of multi-vector energy networks: A systematic review," *Renewable and Sustainable Energy Reviews*, vol. 133, p. 110216, 2020.

[2] V. C. Gungor, D. Sahin, T. Kocak, S. Ergut, C. Buccella, C. Cecati, and G. P. Hancke, "Smart grid technologies: Communication technologies and standards," *IEEE Transactions on Industrial Informatics*, vol. 7, no. 4, pp. 529–539, 2011.

[3] W. Hua, "Smart grid enabling low carbon future power systems towards prosumers era," Ph.D. dissertation, Durham University, 2020.

[4] J. Morley, K. Widdicks, and M. Hazas, "Digitalisation, energy and data demand: The impact of internet traffic on overall and peak electricity consumption," *Energy Research & Social Science*, vol. 38, pp. 128–137, 2018.

[5] A. Millot, A. Krook-Riekkola, and N. Majzi, "Guiding the future energy transition to net-zero emissions: Lessons from exploring the differences between france and sweden," *Energy Policy*, vol. 139, p. 111358, 2020.

[6] W. Hua, J. Jiang, H. Sun, and J. Wu, "A blockchain based peer-to-peer trading framework integrating energy and carbon markets," *Applied Energy*, vol. 279, p. 115539, 2020.

[7] G. Boyle, Peake, Stephen. Renewable energy-power for a sustainable future. No. Ed. 4. Oxford university press, 2018.

[8] J. A. Turner, "A realizable renewable energy future," *Science*, vol. 285, no. 5428, pp. 687–689, 1999.

[9] E. Heylen, G. Deconinck, and D. Van Hertem, "Review and classification of reliability indicators for power systems with a high share of renewable energy sources," *Renewable and Sustainable Energy Reviews*, vol. 97, pp. 554–568, 2018.

[10] B. Muruganantham, R. Gnanadass, and N. Padhy, "Challenges with renewable energy sources and storage in practical distribution systems," *Renewable and Sustainable Energy Reviews*, vol. 73, pp. 125–134, 2017.

[11] W. Hua, J. Jiang, H. Sun, A. M. Tonello, M. Qadrdan, and J. Wu, "Data-driven prosumer-centric energy scheduling using convolutional neural networks," *Applied Energy*, vol. 308, p. 118361, 2022.

[12] J. A. McDowall, "Status and outlook of the energy storage market," in *2007 IEEE Power Engineering Society General Meeting*. IEEE, 2007, pp. 1–3.

[13] M. Hannan, M. Faisal, P. Jern Ker, R. Begum, Z. Dong, and C. Zhang, "Review of optimal methods and algorithms for sizing energy storage systems to achieve decarbonization in microgrid applications," *Renewable and Sustainable Energy Reviews*, vol. 131, p. 110022, 2020.

[14] A. Nasiri, "Integrating energy storage with renewable energy systems," in *2008 34th Annual Conference of IEEE Industrial Electronics*. IEEE, 2008, pp. 17–18.

[15] W. Seward, W. Hua, and M. Qadrdan, "Electricity storage in local energy systems," *Microgrids and Local Energy Systems*, vol. 1, p. 127, 2021.

[16] C. W. Gellings, "The concept of demand-side management for electric utilities," *Proceedings of the IEEE*, vol. 73, no. 10, pp. 1468–1470, 1985.

[17] D. Groppi, A. Pfeifer, D. A. Garcia, G. Krajačić, and N. Duić, "A review on energy storage and demand side management solutions in smart energy islands," *Renewable and Sustainable Energy Reviews*, vol. 135, p. 110183, 2021.

[18] M. Auffhammer, C. Blumstein, and M. Fowlie, "Demand-side management and energy efficiency revisited," *The Energy Journal*, vol. 29, no. 3, 2008.

[19] M. H. Albadi and E. F. El-Saadany, "Demand response in electricity markets: An overview," in *2007 IEEE Power Engineering Society General Meeting*. IEEE, 2007, pp. 1–5.

[20] C. W. Gellings and J. H. Chamberlin, "Demand-side management: Concepts and methods," 1987.

[21] J. A. Short, D. G. Infield, and L. L. Freris, "Stabilization of grid frequency through dynamic demand control," *IEEE Transactions on Power Systems*, vol. 22, no. 3, pp. 1284–1293, 2007.

[22] N. Ruiz, I. Cobelo, and J. Oyarzabal, "A direct load control model for virtual power plant management," *IEEE Transactions on Power Systems*, vol. 24, no. 2, pp. 959–966, 2009.

[23] K. Bhattacharya et al., "Competitive framework for procurement of interruptible load services," *IEEE Transactions on Power Systems*, vol. 18, no. 2, pp. 889–897, 2003.

[24] G. Strbac and D. Kirschen, "Assessing the competitiveness of demand-side bidding," *IEEE Transactions on Power Systems*, vol. 14, no. 1, pp. 120–125, 1999.

[25] W. Hua, J. Jiang, H. Sun, F. Teng, and G. Strbac, "Consumer-centric decarbonization framework using stackelberg game and blockchain," *Applied Energy*, vol. 309, p. 118384, 2022.

[26] E. Celebi and J. D. Fuller, "Time-of-use pricing in electricity markets under different market structures," *IEEE Transactions on Power Systems*, vol. 27, no. 3, pp. 1170–1181, 2012.

[27] S. Kufeoglu, D. Melchiorre, and K. Kotilainen, "Understanding tariff designs and consumer behaviour to employ electric vehicles for secondary purposes in the United Kingdom," *The Electricity Journal*, vol. 32, no. 6, pp. 1–6, 2019.

[28] H. Allcott, "Rethinking real-time electricity pricing," *Resource and Energy Economics*, vol. 33, no. 4, pp. 820–842, 2011.

[29] K. Herter, "Residential implementation of critical-peak pricing of electricity," *Energy Policy*, vol. 35, no. 4, pp. 2121–2130, 2007.

[30] Y. Wang, Q. Chen, T. Hong, and C. Kang, "Review of smart meter data analytics: Applications, methodologies, and challenges," *IEEE Transactions on Smart Grid*, vol. 10, no. 3, pp. 3125–3148, 2019.

[31] D.-M. Han and J.-H. Lim, "Smart home energy management system using ieee 802.15. 4 and zigbee," *IEEE Transactions on Consumer Electronics*, vol. 56, no. 3, pp. 1403–1410, 2010.

CHAPTER 2

Theories of Artificial Intelligence

I n this chapter, we will be introducing the theories of the most common AI approaches related to the applications in energy systems. The focus of this chapter is on the energy-related artificial intelligence (AI) approaches, other AI approaches will not be covered in this chapter. In the meantime, we will try to simplify the illustrations of these theories. For detailed knowledge of the AI, readers are recommended to read refs.[1, 2, 3]. Section 2.1 introduces the backgrounds and development of the AI. Section 2.2 discusses the optimisation problems in energy systems and solution approaches. Section 2.3 describe how to use the game theory to model the decision-making and interactions among stakeholders in energy systems. The support vector machine (SVM) is detailed in Section 2.4 in solving the classification problems for a given dataset. Section 2.5 discusses how to reduce the dimensionality of a high-dimensional dataset by using the principle component analysis (PCA). In Section 2.6, expectation maximisation (EM) algorithm is introduced to solve the maximum likelihood estimation (MLE) problems which are used to estimate the density for a dataset. The Gaussian mixture model (GMM) is introduced in Section 2.7 for estimating the probability density function. Section 2.8 discusses the variational inference which is used to approximate intractable integrals. Section 2.9 describes the implementations of the hidden Markov model (HMM) on the problems of the evaluation, learning, and decoding. Section 2.10 discusses typical architectures of neural networks, and Section 2.11 introduces the reinforcement learning. Section 2.12 concludes this chapter.

2.1 INTRODUCTION

AI could be understood as machines that possess human-like intelligence to act, predict, and make decisions by learning from data [4]. Such intelligence relies on the availability of data and the designed algorithm (such as a programme or a rule). The algorithm processes data to solve problems, typically by identifying certain patterns to decide what to do or predict future outcomes. This procedure seems to give machines an ability to learn from data, and such an ability could significantly improve over time as the data volume increases. In other words, the more data machines get

DOI: 10.1201/9781003170440-2

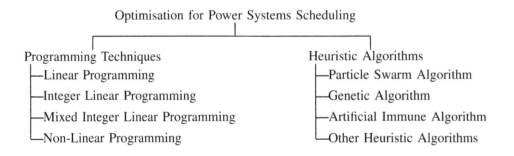

Figure 2.1 Approaches for solving optimisation problems, categorised into programming techniques and heuristic algorithms.

to learn from, the smarter machines become. Like human, AI could have a range of human-like abilities, such as seeing like face recognition in payment systems or smart phones, speaking like Alexa, playing games like AlphaGo, and chatting like Chat-GPT. This book will focus on discussing how the AI could help power systems, or broadly energy systems, to intelligently control relevant assets and make intelligent decisions.

2.2 OPTIMISATION

The optimisation is an essential approach for energy systems in terms of planning, scheduling, and dynamic responses. Under optimisation problems, the objectives of stakeholders in energy systems, e.g., generators, consumers, prosumers, policy makers, power system operators, and market operators, are modelled by predefined formulations and parameters. Solving optimisation problems in energy systems is often constrained by physical limits, e.g., capacities, operating conditions, power rates, ramp rates, voltage/current limits, and security restrictions. By solving the optimisation problems, optimal decisions are yielded, such as optimal size of generation infrastructures [5], optimal control strategies on controllable power units [6], and optimal pricing which reduces peak energy demand [7].

The optimisation approaches can be categorised into programming techniques and heuristic algorithms as presented in Figure 2.1.

The programming techniques consist of the linear programming, non-linear programming, integer linear programming, and mixed integer linear programming. Linear programming is defined as an optimisation problem in which both the objective functions and constraints are the linear functions of decision variables [8]. By contrast, non-linear programming is defined as an optimisation problem in which at least one objective function or constraint is non-linear function of decision variables [9]. Whereas for the integer linear programming problems, only integers, including binary values, can be used as decision variables [10]. For the mixed integer linear programming problems, both integers and non-integers are used as decision variables [11].

The heuristic algorithms consist of particle swarm algorithm, genetic algorithm, artificial immune algorithm, and other heuristic algorithms. These algorithms are primarily used to solve the non-linear programming problems through iteratively searching from the entire solution space which is defined by the constraints of the optimisation problems. The particle swarm algorithm [12] searches from the solution space consisting of particles and moves particles within the space according to the predefined functions which describe the position and velocity of particles, in order to find the optimal solutions. The movement of particles is determined by both the local best known position and global best known position in the searching space. All particles ultimately move towards the best solutions. Learning from the Darwin's Theory of Evolution, the genetic algorithm randomly generates a population of candidate solutions within the solution space. Each generated population is defined as a generation [13]. The value of objective functions for every individual in the population is evaluated and defined as the fitness. After evaluating the fitness for all individuals, most fit individuals are selected and mutated to form a new generation. The population is iteratively evolved towards the best solution in solving the optimisation problems. In the artificial immune algorithm [14], a population of candidate solutions to an optimisation problem is randomly generated, which is similar to the generation of population of the genetic algorithm. The generated population of the artificial immune algorithm is defined as the antigens. The value of objective functions for every antigen in the population is evaluated and defined as the antibody. The antigens are iteratively cloned towards the best solutions.

2.3 GAME THEORY

The game theory serves as an analytical tool for modelling the decision-making and interactions of stakeholders in energy systems. Both cooperative game and non-cooperative game have been implemented into the field of energy systems. The cooperative game enables each stakeholder to at least gain interests through participating in the game instead of acting independently [15]. For the non-cooperative game, each stakeholder seeks to maximise its own interests and ultimately all stakeholders reach to an equilibrium through iterative interactions [16].

Cournot game and Stackelberg game are two classic game-theoretic models for analysing decision-making and interactions of stakeholders in energy systems. The Cournot game describes that stakeholders provide homogeneous products, e.g., electricity, and compete on the amount of products by making individual decisions independently and simultaneously [17]. Whereas, the Stackelberg game is a hierarchical two-level or multi-level sequential decision-making process [18]. For the two-level decision-making, stakeholders are categorised into the leader level which make decisions in the first place and the follower level which make responding decisions after receiving decisions from leaders. For the multi-level decision-making, after each level of followers make responding decisions, they become a leader. This sequential decision-making continues until the last level of followers.

An example of the two-level Stackelberg game in energy systems is that the policy makers (as the leader) determine policy measures in the first place, e.g., carbon pricing or renewable regulations. Receiving the policy signals, generators and consumers (i.e., the followers) make responding decisions, e.g., investing in low-carbon generators or

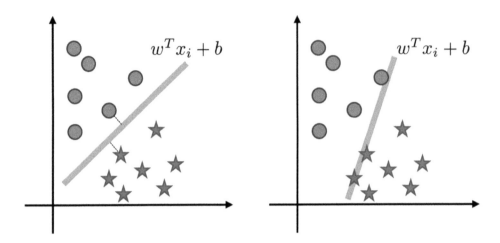

Figure 2.2 Using a hyperplane to classify the dataset into two classes. The hyperplane is denoted as $(w^T x_i + b)$. The figure on the left hand side is an accurate classification whereas the figure on the right hand side is an inaccurate classification. The dashed line indicates the maximum distance between the hyperplane and the closest data point.

reshaping consumption behaviours (readers are referred to [19]). Another example for the multi-level Stackelberg game is when in the energy markets, the policy maker (as the leader) charges the carbon tax from generators, the generators (as the first follower) would increase wholesale prices responding to the increase of generating costs caused by the increased carbon tax. The increase of wholesale prices would cause the increase of retail prices by energy suppliers (i.e., the second follower), which results in consumers (i.e., the last follower) changing their consumption behaviours for reducing the electricity bills as responses (readers are referred to [20]).

2.4 SUPPORT VECTOR MACHINE

As a type of the supervised learning model, the SVM is used for classifying the training dataset. For a training dataset containing N data points as

$$\left\{ (x_i, y_i) \mid_{i=1}^{N}, x_i \in \mathbb{R}^p, y_i \in \{1, -1\} \right\}, \tag{2.1}$$

where x_i is the sample i of the dataset, i.e., the training inputs, and y_i is a binary value indicating the class to which the point x_i belongs, i.e., the training labels.

The aim of the SVM is to find a hyperplane, denoted as $(w^T x_i + b)$, to classify the datasets. As shown in Figure 2.2, the figure on the left hand side is an accurate classification whereas the figure on the right hand side is an inaccurate classification. Hence, if the data point that is closest to the hyperplane, in each category has the maximum distance to the hyperplane, as indicated by the dashed lines in Figure 2.2, the hyperplane is the best one to classify the dataset.

Let a margin function describe the distance between the hyperplane to the closest data point as

$$f_{\mathrm{margin}}\left(w, b\right) := \min_{w,b} l\left(w, b, x_i\right), \forall i = 1, ..., N, \tag{2.2}$$

where $f_{\mathrm{margin}}\left(\cdot\right)$ is the margin function to measure the distance between the hyperplane and the closest data point, w is the weight of the hyperplane, b is the bias of the hyperplane, and $l\left(\cdot\right)$ is the distance function to measure the distance between the hyperplane and the data point x_i which is quantified by

$$l\left(w, b, x_i\right) = \frac{1}{\|w\|} \cdot \left|w^T x_i + b\right|. \tag{2.3}$$

A classifier can be mathematically described as

$$\begin{cases} w^T x_i + b > 0, y_i = 1, \\ w^T x_i + b < 0, y_i = -1, \end{cases} \forall\, i = 1, ..., N. \tag{2.4}$$

The hyperplane $(w^T x_i + b)$ divides a group of data samples x_i into two classes: one is indicated by $y_i = 1$ and another is indicated by $y_i = -1$. Since $(w^T x_i + b)$ and y_i have the same sign, Equation (2.4) can be simplified as:

$$y_i \cdot \left(w^T x_i + b\right) > 0, \forall\, i = 1, ..., N. \tag{2.5}$$

Therefore, the SVM is a classifier which seeks to find the maximum margin as

$$\max_{w,b} f_{\mathrm{margin}}(w, b), \tag{2.6}$$

s.t.

$$y_i \cdot \left(w^T x_i + b\right) > 0, \forall\, i = 1, ..., N. \tag{2.7}$$

Considering Equation (2.3), we have

$$\begin{aligned} &\max_{w,b} f_{\mathrm{margin}}(w, b) \\ &= \max_{w,b} \min_{i} \frac{1}{\|w\|} \cdot \left|w^T x_i + b\right| \\ &= \max_{w,b} \min_{i} \frac{1}{\|w\|} \cdot y_i \cdot \left(w^T x_i + b\right) \\ &= \max_{w,b} \frac{1}{\|w\|} \min_{i} y_i \cdot \left(w^T x_i + b\right). \end{aligned} \tag{2.8}$$

s.t.

$$y_i \cdot \left(w^T x_i + b\right) > 0, \forall\, i = 1, ..., N. \tag{2.9}$$

Given the constraint $y_i \cdot (w^T x_i + b) > 0$, there exists a $\gamma > 0$ to let $\min_i y_i \cdot (w^T x_i + b) = \gamma$, i.e., the margin (the hyperplane to the closest data point) equals to γ. Hence for all data points, their distances to the hyperplane are larger than or equal to γ as

$$y_i \cdot \left(w^T x_i + b \right) \geq \gamma. \tag{2.10}$$

Equation (2.8) can be further simplified as:

$$\max_{w,b} \frac{\gamma}{\|w\|} \tag{2.11}$$

s.t.

$$y_i \cdot \left(w^T x_i + b \right) > \gamma, \forall\, i = 1, ..., N. \tag{2.12}$$

It is noted that when w and b are scaled to $(\lambda \cdot w)$ and $(\lambda \cdot b)$, respectively, γ will be scaled to $(\lambda \cdot \gamma)$, which results in an equivalent optimisation problem that does not affect the results. For simplicity, let $\gamma{=}1$. In addition, the objective function can be written in the minimisation form as:

$$
\begin{aligned}
\max_{w,b} \frac{1}{\|w\|} \\
\equiv \min_{w,b} \frac{1}{2} \|w\| \\
= \min_{w,b} \frac{1}{2} w^T w.
\end{aligned}
\tag{2.13}
$$

Therefore, the SVM approach can be described as solving an optimisation problem as

$$\min_{w,b} \frac{1}{2} w^T w \tag{2.14}$$

s.t.

$$y_i \cdot \left(w^T x_i + b \right) > 1, \forall\, i = 1, ..., N. \tag{2.15}$$

Remark: In addition to the SVM, there are other classification approaches. These approaches can be categorised into the hard margin and soft margin. A primary difference is that for the hard margin, $y_i \in \{0, 1\}$; whereas for the soft margin, $y_i \in [0, 1]$. The hard margin includes the linear discriminant analysis [21] and perceptron [22]. The soft margin includes the generative model [23], e.g., Gaussian discriminant analysis [24], and discriminative model [25] such as logistic regression [26].

2.5 DIMENSIONALITY REDUCTION

This section introduces the dimensionality reduction. The dimensionality reduction is used in the machine learning to address the issue of overfitting, which occurs when a statistical model fits exactly against its training data [27]. The common solutions to address the overfitting issues include: (1) increasing the data volume, (2) regularization, and (3) dimensionality reduction. The motivation of the dimensionality reduction is the curse of dimensionality [28] which is the direct reason of the overfitting issue.

Let's look at the curse of dimensionality. From the mathematical perspective, let us take a simple example of a feature represented by a binary value; the required data to accurately estimate the feature space increases exponentially by two. For more complicated features, much more data is required to accurately estimate the feature spaces.

From the geometric perspective, we consider a hypercube and suprasphere with the same diameter:

- In two-dimension, the hypercube is a square with the area of 1 and the suprasphere is a circle with the area of $\pi \cdot 0.5^2$.

- In three-dimension, the hypercube is a cube with the volume of 1 and the suprasphere is a sphere with the volume of $\frac{4}{3} \cdot 0.5^3$.

Therefore, we can use a general formula to represent the volume of the suprasphere as:

$$V_{\mathrm{s}} = k \cdot 0.5^d, \tag{2.16}$$

where V_{s} is the volume of the suprasphere, k is the coefficient, and d is the dimension of the suprasphere. By contrast, the volume of the hypercube remains 1 irrespective of the increase of the dimension. When the dimension approaches to infinity, i.e., $d \to +\infty$, the volume of the suprasphere would approach to 0. This means that in the high-dimensional space, the suprasphere is empty and the samples are distributed in the space between the outer of suprasphere and inner of hypercube. This results in the difficulty of the classification due to the sparsity and heterogeneity of samples.

The approaches of the dimensionality reduction include (1) feature selection, (2) linear dimensionality reduction, and (3) non-linear dimensionality reduction. The linear dimensionality reduction includes PCA [29] and multi-dimensional scaling [30]. The non-linear dimensionality reduction includes Isomap [31] and locally linear embedding [32]. The book here takes PCA as a typical approach of the dimensionality reduction.

2.5.1 Probabilistic Fundamental

To facilitate the illustration of PCA, this subsection discusses how to represent the mean and covariance of samples in the matrix format. Let n denote the index of samples and p denote the dimensionality of samples. The matrix of samples can be represented as:

$$X = (\mathbf{x}_1, \mathbf{x}_2, ..., \mathbf{x}_N)^T = \begin{pmatrix} \mathbf{x}_1^T \\ \mathbf{x}_2^T \\ \vdots \\ \mathbf{x}_N^T \end{pmatrix} = \begin{pmatrix} x_{11} & x_{12} & \cdots & x_{1p} \\ x_{21} & x_{22} & \cdots & x_{2p} \\ \vdots & \vdots & \ddots & \vdots \\ x_{N1} & x_{N2} & \cdots & x_{Np} \end{pmatrix}_{N \times p}, \tag{2.17}$$

$$\mathbf{x}_i \in \mathbb{R}^p, i = 1, 2, ..., N,$$

For the one-dimensional samples, the mean and covariance of samples can be defined by the Equations (2.18) and (2.19), respectively.

$$\bar{X} = \frac{1}{N} \sum_{i=1}^{N} x_i, \tag{2.18}$$

$$S = \frac{1}{N} \sum_{i=1}^{N} \left(x_i - \bar{X}\right)^2. \tag{2.19}$$

For the p-dimensional samples, the mean and covariance of samples can be defined by the Equations (2.20) and (2.21), respectively.

$$\bar{X}_{p \times 1} = \frac{1}{N} \sum_{i=1}^{N} \mathbf{x}_i, \tag{2.20}$$

$$S_{p \times p} = \frac{1}{N} \sum_{i=1}^{N} \left(\mathbf{x}_i - \bar{X}\right) \left(\mathbf{x}_i - \bar{X}\right)^T. \tag{2.21}$$

Let I_N denote a N-dimensional unit column vector. Based on the Equation (2.17), the Equations (2.20) and (2.21) can be expressed in the matrix format as:

$$\bar{X} = \frac{1}{N} \left(\mathbf{x}_1, \mathbf{x}_2, ..., \mathbf{x}_N\right) I_N = \boxed{\frac{1}{N} X^T I_N}, \tag{2.22}$$

$$
\begin{aligned}
S_{p \times p} &= \frac{1}{N} \left(\mathbf{x}_1 - \bar{X}, \mathbf{x}_2 - \bar{X}, ..., \mathbf{x}_N - \bar{X}\right) \cdot \begin{pmatrix} \left(\mathbf{x}_1 - \bar{X}\right)^T \\ \left(\mathbf{x}_2 - \bar{X}\right)^T \\ \vdots \\ \left(\mathbf{x}_N - \bar{X}\right)^T \end{pmatrix} \\
&= \frac{1}{N} \left[\left(\mathbf{x}_1, \mathbf{x}_2, ..., \mathbf{x}_N\right) - \left(\bar{X}, \bar{X}, ..., \bar{X}\right)\right] \cdot \begin{pmatrix} \left(\mathbf{x}_1 - \bar{X}\right)^T \\ \left(\mathbf{x}_2 - \bar{X}\right)^T \\ \vdots \\ \left(\mathbf{x}_N - \bar{X}\right)^T \end{pmatrix} \\
&= \frac{1}{N} \left(X^T - \bar{X} I_N^T\right) \cdot \begin{pmatrix} \left(\mathbf{x}_1 - \bar{X}\right)^T \\ \left(\mathbf{x}_2 - \bar{X}\right)^T \\ \vdots \\ \left(\mathbf{x}_N - \bar{X}\right)^T \end{pmatrix} \\
&= \frac{1}{N} \left(X^T - \frac{1}{N} X^T I_N I_N^T\right) \cdot \begin{pmatrix} \left(\mathbf{x}_1 - \bar{X}\right)^T \\ \left(\mathbf{x}_2 - \bar{X}\right)^T \\ \vdots \\ \left(\mathbf{x}_N - \bar{X}\right)^T \end{pmatrix} \\
&= \frac{1}{N} X^T \left(I_N - \frac{1}{N} I_N I_N^T\right) \cdot \left(I_N - \frac{1}{N} I_N I_N^T\right)^T X.
\end{aligned}
\tag{2.23}
$$

The term $X^T \left(I_N - \frac{1}{N} I_N I_N^T \right)$ is defined as the centring matrix and is denoted as H. Hence, the Equation (2.23) can be rewritten as:

$$S_{p \times p} = \frac{1}{N} X^T H \cdot H^T X. \tag{2.24}$$

The function of the centring matrix is to subtract mean value in every dimension, so that the samples are distributed around the origin, with the following properties:

- *Property 1*: The transposition of H is itself as

$$H^T = (I_N - \frac{1}{N} I_N I_N^T) = H \tag{2.25}$$

- *Property 2*: The product of H is itself as

$$\begin{aligned}
H^2 &= (I_N - \frac{1}{N} I_N I_N^T)(I_N - \frac{1}{N} I_N I_N^T) \\
&= I_N - \frac{2}{N} I_N I_N^T - \frac{1}{N^2} I_N I_N^T I_N I_N^T \\
&= I_N - \frac{2}{N} I_{N \times N} + \frac{1}{N^2} I_{N \times N} I_{N \times N} \\
&= I_N - (\frac{2}{N})_{N \times N} + \frac{1}{N^2} N_{N \times N} \\
&= I_N - (\frac{2}{N})_{N \times N} + (\frac{1}{N})_{N \times N} \\
&= I_N - (\frac{1}{N})_{N \times N} \\
&= I_N - \frac{1}{N} I_{N \times N} = H,
\end{aligned} \tag{2.26}$$

where $I_{N \times N}$ is an N-dimensional unit square matrix. From this property, we have:

$$S_{p \times p} = \boxed{\frac{1}{N} X^T H X}. \tag{2.27}$$

2.5.2 Principal Component Analysis

PCA aims to transform a set of linear dependent variables into a set of linear independent variables by using the orthogonal transformation, i.e., reconstructing the original feature space [33]. For instance, the variables of an energy consumer include location, energy supplier, energy demand, retail electricity price, and wholesale electricity price. There is correlation between the retail electricity price and wholesale electricity price. PCA can reconstruct the original feature space for transforming the retail electricity price and wholesale electricity price to be linearly independent. Reconstructing the original feature space is achieved by the maximum projection variance or minimum cost of reconstructing.

As shown in Figure 2.3, we have a set of samples in a two-dimensional space with projections from various directions, e.g., projection 1 and projection 2. The

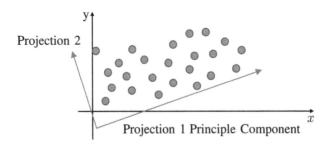

Figure 2.3 Example of finding the principle component in the two-dimensional space. Projection 1 and projection 2 are two instantiated projections of data samples, in which the projection 1 has the most dispersed distribution that the principle component analysis aims to find, since the data samples have the maximum variance from this projection. The projection 1 is defined as the principle component.

projection 1 has the most dispersed distribution that PCA aims to find. At this projection, these samples have the maximum variance. This projection is defined as the principal component. If we want to reduce the dimensionality to q dimensions, we only need to find the first q principle components. The variance represents how much features the data can cover. Hence, finding the maximum projection variance means covering information as much as possible after the projections. By contrast, if all the projections of samples concentrate on a crowded region with overlap, the costs of reconstructing from projections to the original coordination are large. PCA also aims to find the projection with the minimum cost of the reconstitution.

2.5.3   Maximum Projection Variance

We firstly think about how to mathematically represent the maximum projection variance. Assume a sample $\mathbf{x}_i$ normalised by the mean $\bar{X}$ as $(\mathbf{x}_i - \bar{X})$. Given a unit vector $\|\mathbf{u}_1\|=1$, the projection of $(\mathbf{x}_i - \bar{X})$ on the unit vector $\mathbf{u}_1$ is the dot product of $(\mathbf{x}_i - \bar{X})$ and $\mathbf{u}_1$ as

$$
\left(\mathbf{x}_i - \bar{X}\right) \cdot \mathbf{u}_1
$$
$$
= \left\|\mathbf{x}_i - \bar{X}\right\| \cdot \|\mathbf{u}_1\| \cdot \cos\theta
$$
$$
= \left\|\mathbf{x}_i - \bar{X}\right\| \cdot \cos\theta \tag{2.28}
$$
$$
= \left(\mathbf{x}_i - \bar{X}\right)^T \mathbf{u}_1,
$$

s.t.
$$
\mathbf{u}_1^T \mathbf{u}_1 = 1. \tag{2.29}
$$

Therefore, for N samples, the projection of $\left(\mathbf{x}_i - \bar{X}\right)$ on $\mathbf{u}_1$ can be calculated by the square mean length as

$$J = \frac{1}{N} \sum_{i=1}^{N} \left[\left(\mathbf{x}_i - \bar{X}\right)^T \mathbf{u}_1\right]^2,$$

$$= \mathbf{u}_1^T \left[\frac{1}{N} \sum_{i=1}^{N} \left(\mathbf{x}_i - \bar{X}\right)\left(\mathbf{x}_i - \bar{X}\right)^T\right] \mathbf{u}_1,$$

(2.30)

where J is the projection of $\left(\mathbf{x}_i - \bar{X}\right)$ on $\mathbf{u}_1$.

With the Equation (2.21), the Equation (2.30) can be written as:

$$J = \mathbf{u}_1^T S \mathbf{u}_1.$$

(2.31)

Therefore, the problem of finding the maximum projection variance can be expressed as an optimisation problem as:

$$\mathbf{u}_1^* = \arg \max_{\mathbf{u}_1} \mathbf{u}_1^T S \mathbf{u}_1,$$

(2.32)

s.t.

$$\mathbf{u}_1^T \mathbf{u}_1 = 1.$$

(2.33)

This optimisation problem can be solved by the Lagrange multiplier as

$$L(\mathbf{u}_1, \lambda) = \mathbf{u}_1^T S \mathbf{u}_1 + \lambda \left(1 - \mathbf{u}_1^T \mathbf{u}_1\right),$$

(2.34)

$$\frac{\partial L}{\partial \mathbf{u}_1} = 2 \cdot S \cdot \mathbf{u}_1 - 2 \cdot \lambda \cdot \mathbf{u}_1 = 0.$$

(2.35)

We have

$$\boxed{S \cdot \mathbf{u}_1 = \lambda \cdot \mathbf{u}_1},$$

(2.36)

where λ is defined as the Eigenvalue and $\mathbf{u}_1$ is defined as the Eigenvector [34].

In summary, in p dimensions, we have p Eigenvectors $\mathbf{u}_1$, ..., $\mathbf{u}_p$, corresponding to p Eigenvalues λ_1, ..., λ_p. Reducing to q dimensions means taking the first q Eigenvalues λ_1, ..., λ_q.

2.5.4 Minimum Cost of Reconstruction

Next, we think about how to mathematically represent the cost of the reconstruction. As shown in Figure 2.4, $\mathbf{x}_i$ is a sample in the two-dimensional space, $\mathbf{u}_1$ and $\mathbf{u}_2$ are Eigenvectors of the covariance matrix and their Eigenvalues are λ_1 and λ_2, respectively. The coordinate of $\mathbf{x}_i$ on $\mathbf{u}_1$ and $\mathbf{u}_2$ are the product of the original coordinate $(\mathbf{x}_i^T \mathbf{u}_p)$ and the vector $\mathbf{u}_p$ as

$$(\mathbf{x}_i^T \mathbf{u}_1)\mathbf{u}_1 + (\mathbf{x}_i^T \mathbf{u}_2)\mathbf{u}_2.$$

(2.37)

When reducing to single dimension, we have:

$$\mathbf{x}_i' = (\mathbf{x}_i^T \mathbf{u}_1)\mathbf{u}_1.$$

(2.38)

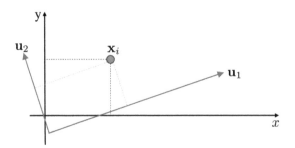

Figure 2.4 Example of the Eigenvector of a sample in the two-dimensional space. $\mathbf{u}_1$ and $\mathbf{u}_2$ are Eigenvectors of the covariance matrix.

From this simple example, we can see that the dimensionality reduction is to reduce the Eigenvector, but the dimension of $\mathbf{x}_i$ remains unchanged.

For the problem of the dimensionality reduction from p dimensions to q dimensions, first, samples are normalised as:

$$\mathbf{x}_i' = \mathbf{x}_i - \bar{X}. \tag{2.39}$$

Second, projecting samples to the Eigenvectors, the new coordinate is:

$$\mathbf{x}_i'' = \sum_{k=1}^{p} (\mathbf{x}_i'^T \mathbf{u}_k) \mathbf{u}_k. \tag{2.40}$$

Third, ranking the Eigenvectors in descending order of Eigenvalues λ and keeping the first q dimension as

$$\hat{\mathbf{x}}_i'' = \sum_{k=1}^{q} (\mathbf{x}_i'^T \mathbf{u}_k) \mathbf{u}_k. \tag{2.41}$$

The minimum cost of the reconstruction is the cost to convert from $\hat{\mathbf{x}}_i''$ back to $\mathbf{x}_i''$, which is denoted as

$$\begin{aligned} J &= \frac{1}{N} \sum_{i=1}^{N} \|\mathbf{x}_i'' - \hat{\mathbf{x}}_i''\|^2 \\ &= \frac{1}{N} \sum_{i=1}^{N} \left\| \sum_{k=q+1}^{p} (\mathbf{x}_i'^T \mathbf{u}_k) \mathbf{u}_k \right\|^2 \end{aligned} \tag{2.42}$$

The 2-norm of a vector is the sum of squares of this vector in every dimension. We have

$$\left\| \sum_{k=q+1}^{p} (\mathbf{x}_i'^T \mathbf{u}_k) \mathbf{u}_k \right\|^2 = \sum_{k=q+1}^{p} (\mathbf{x}_i'^T \mathbf{u}_k)^2. \tag{2.43}$$

Hence,

$$J = \frac{1}{N} \sum_{i=1}^{N} \sum_{k=q+1}^{p} (\mathbf{x}_i'^T \mathbf{u}_k)^2$$

$$= \sum_{k=q+1}^{p} \frac{1}{N} \sum_{i=1}^{N} (\mathbf{x}_i'^T \mathbf{u}_k)^2 \tag{2.44}$$

$$= \sum_{k=q+1}^{p} \frac{1}{N} \sum_{i=1}^{N} \left[(\mathbf{x}_i - \bar{X})^T \mathbf{u}_k \right]^2,$$

With Equation (2.31), we have

$$J = \sum_{k=q+1}^{p} \mathbf{u}_k^T S \mathbf{u}_k. \tag{2.45}$$

Therefore, the minimum cost of the reconstruction can be described as:

$$\mathbf{u}_k^* = \arg \min_{\mathbf{u}_k} \sum_{k=q+1}^{p} \mathbf{u}_k^T S \mathbf{u}_k, \tag{2.46}$$

s.t.

$$\mathbf{u}_k^T \mathbf{u}_k = 1. \tag{2.47}$$

Since $\mathbf{u}_k$ is a linear independent vector. This optimisation problem can be converted into multiple independent optimisation problems as:

$$\mathbf{u}_k^* = \arg \min \mathbf{u}_k^T S \mathbf{u}_k, \tag{2.48}$$

s.t.

$$\mathbf{u}_k^T \mathbf{u}_k = 1. \tag{2.49}$$

This is the same problem as described in the Section 2.5.3.

Therefore, PCA is to find the first q Eigenvectors of the covariance matrix, corresponding to the fist q largest Eigenvalues.

2.6   EXPECTATION MAXIMISATION

MLE is used to estimate the density of a dataset by searching various probability distributions and fitting the parameters [35]. For a simple statistical model, parameters can be estimated as:

$$\theta^* = \arg \max_{\theta} \log P(x|\theta), \tag{2.50}$$

where θ is the parameter to be estimated, x is the sample, and $\log P(x|\theta)$ is the log likelihood.

By contrast, for a complex mixture model, e.g., GMM, it is difficult to estimate parameters using the MLE, since we have no information about the distribution of the dataset and need to use latent variables to generate x as

$$P(x) = \int_z P(x, z) \, \mathrm{d}z, \tag{2.51}$$

where z is the latent variable.

For this reason, we introduce EM which is an algorithm to estimate parameters for the mixture model in the presence of latent variables [36]. The EM algorithm can be described as:

$$\theta^{t+1} = \arg\max_{\theta} \int_z \log P\left(x, z|\theta\right) \cdot P\left(z|x, \theta^t\right) dz, \qquad (2.52)$$

where z is the latent variable, θ^{t+1} is the parameter at the iteration $t+1$ which is a decision variable, θ^t is the parameter at the iteration t which is a constant, $P\left(z|x, \theta^t\right)$ is the posterior probability, and the joint probability $\log P\left(x, z|\theta\right)$ is the complete data. The term $\int_z \log P\left(x, z|\theta\right) \cdot P\left(z|x, \theta^t\right) dz$ can also be denoted in the expectation format as:

$$E_{z|x, \theta^t}\left[\log P\left(x, z|\theta\right)\right]. \qquad (2.53)$$

The procedures of EM algorithm can be described as the following two steps, as indicated by its name:

- 1) *E-step:* Given the posterior probability $P\left(z|x, \theta^t\right)$, calculating the expectation as

$$E_{z|x, \theta^t}\left[\log P\left(x, z|\theta\right)\right]. \qquad (2.54)$$

- 2) *M-step:* Maximising the expectation

$$\theta^{t+1} = \arg\max_{\theta} E_{z|x, \theta^t}\left[\log P\left(x, z|\theta\right)\right]. \qquad (2.55)$$

2.6.1 Convergence of Expectation Maximisation Algorithm

EM algorithm iteratively updates the parameter from the iteration t (θ^t) to the iteration $t+1$ (θ^{t+1}), resulting in an updated log likelihood from $\log P\left(x|\theta^t\right)$ to $\log P\left(x|\theta^{t+1}\right)$. In order to yield the maximum expectation, i.e., the convergence of the EM algorithm, we need to ensure

$$\log P\left(x|\theta^t\right) \leq \log P\left(x|\theta^{t+1}\right). \qquad (2.56)$$

The proof of existing convergence of EM algorithm is given in this subsection. First, we have

$$\log P\left(x|\theta\right) = \log \frac{P\left(x, z|\theta\right)}{P\left(z|x, \theta\right)} \qquad (2.57)$$
$$= \log P\left(x, z|\theta\right) - \log P\left(z|x, \theta\right).$$

In both sides of the equation, calculating the expectation with respect to $P\left(z|x, \theta^t\right)$:

- For left hand, given $\int_z P\left(z|x, \theta^t\right) dz = 1$, we have

$$\text{left hand} = \int_z P\left(z|x, \theta^t\right) \cdot \log P\left(x|\theta\right) dz$$
$$= \log P\left(x|\theta\right) \int_z P\left(z|x, \theta^t\right) dz \qquad (2.58)$$
$$= \log P\left(x|\theta\right)$$

- For right hand, we have

$$\text{right hand} = \int_z P\left(z|x,\theta^t\right) \cdot \log P\left(x,z|\theta\right) dz - \int_z P\left(z|x,\theta^t\right) \cdot \log P\left(z|x,\theta\right) dz. \tag{2.59}$$

Define $\int_z P\left(z|x,\theta^t\right) \cdot \log P\left(x,z|\theta\right) dz$ as $Q\left(\theta,\theta^t\right)$ and $\int_z P\left(z|x,\theta^t\right) \cdot \log P\left(z|x,\theta\right) dz$ as $H\left(\theta,\theta^t\right)$. Hence, in order to prove $\log P\left(x|\theta^t\right) \leq \log P\left(x|\theta^{t+1}\right)$ as in Equation (2.56), we need to prove

$$Q\left(\theta^t,\theta^t\right) - H\left(\theta^t,\theta^t\right) \leq Q\left(\theta^{t+1},\theta^t\right) - H\left(\theta^{t+1},\theta^t\right). \tag{2.60}$$

This proof is given as follows:

- For the $Q\left(\theta,\theta^t\right)$, we have

$$Q\left(\theta^t,\theta^t\right) = \int_z P\left(z|x,\theta^t\right) \cdot \log P\left(x,z|\theta^t\right) dz, \tag{2.61}$$

$$Q\left(\theta^{t+1},\theta^t\right) = \int_z P\left(z|x,\theta^t\right) \cdot \log P\left(x,z|\theta^{t+1}\right) dz. \tag{2.62}$$

Given that

$$\theta^{t+1} = \arg\max_{\theta} \int_z \log P\left(x,z|\theta\right) \cdot P\left(z|x,\theta^t\right) dz, \tag{2.63}$$

we have

$$Q\left(\theta^{t+1},\theta^t\right) \geq Q\left(\theta^t,\theta^t\right). \tag{2.64}$$

- For $H\left(\theta,\theta^t\right)$, we have

$$
\begin{aligned}
& H\left(\theta^{t+1},\theta^t\right) - H\left(\theta^t,\theta^t\right) \\
&= \int_z P\left(z|x,\theta^t\right) \cdot \log P\left(z|x,\theta^{t+1}\right) dz - \int_z P\left(z|x,\theta^t\right) \cdot \log P\left(z|x,\theta^t\right) dz \\
&= \int_z P\left(z|x,\theta^t\right) \cdot \left[\log P\left(z|x,\theta^{t+1}\right) - \log P\left(z|x,\theta^t\right)\right] dz \\
&= \int_z P\left(z|x,\theta^t\right) \cdot \log \frac{P\left(z|x,\theta^{t+1}\right)}{P\left(z|x,\theta^t\right)} dz.
\end{aligned}
\tag{2.65}
$$

To prove $H\left(\theta^{t+1},\theta^t\right) \leq H\left(\theta^t,\theta^t\right)$, one option is to use the Kullback–Leibler (KL) divergence [37], through which the Equation (2.65) can be denoted as:

$$-KL\left(P\left(z|x,\theta^t\right) \| P\left(z|x,\theta^{t+1}\right)\right). \tag{2.66}$$

Given $KL \geq 0$, we have

$$-KL\left(P\left(z|x,\theta^t\right) \| P\left(z|x,\theta^{t+1}\right)\right) \leq 0. \tag{2.67}$$

Hence,

$$H\left(\theta^{t+1},\theta^t\right) \leq H\left(\theta^t,\theta^t\right). \tag{2.68}$$

Alternatively, the Jensen inequality can be used to prove $H\left(\theta^{t+1}, \theta^t\right) \leq H\left(\theta^t, \theta^t\right)$. For a convex function, e.g., the log function, we have

$$E\left[\log x\right] \leq \log E\left[\log x\right]. \tag{2.69}$$

Hence,

$$
\begin{aligned}
H&\left(\theta^{t+1}, \theta^t\right) - H\left(\theta^t, \theta^t\right) \\
&= \int_z P\left(z|x, \theta^t\right) \cdot \log \frac{P\left(z|x, \theta^{t+1}\right)}{P\left(z|x, \theta^t\right)} \mathrm{d}z \\
&\leq \log \int_z P\left(z|x, \theta^t\right) \cdot \frac{P\left(z|x, \theta^{t+1}\right)}{P\left(z|x, \theta^t\right)} \mathrm{d}z \\
&= \log \int_z P\left(z|x, \theta^{t+1}\right) \mathrm{d}z \\
&= \log 1 \\
&= 0,
\end{aligned} \tag{2.70}
$$

which means that $H\left(\theta^{t+1}, \theta^t\right) \leq H\left(\theta^t, \theta^t\right)$.

Therefore, the Equation (2.60) is proved, i.e., $\log P\left(x|\theta^t\right) \leq \log P\left(x|\theta^{t+1}\right)$.

In the following two subsections, more strict proofs of convergence will be provided from the KL divergence and Jensen inequality, respectively.

2.6.2 Kullback–Leibler Divergence

This subsection provides the strict proof for the convergence of EM algorithm from the KL divergence. We have

$$
\begin{aligned}
\log P\left(x|\theta\right) &= \log P\left(x, z|\theta\right) - \log P\left(z|x, \theta\right), \\
&= \log P\left(x, z|\theta\right) - \log q\left(z\right) - \left[\log P\left(z|x, \theta\right) - \log q\left(z\right)\right] \\
&= \log \frac{P\left(x, z|\theta\right)}{q\left(z\right)} - \log \frac{P\left(z|x, \theta\right)}{q\left(z\right)}.
\end{aligned} \tag{2.71}
$$

In both sides of the equation, calculating the expectation with respect to $q\left(z\right)$ as

$$
\begin{aligned}
\text{left hand} &= \int_z q\left(z\right) \cdot \log P\left(x|\theta\right) \mathrm{d}z \\
&= \log P\left(x|\theta\right) \int_z q\left(z\right) \mathrm{d}z \\
&= \log P\left(x|\theta\right)
\end{aligned} \tag{2.72}
$$

$$
\text{right hand} = \int_z q\left(z\right) \cdot \log \frac{P\left(x, z|\theta\right)}{q\left(z\right)} \mathrm{d}z - \int_z q\left(z\right) \cdot \log \frac{P\left(z|x, \theta\right)}{q\left(z\right)} \mathrm{d}z, \tag{2.73}
$$

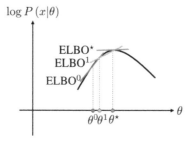

Figure 2.5 Schematic illustration of the relationship between the ELBO and $\log P(x|\theta)$. Given an initial parameter θ^0, EM tries to find the maximal ELBO and then uses this ELBO to update θ^0 to θ^1. This procedure iteratively proceeds until it finds the optimum parameter $\theta^\star$ which yields the maximum $\log P(x|\theta)$.

where the first term $\int_z q(z) \cdot \log \frac{P(x,z|\theta)}{q(z)} dz$ is defined as the evidence lower bound (ELBO) and the second term is the KL divergence as $KL(q(z) \parallel P(z|x,\theta))$. Hence, we have

$$\log P(x|\theta) = \text{ELBO} + KL(q \parallel p). \tag{2.74}$$

Given $KL(q \parallel p) \geq 0$ and the equality holds only if $p = q$, we have

$$\log P(x|\theta) \geq \text{ELBO}. \tag{2.75}$$

The schematic illustration of the relationship between the ELBO and $\log P(x|\theta)$ is presented in Figure 2.5. Given an initial parameter θ^0, EM tries to find the maximal ELBO and then uses this ELBO to update θ^0 to θ^1. This procedure iteratively proceeds until it finds the optimum parameter $\theta^\star$ which yields the maximum $\log P(x|\theta)$ as

$$\theta^* = \arg\max_{\theta} \text{ELBO}$$
$$= \arg\max_{\theta} \int_z q(z) \cdot \log \frac{P(x,z|\theta)}{q(z)} dz \tag{2.76}$$

Using $P(z|x,\theta^t)$ to replace $q(z)$, we have

$$\theta^* = \arg\max_{\theta} \int_z P(z|x,\theta^t) \cdot \log \frac{P(x,z|\theta)}{P(z|x,\theta^t)} dz$$
$$= \arg\max_{\theta} \int_z P(z|x,\theta^t) \cdot [\log P(x,z|\theta) - \log P(z|x,\theta^t)] dz. \tag{2.77}$$

Since θ^t is a constant, this equation can be simplified as

$$\theta^* = \arg\max_{\theta} \int_z P(z|x,\theta^t) \cdot \log P(x,z|\theta) dz. \tag{2.78}$$

2.6.3 Jensen Inequality

This section provides strict proof for the convergence of EM algorithm from the Jensen inequality [38]. According to Jensen inequality, we have

$$
\begin{aligned}
\log P\left(x|\theta\right) &= \int_z P\left(x, z|\theta\right) \mathrm{d}z \\
&= \log \int_z \frac{P\left(x, z|\theta\right)}{q\left(z\right)} q\left(z\right) \mathrm{d}z \\
&= \log E_{q(z)} \left[\frac{P\left(x, z|\theta\right)}{q\left(z\right)}\right] \\
&\geq E_{q(z)} \left[\log \frac{P\left(x, z|\theta\right)}{q\left(z\right)}\right].
\end{aligned}
\tag{2.79}
$$

The equality holds only if

$$
\frac{P\left(x, z|\theta\right)}{q\left(z\right)} = C,
\tag{2.80}
$$

where C is a constant. In this equation, the term $E_{q(z)} \left[\log \frac{P(x,z|\theta)}{q(z)}\right]$ is the ELBO.

Next, we need to prove $q\left(z\right) = P\left(z|x, \theta\right)$. When the equality holds, we have

$$
q\left(z\right) = \frac{1}{C} P\left(x, z|\theta\right)
\tag{2.81}
$$

Calculating expectation with respect to z under the probability distribution of $q\left(z\right)$, we have

$$
\int_z q\left(z\right) \mathrm{d}z = \int_z \frac{1}{C} P\left(x, z|\theta\right) \mathrm{d}z = 1
\tag{2.82}
$$

Hence,

$$
\frac{1}{C} P\left(x|\theta\right) = 1
\tag{2.83}
$$

i.e., $C = P\left(x|\theta\right)$.

Therefore,

$$
q\left(z\right) = \frac{P\left(x, z|\theta\right)}{P\left(x|\theta\right)} = P\left(z|x, \theta\right).
\tag{2.84}
$$

2.6.4 Generalised Expectation Maximisation

This subsection introduces a generalised form of the EM. From previous subsections, we have

$$
\log P\left(x|\theta\right) = ELBO + KL\left(q \parallel p\right),
\tag{2.85}
$$

where

$$
\begin{cases}
ELBO = E_{q(z)} \left[\log \frac{P(x,z|\theta)}{q(z)}\right], \\
KL\left(q \parallel p\right) = \int_z q\left(z\right) \cdot \log \frac{q(z)}{P(z|x,\theta)} \mathrm{d}z.
\end{cases}
\tag{2.86}
$$

Denote ELBO as $L\left(q, \theta\right)$. Given $KL \geq 0$, we have

$$
\log P\left(x|\theta\right) \geq L\left(q, \theta\right).
\tag{2.87}
$$

Although we assume $q(z) = P(z|x, \theta)$, for a complex generative model, $P(z|x, \theta)$ is intractable. Hence, the aim is to let the value of $q(z)$ approach to the value of $P(z|x, \theta)$ through iterative optimisation as:

- First, when θ is fixed, i.e., $\log P(x|\theta)$ is fixed, when $q(z)$ approaches to $P(z|x, \theta)$, the KL decreases and ELBO increases.

- Second, when q^* is fixed, the EM maximises the ELBO through finding the optimal θ.

Therefore, the generalised form of the EM can be described as:

$$
\begin{cases}
\text{E-Step:} & q^{t+1} = \arg\max_q L(q, \theta^t) \\
\text{M-Step:} & \theta^{t+1} = \arg\max_\theta L(q^{t+1}, \theta)
\end{cases}
\tag{2.88}
$$

The ELBO can be further simplified as:

$$
\begin{aligned}
L(q, \theta) &= E_{q(z)}[\log P(x, z) - \log q(z)] \\
&= E_{q(z)}[\log P(x, z)] - E_{q(z)}[\log q(z)] \\
&= E_{q(z)}[\log P(x, z)] + H[q(z)],
\end{aligned}
\tag{2.89}
$$

where $H[q(z)]$ is the entropy of $q(z)$ which is an additional term for the generalised EM. This is because we have no information about the distribution of $q(z)$.

2.7 GAUSSIAN MIXTURE MODEL

This section introduces the GMM which is a mixture of multiple Gaussian distributions [39]. The probability density function (pdf) of a set of samples is shown in Figure 2.6. The pdf indicated by the dashed blue lines is the Gaussian distribution of each sample. The pdf indicated by the green line is the mixture of multiple Gaussian distributions.

2.7.1 Model Introduction

From the geometry perspective, GMM is the weighted sum of multiple Gaussian distributions. Let $N(\mu_k, \Sigma_k)$ denote the Gaussian distribution k, where μ_k is the expectation of the Gaussian distribution k and Σ_k is the variance of the Gaussian distribution k. The pdf of the sample x obtained from the mixture of total K Gaussian distributions can be denoted as:

$$
\sum_{k=1}^{K} \alpha_k N(x|\mu_k, \Sigma_k),
\tag{2.90}
$$

s.t.

$$
\sum_{k=1}^{K} \alpha_k = 1,
\tag{2.91}
$$

where α_k is the weight of the Gaussian distribution k.

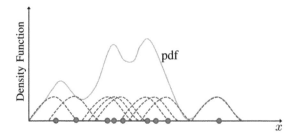

Figure 2.6 Schematic illustration of generating the probability density function using the Gaussian mixture model. The probability density function indicated by the dashed blue lines is the Gaussian distribution of each samples. The probability density function indicated by the green line is the mixture of multiple Gaussian distributions.

As shown in Figure 2.7, there are two Gaussian distributions in a two-dimensional space as indicated by the contour lines. For a sample x, it belongs to both of these two distributions with different probabilities. It is high-probable that x belongs to the Gaussian distribution a and low-probable that x belongs to the Gaussian distribution b. To illustrate these different probabilities, we introduce a new variable z, defined as the latent variable. The latent variable indicates the all possible Gaussian distributions of the variable x, and the probability of each Gaussian distribution as shown in Table 2.1. We have

$$\sum_{k=1}^{K} p_k = 1, \tag{2.92}$$

where p_k is the probability that the variable x belongs to the Gaussian distribution k.

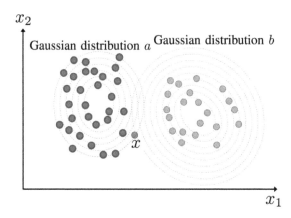

Figure 2.7 Example of two Gaussian distributions in a two-dimensional space. Each Gaussian distribution is indicated by the contour lines. For a sample x, it belongs to both of these two distributions with different probabilities. It is high-probable that x belongs to the Gaussian distribution a and low-probable that x belongs to the Gaussian distribution b.

Table 2.1 Latent variable to describe the probability of mixture Gaussian distributions

z	Gaussian distribution 1	Gaussian distribution 2	...	Gaussian distribution K
Probability	p_1	p_2	...	p_K

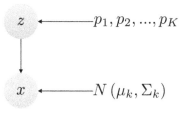

Figure 2.8 The Gaussian mixture model from the probabilistic graphical perspective. The Gaussian mixture model firstly determines the latent variable z with the probability p_k from a set of probabilities $p_1, p_2, ..., p_K$, and then generates x being subject to $N(\mu_k, \Sigma_k)$.

From the perspective of a generative model, GMM firstly selects one of K Gaussian distributions and secondly generates a sample from the selected Gaussian distribution.

From the probabilistic graphical perspective as shown in Figure 2.8, GMM firstly determines the latent variable z with the probability p_k from a set of probabilities $p_1, p_2, ..., p_K$, and then generates x being subject to $N(\mu_k, \Sigma_k)$.

Therefore, α_k from the geometry perspective equals to p_k from the probabilistic graphical perspective, given

$$
\begin{aligned}
P(x) &= \sum_z P(x, z) \\
&= \sum_{k=1}^{K} P(x, z = C_k) \\
&= \sum_{k=1}^{K} P(z = C_k) P(x|z = C_k) \quad = \sum_{k=1}^{K} p_k N(x|\mu_k, \Sigma_k),
\end{aligned}
\tag{2.93}
$$

where C_k is the Gaussian distribution k.

2.7.2 Solution of Gaussian Mixture Model

EM algorithm is an efficient algorithm to solve the GMM. As introduced in Section 2.6, the EM algorithm is given as follows:

$$
\theta^{t+1} = \arg\max E_{z|x,\theta^t}\left[\log P(x, z|\theta)\right],
\tag{2.94}
$$

which consists of the E-step and M-step as detailed in the following subsections.

2.7.2.1 E-Step

First, we will introduce the E-step of the EM algorithm. Let $Q\left(\theta, \theta^t\right)$ denote $E_{z|x,\theta^t}\left[\log P\left(x, z|\theta\right)\right]$, X denote a set of N observed samples, and Z denote a set of N latent variables. We have

$$Q\left(\theta, \theta^t\right) = \int_Z \log P\left(X, Z|\theta\right) \cdot P\left(Z|x, \theta^t\right)$$

$$= \sum_Z \log \prod_{i=1}^N P\left(x_i, z_i|\theta\right) \prod_{i=1}^N P\left(z_i|x_i, \theta^t\right)$$

$$= \sum_Z \sum_{i=1}^N \log P\left(x_i, z_i|\theta\right) \prod_{i=1}^N P\left(z_i|x_i, \theta^t\right) \qquad (2.95)$$

$$= \sum_Z \left[\log P\left(x_1, z_1|\theta\right) + \log P\left(x_2, z_2|\theta\right) + \right.$$

$$\left. \dots + \log P\left(x_N, z_N|\theta\right)\right] \prod_{i=1}^N P\left(z_i|x_i, \theta^t\right)$$

Taking the first term $\log P\left(x_1, z_1|\theta\right)$ as an example. We have

$$\sum_{z_1, z_2, \dots, z_N} \log P\left(x_1, z_1|\theta\right) \prod_{i=1}^N P\left(z_i|x_i, \theta^t\right)$$

$$= \sum_{z_1} \log P\left(x_1, z_1|\theta\right) P\left(z_1|x_1, \theta^t\right) \sum_{z_2, \dots, z_N} \prod_{i=2}^N P\left(z_i|x_i, \theta^t\right)$$

$$= \sum_{z_1} \log P\left(x_1, z_1|\theta\right) P\left(z_1|x_1, \theta^t\right) \sum_{z_2, \dots, z_N} \sum_{z_2} P\left(z_2|x_2, \theta^t\right) \qquad (2.96)$$

$$\sum_{z_3} P\left(z_3|x_3, \theta^t\right) \dots \sum_{z_N} P\left(z_N|x_N, \theta^t\right)$$

Given $\sum_{z_i} P\left(z_i|x_i, \theta^t\right) = 1$, this equation can be simplified as

$$\sum_{z_1} \log P\left(x_1, z_1|\theta\right) \cdot P\left(z_1|x_1, \theta^t\right). \qquad (2.97)$$

Hence,

$$Q\left(\theta, \theta^t\right) = \sum_{z_1} \log P\left(x_1, z_1|\theta\right) \cdot P\left(z_1|x_1, \theta^t\right) + \dots + \sum_{z_N} \log P\left(x_N, z_N|\theta\right) \cdot P\left(z_N|x_N, \theta^t\right)$$

$$= \sum_{i=1}^N \sum_{z_i} \log P\left(x_i, z_i|\theta\right) \cdot P\left(z_i|x_i, \theta^t\right). \tag{2.98}$$

In the GMM, we have

$$P\left(x\right) = \sum_{k=1}^K p_k \cdot N\left(\mu_k, \Sigma_k\right), \qquad (2.99)$$

$$P(x, z) = P(z) P(x|z) = p_z \cdot N(x|\mu_z, \Sigma_z), \tag{2.100}$$

$$P(z|x) = \frac{P(x, z)}{P(x)} = \frac{p_z \cdot N(x|\mu_z, \Sigma_z)}{\sum_{k=1}^{K} p_k \cdot N(x|\mu_k, \Sigma_k)}. \tag{2.101}$$

Therefore, we have

$$Q(\theta, \theta^t) = \sum_{i=1}^{N} \sum_{z_i} \log [p_{z_i} \cdot N(x_i|\mu_{z_i}, \Sigma_{z_i})] \cdot \frac{p_{z_i} \cdot N(x_i|\mu_{z_i}^t, \Sigma_{z_i}^t)}{\sum_{k=1}^{K} p_k^t \cdot N(x_i|\mu_k^t, \Sigma_k^t)} \tag{2.102}$$

2.7.2.2 M-Step

Next, we will introduce the M-step of the EM algorithm. Since the term $\frac{p_{z_i} \cdot N(x_i|\mu_{z_i}^t, \Sigma_{z_i}^t)}{\sum_{k=1}^{K} p_k^t \cdot N(x_i|\mu_k^t, \Sigma_k^t)}$ in Equation (2.102) is a constant, for simplicity, let $P(z_i|x_i, \theta^t)$ denote this term. We have

$$Q(\theta, \theta^t) = \sum_{z_i} \sum_{i=1}^{N} \log [p_{z_i} \cdot N(x_i|\mu_{z_i}, \Sigma_{z_i})] \cdot P(z_i|x_i, \theta^t). \tag{2.103}$$

Expending z_i over $k = 1, ..., K$. We have

$$Q(\theta, \theta^t) = \sum_{k=1}^{K} \sum_{i=1}^{N} \log [p_k \cdot N(x_i|\mu_k, \Sigma_k)] \cdot P(z_k = C_k|x_i, \theta^t). \tag{2.104}$$

Further expanding the log function, we have

$$Q(\theta, \theta^t) = \sum_{k=1}^{K} \sum_{i=1}^{N} [\log p_k + \log N(x_i|\mu_k, \Sigma_k)] \cdot P(z_k = C_k|x_i, \theta^t). \tag{2.105}$$

The optimal parameter θ is given by

$$\theta^{t+1} = \arg \max_{\theta} Q(\theta, \theta^t). \tag{2.106}$$

By solving this optimisation problem, the optimal values of p_k, μ_k, and Σ_k can be obtained.

2.8 VARIATIONAL INFERENCE

Bayesian inference is a statistical inference method including the *exact inference* which can obtain the exact distribution of variables [40] and *approximate inference* which can efficiently obtain the approximated distribution for a complex model through sacrificing the accuracy to some extent [41]. Approximate inference includes the deterministic approximate inference, e.g., variational inference [42], and stochastic approximate inference, e.g., Gibbs sampling [43]. Variational inference is an approach for approximating the intractable integrals which are used in complex statistical models containing observed variables, unknown parameters, and latent variables as described as follows:

$$P(Z|X) = \frac{P(X|Z) \cdot P(Z)}{P(X)}. \tag{2.107}$$

This section will introduce how to obtain the posteriori $P(Z|X)$ by using the variational inference.

Recall that in the Section 2.6, we derived the Equation (2.74) as:

$$\log P(X|\theta) = \text{ELBO} + KL\left[q(Z) \| p(Z|X)\right]. \tag{2.108}$$

Let $L(q)$ denote the ELBO. Given $KL(q \| p) \geq 0$, we need to find a $q(Z)$ to approach $p(Z|X)$, so that $KL\left[q(Z) \| p(Z|X)\right]$ equals to zero.

Given $\log P(X|\theta)$ is fixed, when $L(q)$ approaches to the maximum, $KL[q(Z) \| p(Z|X)] = 0$. Hence, we have

$$q^*(Z) = \arg\max_{q(Z)} L(q) \Rightarrow q^*(Z) \approx P(Z|X). \tag{2.109}$$

Let's disaggregate Z into M independent variables $z_1, z_2, ..., z_M$. We have

$$q(Z) = \prod_{i=1}^{M} q_i(z_i). \tag{2.110}$$

Next, let's fix other variables to calculate one term q_j $(j \in M, j \neq i)$ along. Recall the formulation of ELBO, we have

$$L(q) = \underbrace{\int_Z q(Z)\log P(X,Z)\,\mathrm{d}Z}_{(1)} - \underbrace{\int_Z q(Z)\log q(Z)\,\mathrm{d}Z}_{(2)}, \tag{2.111}$$

in which the terms (1) and (2) can be transformed as follows:

- For the term (1):

$$(1) = \int_{z_1,z_2,...,z_M} \prod_{i=1}^{M} q_i(z_i)\log P(X,Z)\,\mathrm{d}z_1\mathrm{d}z_2...\mathrm{d}z_M$$

$$= \int_{z_j} q_j(z_j)\left[\int_{z_1,...,z_{j-1},z_{j+1},...,z_M} \log P(X,Z)\prod_{i=1,i\neq j}^{M} q_i(z_i)\,\mathrm{d}z_1...\mathrm{d}z_{j-1}\mathrm{d}z_{j+1}...\mathrm{d}z_M\right]\mathrm{d}z_j$$

$$= \int_{z_j} q_j(z_j)\cdot E_{\prod_{i=1,i\neq j}^{M} q_i(z_i)}\left[\log P(X,Z)\right]\mathrm{d}z_j. \tag{2.112}$$

Rewriting (1) in the log form by using $\log \hat{p}(X,z_j)$ to replace $E_{\prod_{i=1,i\neq j}^{M} q_i(z_i)}$ $\left[\log P(X,Z)\right]$ as

$$(1) = \int_{z_j} q_j(z_j)\cdot E_{\prod_{i=1,i\neq j}^{M} q_i(z_i)}\left[\log P(X,Z)\right]\mathrm{d}z_j$$

$$= \int_{z_j} q_j(z_j)\cdot \log \hat{p}(X,z_j)\,\mathrm{d}z_j \tag{2.113}$$

- For the term (2):

$$(2) = \int_Z \prod_{i=1}^{M} q_i(z_i) \cdot \log \prod_{i=1}^{M} q_i(z_i) \, dZ$$

$$= \int_Z \prod_{i=1}^{M} q_i(z_i) \cdot \sum_{i=1}^{M} \log q_i(z_i) \, dZ \qquad (2.114)$$

$$= \int_Z \prod_{i=1}^{M} q_i(z_i) \cdot (\log q_1(z_1) + \log q_2(z_2) + ... + \log q_M(z_M)) \, dZ$$

Taking any one term from Equation (2.114) as an example, e.g., $\int_Z \prod_{i=1}^{M} q_i(z_i) \cdot \log q_1(z_1) \, dZ$, we have

$$\int_Z \prod_{i=1}^{M} q_i(z_i) \cdot \log q_1(z_1) \, dZ$$

$$= \int_Z q_1(z_1) \cdot q_2(z_2) ... q_M(z_M) \cdot \log q_1(z_1) \, dZ$$

$$= \int_{z_1, z_2, ..., z_M} q_1(z_1) \cdot q_2(z_2) ... q_M(z_M) \cdot \log q_1(z_1) \, dz_1 \cdot dz_2 ... dz_M \qquad (2.115)$$

$$= \int_{z_1} q_1(z_1) \cdot \log q_1(z_1) \, dz_1 \int_{z_2} q_2(z_2) \, dz_2 ... \int_{z_M} q_M(z_M) \, dz_M.$$

Given $\int_{z_i} q_i(z_i) \, dz_i = 1$, we have

$$\int_Z \prod_{i=1}^{M} q_i(z_i) \cdot \log q_1(z_1) \, dZ = \int_{z_1} q_1(z_1) \cdot \log q_1(z_1) \, dz_1. \qquad (2.116)$$

Hence,

$$(2) = \sum_{i=1}^{M} \int_{z_i} q_i(z_i) \cdot \log q_i(z_i) \, dz_i. \qquad (2.117)$$

Taking the jth $(j \in M, j \neq i)$ term and denote other terms as a constant C. Hence,

$$(2) = \int_{z_j} q_j(z_j) \cdot \log q_j(z_j) \, dz_j + C. \qquad (2.118)$$

Therefore, we have

$$L(q) = (1) - (2)$$

$$= \int_{z_j} q_j(z_j) \cdot \log \hat{p}(X, z_j) \, dz_j - \int_{z_j} q_j(z_j) \cdot \log q_j(z_j) \, dz_j - C \qquad (2.119)$$

$$= \int_{z_j} q_j(z_j) \cdot \log \frac{\hat{p}(X, z_j)}{q_j(z_j)} \, dz_j - C.$$

By omitting the constant term, we have

$$L(q) = \int_{z_j} q_j(z_j) \cdot \log \frac{\hat{p}(X, z_j)}{q_j(z_j)} \, dz_j$$

$$= KL(q_j(z_j) \parallel \hat{p}(X, z_j)) \qquad (2.120)$$

$$\leq 0.$$

Therefore, when $q_j(z_j) = \hat{p}(X, z_j)$, the equality of Equation (2.120) holds.

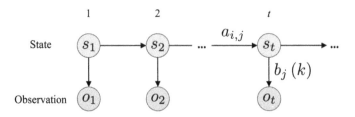

Figure 2.9 Schematic illustration of the hidden Markov model. The model consists of two parts. The first part is the time series state variable, denoted as $S = s_1, s_2, ..., s_t, ...$, with the space of $Q = \{q_1, q_2, ..., q_N\}$. The transition between states is modelled by a transition probability matrix $A = a_{i,j}, a_{i,j} = P(s_{t+1} = q_j|s_t = q_i)$. The second part, which describes the underlying dynamics of systems, is the observation variable, denoted as $O = o_1, o_2, ..., o_t, ...$, with the space of $V = \{v_1, v_2, ..., v_N\}$. The generation from the state to observation is modelled by a emission probability matrix $B = b_j(k), b_j(k) = P(o_t = v_k|s_t = q_j)$.

2.9   HIDDEN MARKOV MODEL

HMM is a statistical Markov model which consists of following two parts of the statistical process [44] as shown in Figure 2.9:

- *State variable:* The first part is the time series state variable, denoted as $S = s_1, s_2, ..., s_t, ...$, with the space of $Q = \{q_1, q_2, ..., q_N\}$. The transition between states is modelled by a transition probability matrix $A = a_{i,j}, a_{i,j} = P(s_{t+1} = q_j|s_t = q_i)$.

- *Observation variable:* The second part which describes the underlying dynamics of systems is the observation variable, denoted as $O = o_1, o_2, ..., o_t, ...$, with the space of $V = \{v_1, v_2, ..., v_N\}$. The generation from the state to observation is modelled by a emission probability matrix $B = b_j(k), b_j(k) = P(o_t = v_k|s_t = q_j)$.

HMM is subject to the following two assumptions:

- *Assumption 1:* The current state is only dependent on the previous one state as

$$P(s_{t+1}|s_t, s_{t-1}, ..., s_1, o_t, o_{t-1}, ..., o_1) = P(s_{t+1}|s_t). \qquad (2.121)$$

- *Assumption 2:* The current observation is only dependent on the current state as

$$P(o_t|s_t, s_{t-1}, ..., s_1) = P(o_t|s_t). \qquad (2.122)$$

This model is applicable for three tasks: evaluation, learning, and decoding, which will be detailed in the following subsections.

2.9.1 Evaluation

The evaluation task aims to compute the probability of observation by given model parameters, i.e., $P(O|\lambda)$, where λ is the model parameters. Let's take the state variable into the consideration by using the sum of the state variable, and then use the conditional probability formula as:

$$
\begin{aligned}
P(O|\lambda) &= \sum_S P(O, S|\lambda) \\
&= \sum_S P(O|S, \lambda) \cdot P(S|\lambda).
\end{aligned}
\tag{2.123}
$$

Therefore, the calculation of $P(O|\lambda)$ can be converted to the calculation of $P(O|S, \lambda)$ and $P(S|\lambda)$ as follows:

- *Calculating $P(S|\lambda)$:*

  The expansion of $P(S|\lambda)$ is

$$
\begin{aligned}
P(S|\lambda) &= P(s_1, s_2, ..., s_T|\lambda) \\
&= P(s_T|s_1, s_2, ..., s_{T-1}, \lambda) \cdot P(s_1, s_2, ..., s_{T-1}|\lambda).
\end{aligned}
\tag{2.124}
$$

  Based on the *Assumption 1*, we have

$$
P(s_T|s_1, s_2, ..., s_{T-1}, \lambda) = P(s_T|s_{T-1}, \lambda).
\tag{2.125}
$$

  Analogously, we have

$$
\begin{aligned}
P(s_1, s_2, ..., s_{T-1}|\lambda) &= P(s_{T-1}|s_1, s_2, ..., s_{T-2}, \lambda) \cdot P(s_1, s_2, ..., s_{T-2}|\lambda) \\
&= P(s_{T-1}|s_{T-2}, \lambda) \cdot P(s_1, s_2, ..., s_{T-2}|\lambda).
\end{aligned}
\tag{2.126}
$$

  Through recursive iterations, we have

$$
\begin{aligned}
P(S|\lambda) &= P(s_T|s_{T-1}, \lambda) \cdot P(s_{T-1}|s_{T-2}, \lambda) ...P(s_2|s_1, \lambda) \cdot P(s_1, \lambda) \\
&= \pi(a_{s_1}) \cdot \prod_{t=2}^{T} a_{s_{t-1}, s_t},
\end{aligned}
\tag{2.127}
$$

  where $\pi(a_{s_1})$ is the probability distribution of the initial state s_1.

- *Calculating $P(O|S, \lambda)$:*

  The expansion of $P(O|S, \lambda)$ is

$$
\begin{aligned}
P(O|S, \lambda) =& P(o_T, o_{T-1}, ..., o_1|s_T, s_{T-1}, ..., s_1, \lambda) \\
=& P(o_T|o_{T-1}, ..., o_1, s_T, s_{T-1}, ..., s_1, \lambda) \cdot \\
& P(o_{T-1}, ..., o_1|s_T, s_{T-1}, ..., s_1, \lambda).
\end{aligned}
\tag{2.128}
$$

  Based on the *Assumption 2*, we have

$$
P(o_T|o_{T-1}, ..., o_1, s_T, s_{T-1}, ..., s_1, \lambda) = P(o_T|s_T, \lambda).
\tag{2.129}
$$

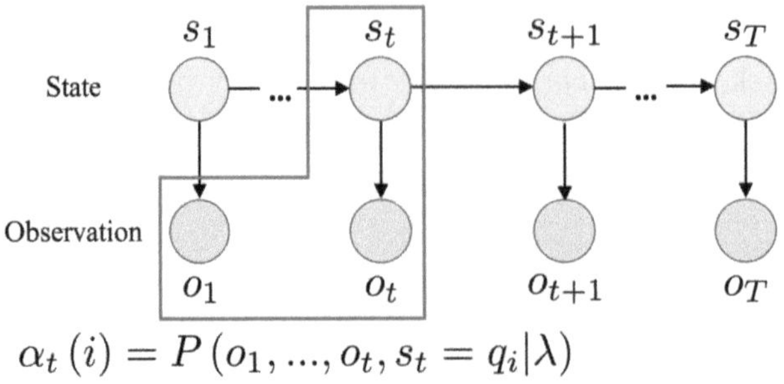

$$\alpha_t\left(i\right) = P\left(o_1, ..., o_t, s_t = q_i | \lambda\right)$$

Figure 2.10 Schematic illustration of using the forward algorithm to solve the evaluation problem of the hidden Markov model. The part in the red box contains the observations in the first t steps and the state in the t-th step, denoted as $\alpha_t\left(i\right)$.

Through recursive iterations, we have

$$P\left(O|S, \lambda\right) = P\left(o_T | s_T, \lambda\right) \cdot P\left(o_{T-1} | s_{T-1}, \lambda\right) ... P\left(o_1 | s_1, \lambda\right)$$
$$= \prod_{t=1}^{T} b_{s_t}\left(o_t\right). \tag{2.130}$$

Therefore, by calculating $P\left(S|\lambda\right)$ and $P\left(O|S, \lambda\right)$, $P\left(O|\lambda\right)$ can be calculated as

$$P\left(O|\lambda\right) = \sum_{S}\left[\pi\left(a_{s_1}\right) \cdot \prod_{t=2}^{T} a_{s_{t-1}, s_t} \cdot \prod_{t=1}^{T} b_{s_t}\left(o_t\right)\right]. \tag{2.131}$$

In Equation (2.131), the sum of S can be expanded as

$$P\left(O|\lambda\right) = \sum_{s_1}\sum_{s_2}\cdots\sum_{s_T}\left[\pi\left(a_{s_1}\right) \cdot \prod_{t=2}^{T} a_{s_{t-1}, s_t} \cdot \prod_{t=1}^{T} b_{s_t}\left(o_t\right)\right], \tag{2.132}$$

which means that there are T-dimensional s and each s has N possible values. Hence, the computational complexity of calculating Equation (2.131) is $O\left(N^T\right)$. The forward algorithm [45] is an efficient algorithm to solve this problem by reducing the computational complexity from $O\left(N^T\right)$ to $O\left(T \cdot N^2\right)$, which is introduced in subsequent text.

As presented in Figure 2.10, let $\alpha_t\left(i\right)$ denote the part in the red box containing the observations in the first t steps and the state in the t-th step. We have

$$\alpha_t\left(i\right) = P\left(o_1, ..., o_t, s_t = q_i | \lambda\right), \tag{2.133}$$

$$\alpha_T\left(i\right) = P\left(O, s_T = q_i | \lambda\right). \tag{2.134}$$

Hence,

$$P\left(O|\lambda\right) = \sum_{s=1}^{N} P\left(O, s_T = q_i | \lambda\right)$$
$$= \sum_{s=1}^{N} \alpha_T\left(i\right). \tag{2.135}$$

For the next time step, we have

$$
\begin{aligned}
\alpha_{t+1}\left(j\right) =& P\left(o_1, ..., o_{t+1}, s_{t+1} = q_j | \lambda\right) \\
=& \sum_{s=1}^{N} P\left(o_1, ..., o_t, o_{t+1}, s_{t+1} = q_j, s_t = q_i | \lambda\right) \\
=& \sum_{s=1}^{N} P\left(o_{t+1} | o_1, ..., o_t, s_t = q_i, s_{t+1} = q_j, \lambda\right) \cdot P\left(o_1, ..., o_t, s_t = q_i, s_{t+1} = q_j | \lambda\right) \\
=& \sum_{s=1}^{N} P\left(o_{t+1} | o_1, ..., o_t, s_t = q_i, s_{t+1} = q_j, \lambda\right) \cdot P\left(s_{t+1} = q_j | o_1, ..., o_t, s_t = q_i, \lambda\right) \\
& \cdot P\left(o_1, ..., o_t, s_t = q_i | \lambda\right) \\
=& \sum_{s=1}^{N} P\left(o_{t+1} | s_{t+1} = q_j, \lambda\right) \cdot P\left(s_{t+1} = q_j | s_t = q_i, \lambda\right) \cdot \alpha_t\left(i\right),
\end{aligned}
$$

$$(2.136)$$

in which the last step uses *Assumption 1* and *Assumption 2* to simplify the equation. This results in a recursive relationship between $\alpha_{t+1}\left(j\right)$ and $\alpha_t\left(i\right)$. Using the transition probability matrix and emission probability matrix to replace the $P\left(o_{t+1} | s_{t+1} = q_j, \lambda\right)$ and $P\left(s_{t+1} = q_j | s_t = q_i, \lambda\right)$ in Equation (2.136), respectively, we have

$$
\alpha_{t+1}\left(j\right) = \sum_{s=1}^{N} b_j\left(O_{t+1}\right) \cdot a_{i,j} \cdot \alpha_t\left(i\right),
\tag{2.137}
$$

through which the probability $P\left(O | \lambda\right)$ can be obtained.

2.9.2 Learning

The learning problem of the HMM can de described as an MLE problem as

$$
\lambda^\star = \arg\max_{\lambda} P\left(O | \lambda\right),
\tag{2.138}
$$

where the parameter $\lambda = \left(\pi, A, B\right)$.

Recall that in Section 2.6 the EM algorithm is an efficient algorithm to solve the MLE problem. On the context of the HMM, the EM algorithm can be described as

$$
\lambda^{t+1} = \arg\max_{\lambda} \sum_{S} \log P\left(O, S | \lambda\right) \cdot P\left(S | O, \lambda^t\right).
\tag{2.139}
$$

We have

$$
P\left(S | O, \lambda^t\right) = \frac{P\left(S, O | \lambda^t\right)}{P\left(O | \lambda^t\right)}.
\tag{2.140}
$$

Since λ^t is a known input from the last time step and the observation O is given, the term $P\left(O | \lambda^t\right)$ is a constant which is independent of the maximisation problem. Equation (2.139) can be further simplified as

$$
\lambda^{t+1} = \arg\max_{\lambda} \sum_{S} \log P\left(O, S | \lambda\right) \cdot P\left(O, S | \lambda^t\right).
\tag{2.141}
$$

Let $Q(\lambda, \lambda^t)$ denote $\sum_S \log P(O, S|\lambda) \cdot P(O, S|\lambda^t)$.

Recall in Equation (2.132), we have

$$P(O|\lambda) = \sum_S P(O, S|\lambda)$$

$$= \sum_{s_1} \sum_{s_2} \cdots \sum_{s_T} \left[\pi(a_{s_1}) \cdot \prod_{t=2}^T a_{s_{t-1}, s_t} \cdot \prod_{t=1}^T b_{s_t}(o_t) \right]. \tag{2.142}$$

Hence,

$$P(O, S|\lambda) = \pi(a_{s_1}) \cdot \prod_{t=2}^T a_{s_{t-1}, s_t} \cdot \prod_{t=1}^T b_{s_t}(o_t). \tag{2.143}$$

Taking Equation (2.143) to $Q(\lambda, \lambda^t)$ and using π_{s_1} to denote $\pi(a_{s_1})$, we have

$$Q(\lambda, \lambda^t) = \sum_S \left[\left(\log \pi_{s_1} + \sum_{t=2}^T \log a_{s_{t-1}, s_t} + \sum_{t=1}^T \log b_{s_t}(o_t) \right) \cdot P(O, S|\lambda^t) \right]. \tag{2.144}$$

We take solving the parameter π as an example to demonstrate the solutions. Since we aim to solve the parameter π, the terms $\sum_{t=2}^T \log a_{s_{t-1}, s_t}$ and $\sum_{t=1}^T \log b_{s_t}(o_t)$ can be omitted. We have

$$\pi^{t+1} = \arg \max_\pi Q(\lambda, \lambda^t)$$

$$= \arg \max_\pi \sum_S \left[\log \pi_{s_1} \cdot P(O, S|\lambda^t) \right]$$

$$= \arg \max_\pi \sum_{s_1} \cdots \sum_{s_T} \left[\log \pi_{s_1} \cdot P(O, s_1, ..., s_T|\lambda^t) \right] \tag{2.145}$$

$$= \arg \max_\pi \sum_{s_1} \left[\log \pi_{s_1} \cdot P(O, s_1|\lambda^t) \right].$$

Given s_1 has N possible values $\{q_1, q_2, ..., q_N\}$ and π_{s_1} is the probability when $s_1 = q_i$, use s to replace s_1. We have

$$\arg \max_\pi \sum_{s_1} \left[\log \pi_{s_1} \cdot P(O, s_1|\lambda^t) \right]$$

$$= \arg \max_\pi \sum_{s=1}^N \left[\log \pi_s \cdot P(O, s_1 = q_i|\lambda^t) \right], \tag{2.146}$$

s.t.

$$\sum_{s=1}^N \pi_s = 1. \tag{2.147}$$

This optimisation problem can be solved by the Lagrange multiplier as

$$L(\pi, \eta) = \sum_{s=1}^N \left[\log \pi_s \cdot P(O, s_1 = q_i|\lambda^t) \right] + \eta \cdot \left(\sum_{s=1}^N \pi_s - 1 \right). \tag{2.148}$$

$$\frac{\partial L}{\partial \pi_s} = \frac{1}{\pi_s} P(O, s_1 = q_i|\lambda^t) + \eta = 0. \tag{2.149}$$

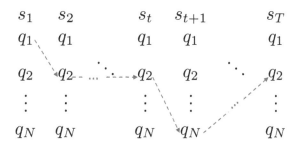

Figure 2.11 Schematic illustration of solving the decoding problem of the hidden Markov model. The aim of the decoding problem is to find a sequence $\{s_1, s_2, ..., s_T\}$ with the highest probability as indicated by the dashed red arrow. In each time step t, there are N potential values.

Hence,

$$P\left(O, s_1 = q_i | \lambda^t\right) + \eta \cdot \pi_s = 0. \tag{2.150}$$

$$\sum_{s=1}^{N} \left[P\left(O, s_1 = q_i | \lambda^t\right) + \eta \cdot \pi_s\right] = 0. \tag{2.151}$$

$$P\left(O|\lambda^t\right) + \eta = 0. \tag{2.152}$$

$$\eta = -P\left(O|\lambda^t\right) \tag{2.153}$$

Taking η to Equation (2.150), we have

$$\pi_s^{t+1} = \frac{P\left(O, s_1 = q_i | \lambda^t\right)}{P\left(O|\lambda^t\right)}. \tag{2.154}$$

This is the process to solve the learning problem of the HMM. The solution algorithm is defined as Baum–Welch algorithm [46].

2.9.3 Decoding

The decoding problem of the HMM can be described as

$$S^\star = \arg\max_S P\left(S|O, \lambda\right). \tag{2.155}$$

As shown in Figure 2.11, the aim of the decoding problem is to find a sequence $\{s_1, s_2, ..., s_T\}$ with the highest probability as indicated by the dashed red arrow. In each time step t, there are N potential values. Hence, the total possible combination of the sequence is N^T.

Let $\delta_t(i)$ denote the maximum probability of choosing the value q_i at the time step t as

$$\delta_t(i) = \max_{s_1, s_2, ..., s_{t-1}} P\left(o_1, o_2, ..., o_t, s_1, ..., s_{t-1}, s_t = q_i\right). \tag{2.156}$$

It is noted that Equation (2.156) only fixes the value s_t at the time step t, before which the state is random being subject to the sequence with the highest probability.

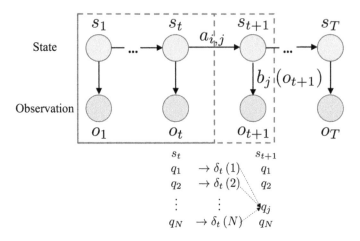

Figure 2.12 Schematic illustration of the recursive relationship between δ_{t+1} and δ_t.

To solve the decoding problem, we need to find the recursive relationship between $\delta_{t+1}(j)$ and $\delta_t(i)$, as indicated in Figure 2.12 from the dashed red box to dashed blue box. δ_{t+1} can be expressed as:

$$\delta_{t+1}(j) = \max_{s_1, s_2, \ldots, s_t} P(o_1, o_2, \ldots, o_t, o_{t+1}, s_1, \ldots, s_t, s_{t+1} = q_j). \tag{2.157}$$

This recursive relationship can be expressed as:

$$\delta_{t+1}(j) = \max_{1 \leq i \leq N} \delta_t(i) \cdot a_{i,j} \cdot b_j(o_{t+1}). \tag{2.158}$$

To record the path of maximum probability from the step 1 to the step T, let $\varphi_{t+1}(j)$ denote the selected index i moving to $\delta_{t+1}(j)$. We have

$$\varphi_{t+1}(j) = \arg \max_{1 \leq i \leq N} \delta_t(i) \cdot a_{i,j} \cdot b_j(o_{t+1}). \tag{2.159}$$

Therefore, the optimal solution of the decoding problem is $\{\varphi_1, \varphi_2, \ldots, \varphi_T\}$. This solving algorithm is defined as the Viterbi algorithm [47].

2.10 FEEDFORWARD NEURAL NETWORKS

As one of typical machine learning models, neural networks are inspired by human brains [48]. As presented in Figure 2.13, the neural networks take the dataset as inputs and process the dataset by extracting key features through deep hidden layers of neurons. The extracted features are mapped to the outputs. The aim of training neural networks is to make the predicted outputs to be close to training labels, also defined as the ground truth. During the training process, the weights and bias of these neurons are iteratively updated to accurately predict desired outputs through the back-propagation [13].

Neural networks are in particular suitable for mapping the non-linear relationship between the inputs and outputs. The activation functions are key to the non-linearity.

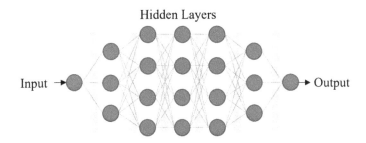

Figure 2.13 Architecture of neural networks. The neural networks take the dataset as inputs and process the dataset by extracting key features through deep hidden layers of neurons. The extracted features are mapped to the outputs.

Common non-linear activation functions include the Sigmoid [49], Tanh [50], ReLU [13], and Softmax [13]. These are expressed as follows:

$$\text{Sigmoid: } f(x) = \frac{1}{1 + e^{-x}}, \tag{2.160}$$

$$\text{Tanh: } f(x) = \frac{e^x - e^{-x}}{e^x + e^{-x}}, \tag{2.161}$$

$$\text{ReLU: } f(x) = \max(0, x), \tag{2.162}$$

$$\text{Softmax: } f(x_i) = \frac{\exp(x_i)}{\sum_i \exp(x_i)}, \tag{2.163}$$

where x is the input sample, i is the index of the sample, and $f(x)$ is the returned output.

Typical neural networks include the recurrent neural networks [13], long short-term memory of recurrent neural networks [13], and convolutional neural networks [13]. These are detailed in the following subsections.

2.10.1 Recurrent Neural Network

For complicated machine learning problems, it is useful to look at the data and find certain patterns, e.g., data organised in a sequencial order or regular circle. This means that the input data is in the form of vectors which are the computers' native language. For instance, in power systems, the daily load profile in every hour can be organised as a vector with 24 elements. With one vector as an input, we can predict the daily power profile for tomorrow which is also a 24-element vector. Next, if there is a case in which we have the data of daily load profile for Monday, but do not have the data of daily load profile for Tuesday, how would we predict the daily load profile for Wednesday? The solution is that we could use the data for Monday to predict the daily load profile for Tuesday, and then use the prediction of the daily load profile on Tuesday as an input. The neural networks would return the prediction of the daily load profile on Wednesday as the output. This is a simple example to show the mechanism behind the recurrent neural networks as shown in Figure 2.14.

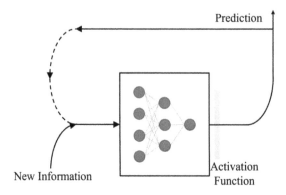

Figure 2.14 Mechanism of the recurrent neural networks. The output prediction is fed back as a new input of neural networks to predict outputs based on the data with recurring patterns.

2.10.2 Long Short-Term Memory

One drawback of the recurrent neural networks is that each prediction only looks back one time step, which is a short-term memory. However, the information from further back would be equally valuable for the accuracy of predictions. To overcome this drawback, we expand the recurrent neural networks by adding the memory to remember the extracted features of many back steps. The mechanism of the long short-term memory of the recurrent neural network is presented in Figure 2.15. The plus junction indicates the sum of element-by-element of two vectors with the same shape. Analogously, the times junction indicates the multiplication of element-by-element of two vectors with the same shape.

- *Prediction:* The new information is passed through the neural networks to get predictions.

- *Forgetting:* The predictions are further passed through the plus junction, and in the meantime, the copy of the predictions is held on for the next time step when the next predictions are yielded by neural networks. Some of predictions are forgotten while other predictions are remembered which are added back into the next predictions by the plus junction. There are separate neural networks which take the sum of prediction and memories as the input, and are trained to learn when to forget what.

- *Selection:* When combining the prediction with memories, the memories are not necessarily released as new predictions. Hence, the selection gate acts as a filter to achieve this function through another separate neural networks.

- *Ignoring:* The ignoring gate ignores possible predictions, which sets aside the elements which are not of immediate relevance.

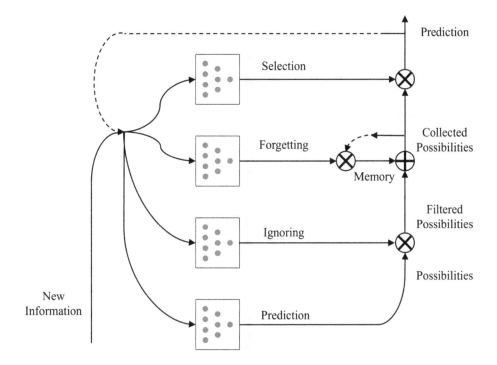

Figure 2.15 Mechanism of the long short-term memory of recurrent neural networks. The new information is passed through the neural networks to get predictions. The predictions are further passed through the plus junction, and in the meantime, the copy of the predictions is held on for the next time step when the next predictions are yielded by neural networks. Some of predictions are forgotten while other predictions are remembered which are added back into the next predictions by the plus junction. There are separate neural networks which take the sum of prediction and memories as the input, and are trained to learn when to forget what. When combining the prediction with memories, the memories are not necessarily released as new predictions. Hence, the selection gate acts as a filter to achieve this function through another separate neural networks. The ignoring gate ignores possible predictions, which sets aside the elements which are not of immediate relevance.

2.10.3 Convolutional Neural Network

The high-dimensional inputs would incur high computational burdens for conventional neural networks. To overcome this issue, the convolutional neural networks can improve the computational efficiency and extract key features from high-dimensional inputs. General structure of the convolutional neural networks is presented in Figure 2.16. In convolutional layers, multiple filters slide through the input to capture key feature representations as

$$\Phi = f^{\mathrm{a}}\left(W \otimes x + b\right), \tag{2.164}$$

where Φ is the output feature representation, $f^{\mathrm{a}}\left(\cdot\right)$ is the activation function, W is the weight of a filter, and b is the bias of a filter.

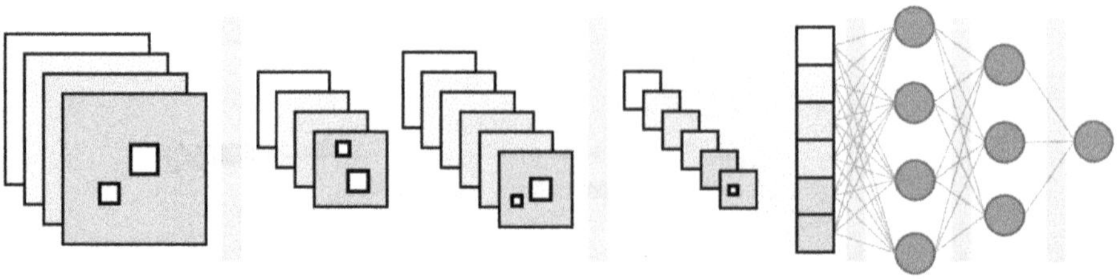

Convolution 1   Pool 1   Convolution 2   Pool 2   Flatten   Fully-Connected Outputs

Figure 2.16 Architecture of convolutional neural networks. In convolutional layers, multiple filters slide through the input to capture key feature representations. The pooling layer follows the convolutional layer to reduce the size of extracted features and keep key features. All extracted features from multiple filters are stacked to form a global feature map. The global feature map is subsequently flattened and processed by fully connected layers to return to the outputs.

The pooling layer follows the convolutional layer to reduce the size of extracted features and keep key features. All extracted features from multiple filters are stacked to form a global feature map. The global feature map is subsequently flattened and processed by fully connected layers to return to the outputs.

2.11 REINFORCEMENT LEARNING

Under the mode of reinforcement learning, the agent seeks to take actions in an environment for maximising the cumulative reward [51] as shown in Figure 2.17. This subsection introduces the reinforcement learning and relevant key concepts.

To help readers understand the reinforcement learning, key terminologies are introduced as follows:

- *State, action, policy function, and reward:* Given an observed state, an agent takes an action by the policy function as

$$\pi(s, a) = P(A = a | S = s), \tag{2.165}$$

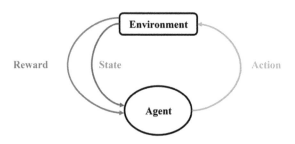

Figure 2.17 Flowchart of reinforcement learning. Under the mode of reinforcement learning, the agent seeks to take actions in an environment for maximising the cumulative reward.

where $\pi\left(\cdot\right)$ is the policy function, $\pi \in [0,1]$, A and S are random variables representing the action and state, respectively, and a and s are specific values taken by the random variables A and S, respectively. Once an agent takes an action, the environment will return a reward, denoted as R.

At the current state s, once an agent takes an action a, the environment will give a new state s', which is defined as the state transition as

$$p\left(s'|s,a\right) = P\left(S' = s'|S = s, A = a\right), \qquad (2.166)$$

where $p\left(\bullet\right)$ is the state transition function.

The state, action, and reward form a trajectory as $s_1, a_1, r_1, s_2, a_2, r_2, ..., s_T, a_T, r_T$.

- *Reward and return:* Return (also defined as the cumulative future reward) accounts for cumulative rewards in future. Considering that the future reward is less valuable than the current reward, the future reward should be discounted as

$$U_t = \gamma \cdot R_t + \gamma^2 \cdot R_{t+1} + ..., \qquad (2.167)$$

where U_t is the return at the time step t, and $\gamma \in [0,1]$ is the discounted rate. At the time step t, all the future rewards are random and therefore the return U_t is random. There are two sources of randomness in the return:

1) Given the state s, the action has the randomness which comes from the policy function, i.e., $\pi\left(a|s\right) = P\left(A = a|S = s\right)$;

2) Given the state s and action a, the new state has the randomness which comes from the state transition function, i.e., $p\left(s'|s,a\right) = P\left(S' = s'|S = s, A = a\right)$. This is because for any time step i in future, $i \geq t$, the reward R_i depends on S_i and A_i.

Therefore, the return U_t depends on all future actions from time t, i.e., $A_t, A_{t+1}, A_{t+2}, ...$ and all future states from time t, i.e., $S_t, S_{t+1}, S_{t+2}, ...$

- *Action-value function:* Since U_t is a random variable, the future situation at time step t is not known. To evaluate the future situations, we can calculate the expectation of U_t, defined as the action-value function as

$$Q_\pi\left(s_t, a_t\right) = E\left[U_t|S_t = s_t, A_t = a_t\right], \qquad (2.168)$$

where $Q_\pi\left(\cdot\right)$ is the action-value function, which depends on the current state s_t, action a_t, and the policy function π. The action-value function indicates that using the policy π, how good it is to take action a_t given the current state s_t.

Given various policy functions, the policy function that can yield the maximum value of the action-value function is the best policy function. The corresponding maximum value of the action-value function is defined as the optimal action-value function as

$$Q^\star (s_t, a_t) = \max_\pi Q_\pi (s_t, a_t) . \tag{2.169}$$

Whatever policy function π is used, the result of taking a_t at state s_t cannot be better than $Q^\star (s_t, a_t)$.

- *State-value function:* State-value function is the expectation of the action-value function with respect to action A as

$$V_\pi (s_t) = E_A [Q_\pi (s_t, A)],$$
$$= \begin{cases} \sum_a \pi (a|s_t) \cdot Q_\pi (s_t, a), & \text{if action is discrete,} \\ \int_a \pi (a|s_t) \cdot Q_\pi (s_t, a) \, \mathrm{d}a, & \text{if action is contineous,} \end{cases} \tag{2.170}$$

where $V_\pi (\cdot)$ is the state-value function. On the one hand, the state-value function indicates how good the current situation is. On the other hand, the average value of the state-value function, i.e., $E_S [V_\pi (S)]$ evaluates how good the policy π is.

Based on these terminologies, an agent has two options to take an action:

1) If the agent has a good policy $\pi (a|s)$, given an observed state s_t, an action a_t can be taken by randomly sampling from the policy function;

2) If the agent has an optimal action-value function $Q^\star (s_t, a_t)$, given an observed state s_t, an action a_t can be taken by maximising the value of the action-value function.

2.12 CHAPTER SUMMARY

This chapter introduced the fundamental theories of AI as the preliminary for applying the AI into smart energy systems. The most common AI approaches have been described, based on the background and development of the current AI technologies and interactions with energy systems. These approaches include the optimisation, game theory, SVM, PCA for dimensionality reduction, EM for MLE, GMM, variational inference, HMM, neural networks, and reinforcement learning. In the next chapter, the fundamental theories and key functions of the Blockchain technologies will be introduced.

Bibliography

[1] P. H. Winston, *Artificial Intelligence.* Addison-Wesley Longman Publishing Co., Inc., 1992.

[2] A. Burkov, *The Hundred-Page Machine Learning Book.* Andriy Burkov Quebec City, QC, Canada, 2019, vol. 1.

[3] S. Gollapudi, *Practical Machine Learning.* Packt Publishing Ltd, 2016.

[4] J. McCarthy, "What is artificial intelligence?" 2007.

[5] H. Wei, Y. Zhang, Y. Wang, W. Hua, R. Jing, and Y. Zhou, "Planning integrated energy systems coupling v2g as a flexible storage," *Energy*, vol. 239, p. 122215, 2022.

[6] W. Hua, D. Li, H. Sun, and P. Matthews, "Unit commitment in achieving low carbon smart grid environment with virtual power plant," in *2017 International Smart Cities Conference (ISC2)*. IEEE, 2017, pp. 1–6.

[7] W. Hua, H. Sun, H. Xiao, and W. Pei, "Stackelberg game-theoretic strategies for virtual power plant and associated market scheduling under smart grid communication environment," in *2018 IEEE International Conference on Communications, Control, and Computing Technologies for Smart Grids (SmartGrid-Comm)*, 2018, pp. 1–6.

[8] G. B. Dantzig, *Linear Programming and Extensions*. Princeton university press, 1998, vol. 48.

[9] M. S. Bazaraa, H. D. Sherali, and C. M. Shetty, *Nonlinear Programming: Theory and Algorithms*. John Wiley & Sons, 2013.

[10] A. Schrijver, *Theory of Linear and Integer Programming*. John Wiley & Sons, 1998.

[11] H. P. Williams, *Logic and Integer Programming*. Springer, 2009.

[12] Y. Shi et al., "Particle swarm optimization: Developments, applications and resources," in *Proceedings of the 2001 Congress on Evolutionary Computation (IEEE Cat. No. 01TH8546)*, vol. 1. IEEE, 2001, pp. 81–86.

[13] K. Deb, A. Pratap, S. Agarwal, and T. Meyarivan, "A fast and elitist multiobjective genetic algorithm: Nsga-ii," *IEEE Transactions on Evolutionary Computation*, vol. 6, no. 2, pp. 182–197, 2002.

[14] D. Dasgupta, *Artificial Immune Systems and their Applications*. Springer Science & Business Media, 2012.

[15] R. Branzei, D. Dimitrov, and S. Tijs, *Models in Cooperative Game Theory*. Springer Science & Business Media, 2008, vol. 556.

[16] Nash Jr, John. "Non-cooperative games." Essays on Game Theory. Edward Elgar Publishing, 1996. 22-33.

[17] H. R. Varian, *Intermediate Microeconomics: A Modern Approach: Ninth International Student Edition*. WW Norton & Company, 2014.

[18] H. Von Stackelberg, *Market Structure and Equilibrium*. Springer Science & Business Media, 2010.

[19] W. Hua, D. Li, H. Sun, and P. Matthews, "Stackelberg game-theoretic model for low carbon energy market scheduling." *IET Smart Grid*, vol. 3, no. 1, pp. 31–41, 2020.

[20] W. Hua, J. Jiang, H. Sun, F. Teng, and G. Strbac, "Consumer-centric decarbonization framework using stackelberg game and blockchain," *Applied Energy*, vol. 309, p. 118384, 2022.

[21] S. Balakrishnama and A. Ganapathiraju, "Linear discriminant analysis: A brief tutorial," *Institute for Signal and Information Processing*, vol. 18, no. 1998, pp. 1–8, 1998.

[22] S. I. Gallant *et al.*, "Perceptron-based learning algorithms," *IEEE Transactions on Neural Networks*, vol. 1, no. 2, pp. 179–191, 1990.

[23] M. C. Wittrock, "A generative model of mathematics learning." *Journal for Research in Mathematics Education* (1974): 181–196.

[24] T. Hastie and R. Tibshirani, "Discriminant analysis by gaussian mixtures," *Journal of the Royal Statistical Society: Series B (Methodological)*, vol. 58, no. 1, pp. 155–176, 1996.

[25] G. Bhat, M. Danelljan, L. V. Gool, and R. Timofte, "Learning discriminative model prediction for tracking," in *Proceedings of the IEEE/CVF International Conference on Computer Vision*, 2019, pp. 6182–6191.

[26] R. E. Wright, "Logistic regression." 1995.

[27] D. M. Hawkins, "The problem of overfitting," *Journal of Chemical Information and Computer Sciences*, vol. 44, no. 1, pp. 1–12, 2004.

[28] M. Köppen, "The curse of dimensionality," in *5th Online World Conference on Soft Computing in Industrial Applications (WSC5)*, vol. 1, 2000, pp. 4–8.

[29] H. Abdi and L. J. Williams, "Principal component analysis," *Wiley Interdisciplinary Reviews: Computational Statistics*, vol. 2, no. 4, pp. 433–459, 2010.

[30] J. B. Kruskal, *Multidimensional Scaling*. Sage, 1978, no. 11.

[31] M. Balasubramanian and E. L. Schwartz, "The isomap algorithm and topological stability," *Science*, vol. 295, no. 5552, pp. 7–7, 2002.

[32] S. T. Roweis and L. K. Saul, "Nonlinear dimensionality reduction by locally linear embedding," *Science*, vol. 290, no. 5500, pp. 2323–2326, 2000.

[33] S. Wold, K. Esbensen, and P. Geladi, "Principal component analysis," *Chemometrics and Intelligent Laboratory Systems*, vol. 2, no. 1-3, pp. 37–52, 1987.

[34] L. I. Smith, "A tutorial on principal components analysis," 2002.

[35] I. J. Myung, "Tutorial on maximum likelihood estimation," *Journal of Mathematical Psychology*, vol. 47, no. 1, pp. 90–100, 2003.

[36] B. North and A. Blake, "Learning dynamical models using expectation-maximisation," in *Sixth International Conference on Computer Vision (IEEE Cat. No. 98CH36271)*. IEEE, 1998, pp. 384–389.

[37] T. Van Erven and P. Harremos, "Rényi divergence and Kullback-Leibler divergence," *IEEE Transactions on Information Theory*, vol. 60, no. 7, pp. 3797–3820, 2014.

[38] E. J. McShane, "Jensen's inequality," *Bulletin of the American Mathematical Society*, vol. 43, no. 8, pp. 521–527, 1937.

[39] D. A. Reynolds, "Gaussian mixture models." *Encyclopedia of Biometrics*, vol. 741, no. 659-663, 2009.

[40] L. Munoz-Gonzalez, D. Sgandurra, M. Barrere, and E. C. Lupu, "Exact inference techniques for the analysis of bayesian attack graphs," *IEEE Transactions on Dependable and Secure Computing*, vol. 16, no. 2, pp. 231–244, 2017.

[41] T. Heskes and O. Zoeter, "Expectation propogation for approximate inference in dynamic bayesian networks," *arXiv preprint arXiv:1301.0572*, 2012.

[42] D. M. Blei, A. Kucukelbir, and J. D. McAuliffe, "Variational inference: A review for statisticians," *Journal of the American Statistical Association*, vol. 112, no. 518, pp. 859–877, 2017.

[43] E. I. George and R. E. McCulloch, "Variable selection via gibbs sampling," *Journal of the American Statistical Association*, vol. 88, no. 423, pp. 881–889, 1993.

[44] S. Fine, Y. Singer, and N. Tishby, "The hierarchical hidden markov model: Analysis and applications," *Machine Learning*, vol. 32, no. 1, pp. 41–62, 1998.

[45] S.-Z. Yu and H. Kobayashi, "An efficient forward-backward algorithm for an explicit-duration hidden markov model," *IEEE Signal Processing Letters*, vol. 10, no. 1, pp. 11–14, 2003.

[46] S. Tu, "Derivation of baum-welch algorithm for hidden markov models," *URL: https://people. eecs. berkeley. edu/Ÿ stephentu/writeups/hmm-baum-welch-derivation. pdf*, 2015.

[47] H.-L. Lou, "Implementing the viterbi algorithm," *IEEE Signal Processing Magazine*, vol. 12, no. 5, pp. 42–52, 1995.

[48] S.-C. Wang, "Artificial neural network," in *Interdisciplinary Computing in Java Programming*. Springer, 2003, pp. 81–100.

[49] A. C. Marreiros, J. Daunizeau, S. J. Kiebel, and K. J. Friston, "Population dynamics: Variance and the sigmoid activation function," *Neuroimage*, vol. 42, no. 1, pp. 147–157, 2008.

[50] B. Karlik and A. V. Olgac, "Performance analysis of various activation functions in generalized mlp architectures of neural networks," *International Journal of Artificial Intelligence and Expert Systems*, vol. 1, no. 4, pp. 111–122, 2011.

[51] L. P. Kaelbling, M. L. Littman, and A. W. Moore, "Reinforcement learning: A survey," *Journal of Artificial Intelligence Research*, vol. 4, pp. 237–285, 1996.

CHAPTER 3

Theories of Blockchain Technologies

This chapter introduces fundamental theories of the Blockchain technologies, with the focus on the cryptocurrency and smart contracts. Key functions of Blockchain technologies related to the energy field are highlighted. Section 3.1 introduces an overview of Blockchain technology and its key properties. The Blockchain-based cryptocurrency is introduced in Section 3.2, including the cryptography theory supporting the operation of Blockchain networks in Sub-Section 3.2.1, structures of individual blocks and Blockchain networks in Sub-Section 3.2.2, consensus of recording transactions in Blockchain networks in Sub-Section 3.2.3, information contained in individual block in Sub-Section 3.2.4, difficulty for maintaining block time in Sub-Section 3.2.5, node types of Blockchain networks in Sub-Section 3.2.6, and communications of Blockchain networks in Sub-Section 3.2.7. Blockchain-based smart contracts are introduced in Section 3.3, including account types of Ethereum Blockchain networks in Sub-Section 3.3.1, data structure of individual block of Ethereum Blockchain networks in Sub-Section 3.3.2, and programmes of smart contracts in Sub-Section 3.3.3. Section 3.4 concludes this chapter.

3.1 AN OVERVIEW OF BLOCKCHAIN

A Blockchain is a distributed ledger system for storing information that is managed by a decentralised community in a peer-to-peer manner through proper incentivisation [1, 2]. As such, the terms Blockchain and distributed ledger can be used interchangeably, although the former primarily describes how different records (i.e., blocks) are put together to form a chain, whereas the latter mainly focuses on how data is replicated, shared, and synchronised across multiple nodes, thus making recorded data immutable. Typically, Blockchain has the following properties [3]:

- *Duplicability:* Recorded data of Blockchain is shared across a peer-to-peer network (public or private) where each network node has a copy of the recorded data.

- *Consensus:* A decentralised consensus is achieved when creating a new block in the Blockchain.

- *Immutability:* Without the approval of the majority of network nodes, it is impossible to alter any data in the block.

- *Verifiability:* Recorded data of a Blockchain is verifiable to establish the digital trust throughout the network.

3.2 BLOCKCHAIN-BASED CRYPTOCURRENCY

The cryptocurrency, such as the Bitcoin, is recognised as the first generation of the Blockchain technologies [4]. The cryptocurrency is not only a trendsetter, but supports a fully decentralised peer-to-peer network. The fundamental theories, e.g., cryptography theory, and designs, e.g., block structures and node types, provide huge potential for supporting peer-to-peer energy trading in local energy markets by ensuring the security, trustworthiness, and privacy of prosumers or consumers. This section will introduce key functions of Blockchain-based cryptocurrency relevant to energy trading.

3.2.1 Cryptography Theory

Although the Blockchain uses the term of *crypto*, it is still open and accessible to all users, which means that the information stored in the Blockchain networks, e.g., the account address, balances, and transactions, can be accessed by all users. Rather, the actual meaning behind the *crypto* refers to the foundation of the cryptocurrency, i.e., the cryptography theory. In the cryptography theory, two important functions are exploited by the cryptocurrency, which are the *cryptographic hash function* [5] and *signature* [6]. In the following text, how these two functions support the operation of the cryptocurrency will be discussed.

3.2.1.1 Cryptographic Hash Function

A hash function can map an arbitrary-size data to fixed-size values, defined as the hash values or digest [7]. The hash values are usually used to index a fixed-size table, called a hash table. The cryptographic hash function has two properties: the *collision resistance* and *hiding* [1].

To understand the property of the collision resistance, first, let us know what is collision. In the cryptography theory, a collision is defined as a cryptographic hash that tries to find two different inputs, e.g., x and y, $x \neq y$, such that these two inputs can produce the same hash value, i.e., $hash(x) = hash(y)$ [8]. In the hash table, the input space is much larger than the output space, which results in the collision to be unavoidable. For instance, in the case of the Secure Hash Algorithm (SHA)-256 with a 256-bit (32 bytes) hash value, the input space is infinite whereas the output space would be 2^{256} [9].

Next, with the knowledge of the collision, the property of the collision resistance means that it is highly unlikely to artificially create a collision attack. This means given an input x, there is not any efficient approach to artificially find another input y, such that $hash(x) = hash(y)$, unless traversing every feasible value in the input space [10]. For a large input space, e.g., the input space of the SHA-256, this traversal would be extremely computationally complex. One important application of the property of the collision resistance is to prevent the message tampering [1]. When a message a is encrypted by the cryptographic hash function $hash(a)$, the tampered message a' would result in a different hash value $hash(a')$, such that $hash(a) \neq hash(a')$.

With the property of hiding, the cryptographic hash functions can hide the input messages by only saving the hash value of the original input messages, instead of saving the input messages themselves [11]. This is because the process of producing the hash value is irreversible, which means that given an input x, the cryptographic hash functions can produce the hash value $hash(x)$. However, given a hash value $hash(x)$, it is impossible to trace any information about the input x, unless traversing every feasible value in the input space to find out which input can produce the hash value $hash(x)$. Again, this would be extremely computationally complex. One important condition for the property of hiding is the input space should be large enough and the input values should be uniformly distributed within this space.

The Blockchain networks are protected by solving the puzzle using the SHA-256 which is secured by the properties of the collision resistance and hiding. As shown in Figure 3.1, the block header includes the version number, Merkle root, previous block hash, timestamp, difficulty target, and nonce. The details of these domains and the block body will be introduced in the following subsections. Here, we focus on the domain of the nonce. In cryptography, nonce is an arbitrary number which would be used just once in a cryptographic communication, i.e., the number once [12]. The input of the SHA-256 is the values in all the domains of the block header, and the output of the SHA-256 is the fixed-length hash value which is the unique identity of a block [13]. The mechanism behind the block mining is to keep trying different nonces until the hash value satisfies the certain targets, e.g., a targeted value with k times of 0 as shown in Figure 3.2. The process of the puzzle solving has the feature of being *moderately hard to solve during the block mining but easy to verify during the validation* [14]. This feature indicates that the only way to find the satisfied nonce is through exhaustively trying. However, once the satisfied nonce is found, the verification can be easily proceeded by executing one time SHA-256.

3.2.1.2 Signature

In the Blockchain networks, a user can set up an account by creating a key pair. This key pair consists of the private key and public key [15]. The concepts of the private key and public key originated from the asymmetric cryptography [16].

An example of the asymmetric cryptography is explained here: Alice has sent an encrypted message to Bob. If Bob wants to read the content of the message, he has to decrypt this message with the same key which Alice has used to encrypt the message. However, there is not any secure way for Alice to send the key to Bob in a plain-text manner. To overcome this issue, in the asymmetric cryptography, instead of using one single key for both the encryption and decryption [17], we use a key pair by which

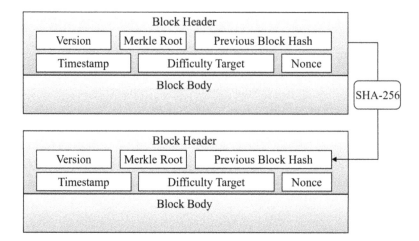

Figure 3.1 Schematic illustration of the Blockchain networks secured by the SHA-256. The input of the SHA-256 is the values in all the domains of the block header, and the output of the SHA-256 is the fixed-length hash value which is the unique identity of a block.

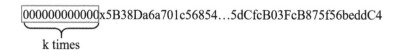

Figure 3.2 Example of a target for the block mining. The miners keep trying different nonces until the hash value satisfies the targeted value with k times of 0.

the public key is used to crypt the message and the private key is used to decrypt the message. For the same example, Alice can use Bob's public key to encrypt the message. When Bob receives the encrypted message, he can use his own private key to decrypt. It should be noted that both the public key and private key belong to Bob, i.e., the message receiver. The advantage of this public–private key pair is that the message receiver only needs to share the public key to the sender while keeping the security of the private key.

In the Blockchain networks, both the private key and public key are generated by the SHA-256, by which the private key is used to generate the public key and the public key is used to generate the public key hash, i.e., the account address (also called the wallet) [18]. The SHA-256 with a 256-bit hash value can provide a good source of randomness for generating a unique key pair for every user. When an user initiates a transaction, this user only needs to sign this transaction with the private key. Other users can use the public key of this initiator of transaction to verify the signature.

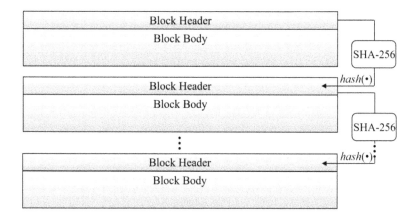

Figure 3.3 Schematic illustration of the hash pointer and data structure of the Blockchain. The blocks are chronologically chained by involving the hash of the previous block into the current block, forming a Blockchain. The hash pointer indicates the position of a block in the Blockchain.

3.2.2 Data Structure

With the knowledge of the key cryptography theory for the cryptocurrency, the following text will introduce how this theory can be exploited to support the data structure of the Blockchain networks from the following two aspects:

- The link between two blocks, i.e., the chaining features;

- The data structure inside a block, i.e., the Merkle tree.

3.2.2.1 Chaining Features

In the Blockchain networks, all the transactions are structured as publicly available ledgers, i.e., blocks. The blocks are chronologically chained by involving the hash of the previous block into the current block, forming a Blockchain [19] as shown in Figure 3.3. Recall that in Sub-Section 3.2.1.1, this chaining feature is secured by the cryptographic hash function, e.g., SHA-256. Hence, the Blockchain can be taken as a linked list and the cryptographic hash function can be taken as a hash pointer which indicates the position of a block in this list.

The advantage of using the hash pointer is that the hash pointer can not only indicate the position of a block, but also guarantee the tamper resistance [20]. This is because if a malicious node tries to tamper with the information stored in one block, it would result in a different hash value which does not match the input of the next block and the following blocks. If a malicious node tries to tamper with all the following blocks, it would be extremely computationally difficult. With this advantage, we can verify whether the information in anywhere of the Blockchain is tampered by only verifying the hash value of the current block.

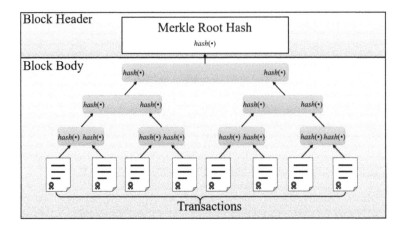

Figure 3.4 Schematic illustration of the data structure inside a block. The data is stored in the form of a Merkle tree. Transactions are structured in the bottom as the leaf nodes. Each leaf node is linked to its parent node through the hash pointer, until reaching the root node. The cryptographic hash of the root node is called the Merkle root hash which is stored in the block header.

3.2.2.2 Merkle Tree

In each block, the transactions are stored in the form of a Merkle tree [21] as shown in Figure 3.4 [22]. In a Merkle tree, the data, i.e. transactions, are structured in the bottom as the leaf nodes. Each leaf node is linked to its parent node through the hash pointer, until reaching the root node. The cryptographic hash of the root node is called the Merkle root hash which is stored in the block header. Merkle tree enables the secure and efficient verification of the transactions stored in a block [23]. Analogous to the advantage of using the hash pointer, we can verify whether the transactions in a block are tampered by only verifying the root hash.

The nodes of the Blockchain networks can be categorised as *full nodes* and *light nodes* [24]. Full nodes store the entire Merkle tree of every block including both the block header and block body. Light nodes only need to store the block header of every block. As shown in Figure 3.5, when a light node needs to verify whether a transaction is stored in a block, this light node can calculate the hash of every node related to this transaction and request the full node for the Merkle proof, such that this light node can verify whether the calculated root hash equals to the root hash stored in the block header.

3.2.3 Consensus

Although the cryptocurrency is secured by the owner's signature using the public–private key pair as introduced in Sub-Section 3.2.1.2, as a digital cash rather than a physical cash, it is still duplicable. When we use this cryptocurrency, e.g., for energy trading, the thing which we do not want to happen is that an energy buyer a spends one unit of the cryptocurrency to buy one unit of energy from an energy seller b, and simultaneously the energy buyer a uses the same cryptocurrency (duplicated) to buy

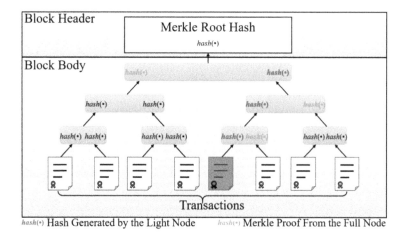

Block Header

Merkle Root Hash
hash(•)

Block Body

Transactions

hash(•) Hash Generated by the Light Node   *hash(•)* Merkle Proof From the Full Node

Figure 3.5 Merkle proof for a light node to verify a transaction. When a light node needs to verify whether a transaction (indicated by the orange colour) is stored in a block, this light node can calculate the hash of every node related to this transaction (indicated by the red colour) and request the full node for the Merkle proof (indicated by the green colour), such that this light node can verify whether the calculated root hash equals to the root hash stored in the block header.

the one unit of energy from another energy seller c, which means the buyer a uses only one unit of cryptocurrency to buy two units of energy. This is defined as the double-spending attack in the field of the cryptocurrency [25].

To overcome this double-spending attack, one solution is to find a trustworthy third party, e.g., banks, to record every transaction for every customer. Nonetheless, in some contexts, it is challenging to find such a trustworthy third party. For instance, if we want to design a cryptocurrency for the decentralised peer-to-peer energy trading, we need to find a reliable peer to verify every transaction, and decide the total amount of the issued currency and time to issue this currency.

In the Bitcoin-centric Blockchain networks, every transaction is recorded by a data structure called the *unspent transaction output* (UTXO) [26] collectively verified by every node in the networks through using the *proof-of-work* (PoW) [27] as a consensus. The trustworthy third party, total amount of the issued currency, and time to issue this currency are determined by the block mining.

3.2.3.1 Unspent Transaction Output

The Bitcoin system is a transaction-based ledger [28]. Different from the account-based ledger by which the account information, e.g., account balance, is recorded in the ledger, the transaction-based ledger only records the transaction information, e.g., the transaction amount, sender, and receiver. The account balance of the transaction-based ledger needs to be audited based on all the inputs and outputs of the related transactions. This transaction-based ledger is defined as the UTXO which is verified by every node in the Blockchain networks.

To help readers understand the operation of the UTXO, an example is provided in Figure 3.6, by which the transactions are chronologically indexed and the UTXO is dynamically updated with the transactions. The details of each transaction are explained as follows:

- In the transaction 1, the sender a sends 15 bitcoin (BTC) (received from previous transactions) to the receivers b (7 BTC) and c (8 BTC), which is signed by the sender a.

- In the transaction 2, the sender c sends 8 BTC (received from the sender a in the transaction 1) to the receivers b (3 BTC) and d (5 BTC), which is signed by the sender c.

- In the transaction 3, the sender b sends 10 BTC (received from the sender a in the transaction 1 and the sender c in the transaction 2) to the receiver e, which is signed by the sender b.

The UTXO correspondingly records the inputs and outputs of every transaction. When the received BTC is spent, the corresponding previous record would be removed and new record would be added in the UTXO. For instance, when b sends all 10 BTC to e, the records of $a(7){\rightarrow}b$ and $c(3){\rightarrow}b$ would be removed and the record $b(10){\rightarrow}e$ would be added. In every state, the total inputs of the UTXO, e.g., $a(7){\rightarrow}b$ and $c(3){\rightarrow}b$, equal to the total outputs of the UTXO, e.g., $b(10){\rightarrow}e$, plus a transaction fee paid to the block miner.

In the UTXO, every input needs to indicate the hash value of the source transaction, e.g., $hash$(Transaction 1), and the index in this source transaction, e.g. $a(7){\rightarrow}b$, which enables every source to be tractable. Therefore, through collectively verifying the UTXO by every node in the Blockchain networks, the double-spending attack can be prevented.

3.2.3.2 Proof-of-Work

After we have the UTXO to record every transaction in the Bitcoin-centric Blockchain networks, the next question is who will decide which transactions are enclosed in a block and sort the order of these transactions. Obviously, if the answer is every node in the Blockchain networks, the consistency of the enclosed transactions in blocks cannot be guaranteed. An alternative solution is that one node encloses certain transactions in a certain order into a new block, and other nodes verify the enclosed transactions. If these transactions are approved by more than 50% of nodes, this new validated block can be added to the Blockchain.

Another issue that was discussed in Section 3.2.1.2 was that any user can easily set up an account by creating a public–private key pair. If a user sets up accounts as much as exceeding 50% of total accounts in the Blockchain networks, this user would dominate the verification and determine which transactions can be enclosed in a block.

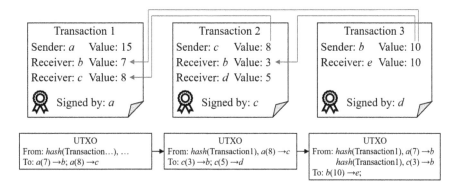

Figure 3.6 Schematic illustration for updating the UTXO with the proceeding of transactions. In the transaction 1, the sender a sends 15 bitcoin (BTC) (received from previous transactions) to the receivers b (7 BTC) and c (8 BTC), which is signed by the sender a. In the transaction 2, the sender c sends 8 BTC (received from the sender a in the transaction 1) to the receivers b (3 BTC) and d (5 BTC), which is signed by the sender c. In the transaction 3, the sender b sends 10 BTC (received from the sender a in the transaction 1 and the sender c in the transaction 2) to the receiver e, which is signed by the sender b. The UTXO correspondingly records the inputs and outputs of every transaction. In the UTXO, every input needs to indicate the hash value of the source transaction, e.g., $hash$(Transaction 1), and the index in this source transaction, e.g. $a(7) \twoheadrightarrow b$, which enables every source to be tractable.

Instead of using the number of nodes to verify new blocks, the Bitcoin-centric Blockchain networks use the computing power for generating new blocks and verifications. Every node can enclose certain transactions in a certain order individually, and then it keeps trying different nonces until the hash value satisfies the certain target, i.e., the block mining as introduced in Sub-Section 3.2.1.1. If the mined block can be approved by more than 50% of the computing power in the Blockchain networks, this block will be added to the Blockchain. The consensus of using computing power for block mining is defined as the PoW [27].

3.2.4 Block Structure

A block, i.e., a ledger, which contains a number of enclosed transactions is a basic unit of a Blockchain. The SHA-256 algorithm imports the previous block header hash as the input. This guarantees the tamper resistance of transactions in the previous block. A malicious node cannot modify transactions without modifying the block header. In this section, the structure of a block including both the *block header* and the *block body* will be introduced. An example of information contained in a mined block [29] is presented in Table 3.1, with details explained as follows:

- *Hash:* The hash is a unique identity for a specific block.

- *Timestamp:* Time is the Unix epoch time when the block miner processed the

Table 3.1 Example of information contained in a mined block [29]

Block	679036
Hash	0000000000000000000645190b239347e8696fa9963 4b80f8d97eaca5df2e2de
Confirmations	1
Timestamp	2021-04-13 10:35
Height	679036
Miner	BTC.TOP
Number of transactions	2,732
Difficulty	23,137,439,666,472.05
Merkle root	b175ffb2eec75fc35b2c5b631f2a56f6c27499bc24cb 36874c2193eae7c05eb3
Version	0x20000000
Bits	386,673,224
Weight	3,993,175 WU
Size	1,314,298 bytes
Nonce	1,412,495,669
Transaction volume	33103.78524111 BTC
Block reward	6.25000000 BTC
Fee reward	0.66344754 BTC

block header. Full nodes will only accept the blocks from miners within two hours in the future.

- *Height:* This accounts for the number of blocks which have already been connected to the Blockchain networks.

- *Miner:* The miner records the address of the node which confirms the transactions in a block.

- *Number of transactions:* The number of transactions is the number of all transactions contained in a block.

- *Difficulty:* it is a value defining the complexity for a miner to find a valid hash for a block.

- *Merkle root hash:* All transactions in a block are structured into a Merkle tree, and the hash value of the Merkle root is stored in the block header to guarantee the tamper resistance of transactions in the current block.

- *Version number:* The version number helps each node in the Blockchain networks identify the rules for validating the blocks.

- *nBits:* The nBits is generated by encoding the targeted threshold. For a valid block, the hash of this block must be less than or equal to the targeted threshold.

- *Weight:* It is a measurement to compare the size of different transactions in proportion to the block size limit.

- *Size:* The size quantifies the total size of a block.

- *Nonce:* The nonce is an random number adjusted by the block miners, so that the hash of the block header is less than or equal to the targeted threshold.

- *Transaction volume:* The transaction volume measures the total amount of currency transacted in a block.

- *Block reward:* The block reward is a static reward to the miner who has successfully find the puzzle.

- *Fee reward:* The fee reward is the total amount of transaction fees rewarding to the miner who has successfully found the puzzle.

3.2.5 Difficulty

In Bitcoin networks, the average time for mining a new block, i.e., the block time, is 10 minutes [30] which theoretically decreases with the increase in number of participating miners and their computational power. However, the Bitcoin networks are able to still maintain this average time by regularly adjusting the difficulty of the block mining. This subsection will introduce the reasons and mechanisms for adjusting the difficulty to maintain this average mining time.

3.2.5.1 *Reasons for Adjusting Difficulty*

The objective of the block mining is to find a proper nonce, so that the hash value of the block header is less than or equal to the predefined target [31] as

$$hash(\text{block header}) \leq \text{target}_{\text{define}}, \tag{3.1}$$

where $\text{target}_{\text{define}}$ is the predefined target value. A larger value of $\text{target}_{\text{define}}$ indicates that the nonce is easier to be found, i.e., a lower value of the difficulty; whereas a smaller value of $\text{target}_{\text{define}}$ indicates that the nonce is harder to be found, i.e., a higher value of the difficulty. Hence, the value of the target is inversely proportional to the value of the difficulty as

$$d = \frac{\text{target}|_{d=1}}{\text{target}_{\text{define}}}, \tag{3.2}$$

where d is the value of the difficulty and $\text{target}|_{d=1}$ is the value of target when the value of difficulty equals to one.

Ideally, the increase of the computational power enables miners to more efficiently find the proper nonce, and thus reduces the block time, which means more transactions can be enclosed into the Blockchain networks. Nonetheless, due to the delay of broadcast over Blockchain networks, different nodes will receive the information of a new mined block at different time. As shown in Figure 3.7 the node a firstly receives the latest update of the new mined block x, and follows this new block to mine the next block $x + 1$. At the same time, the node b does not receive this update and

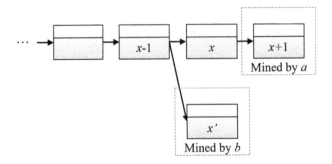

Figure 3.7 Blockchain forking due to the communication delay. The node a firstly receives the latest update of new mined block x, and follows this new block to mine the next block $x + 1$. At the same time, the node b does not receive this update and follows the previous block $x - 1$ to mine the next block x'.

follows the previous block $x - 1$ to mine the next block x'. This inconsistency results in the forking of the Blockchain. Due to multiple forks, it is difficult for all nodes to reach the consensus. Therefore, the Bitcoin networks maintain the average block time at 10 minutes to minimise the forking risk caused by the delay of broadcast [32].

3.2.5.2 Mechanism for Adjusting Difficulty

In the Bitcoin protocol, the system adjusts the difficulty in every 2016 blocks [33]. Given the average block time is 10 minutes, the time interval for adjusting the difficulty is 14 days. The mechanism for adjusting the difficulty can be expressed as

$$\text{target'} = \text{target} \cdot \frac{t_{\text{mine}}}{2016 \cdot 10 \text{ min}}, \tag{3.3}$$

where target' is the adjusted target value, t_{mine} is the actual time for generating the most recent 2016 blocks. If the actual time is longer than the expected time, i.e., $2016 \cdot 10$ min, the target value will be adjusted to increase whereas the difficulty value will decrease according to the Equation (3.2). If the actual time is shorter than the expected time, the target value will be adjusted to decrease whereas the difficulty value will increase. In the Bitcoin protocol, there is a constraint for adjusting the difficulty as

$$\frac{\text{target}}{4} \leq \text{target'} \leq 4 \cdot \text{target}. \tag{3.4}$$

The updated target value is encoded into the nBits domain of the block header [34]. If a miner generates a new block without updating the nBits domain, this mined block cannot be validated by other full nodes, which ensures every node to update the adjusted difficulty values.

3.2.6 Node Types

The participants of Blockchain networks can operate either full nodes or light nodes depending on their storage and computational capabilities.

The full nodes keep the online states all the time to verify every new mined block. They update the entire Blockchain information locally and upload the updated information by mining new blocks. If a new block is successfully mined, other nodes will synchronise with this new block. Full nodes need to verify the validity of every transaction through maintaining the UTXO in their storage. After validating these transactions, the successful miner has the right to decide which transactions to be enclosed into a block and which Blockchain is validate when the Blockchain forks.

Light nodes are online only when they participate in transactions. They only update the block header information and store the transactions related to them. These related transactions can be validated by requesting the Merkle proof from full nodes. They are unable to validate the mined blocks and the Blockchain except verifying the difficulty from the block header.

For the storage capability [35], full nodes need to store the entire Blockchain including both the block header and block body of every block. By contrast, light nodes only need to store the block header. For the computational capabilities [36], full nodes need to verify every block and mine new blocks through solving the mining puzzle. Whereas, light nodes only need to verify their related transactions. Therefore, operating light nodes can dramatically reduce the storage and computational requirements.

3.2.7 Networks

In Blockchain networks, the clients broadcast transactions to the networks, such that full nodes can enclose these transactions into blocks, forming the Blockchain. The broadcast is proceeded on the peer-to-peer overlay networks, under which every node has equal accessibility and communicates with each other by the TCP protocol [37]. Blockchain networks have the features of simplicity and robustness. Every node maintains a random set of neighbouring nodes through which the messages are passed by the flooding broadcast [38]. Once a node receives a message, this node will broadcast it to neighbouring nodes and mark these neighbouring nodes as broadcasted. It is to be noted that these neighbouring nodes are randomly selected without considering the topological structure and geographical locations. Although this design enhances the robustness of the Blockchain networks, the communicating efficiency is compromised.

3.3   BLOCKCHAIN-BASED SMART CONTRACTS

Bitcoin and Ethereum are two primary types of cryptocurrency [39], which are supported by the Blockchain technologies. Bitcoin is recognised as the first generation of the Blockchain technology which decentralises the currency [40]. Ethereum, developed by Vitalik Buterin in 2014 [41], is recognised as the second generation of the Blockchain technology which not only overcomes a number of drawbacks of the Bitcoin, e.g., saving the time of block mining and resisting the mining centralisation, but also supports the operation of the smart contracts [42].

The smart contracts, a term coined by Nick Szabo in 1994 [43], are a programmed transactional protocol with the immutable, verifiable, secure, and replicable features [44]. A general form of the smart contracts is 'A customer a deposits the currency b into the smart contracts for purchasing the seller c' s product. When confirming that a receives the product from c, the smart contracts pay the deposited currency b to c.' Incorporating the smart contracts into the Ethereum enables the Blockchain to evolve from an application to be a platform, under which every customer can encode the standardised terms of contracts.

3.3.1 Account

Recall that as per Sub-Section 3.2.3.1, the Bitcoin system is a transaction-based ledger which only records the inputs and outputs of transactions by using the UTXO. This design increases the complexity for customers for the following three reasons:

- The balance of a customer's account needs to be audited from the UTXO, instead of being recorded in the Blockchain.

- A customer needs to include the source of the transferred currency, i.e., from which sender in which transaction.

- The account balance cannot be split when a customer transfers the currency to others. For instance, the customer a's account balance is 7 Ethers, if a wants to pay 4 Ethers to another customer b, a has to either pay the rest 3 Ethers to the block miner as the transaction fee, or transfer the rest 3 Ethers to a's another account.

Ethereum system evolves to be an user-friendly account-based ledger which records the account balance. The only requirement for the transaction is that a customer has enough balance to transfer, and the account balance can also be split. This account-based ledger can naturally prevent the double-spending attack, since if a malicious node replicates the transaction, the Blockchain will deduct the corresponding amounts twice from the balances of this malicious node.

Nonetheless, this account based ledger has the risk of the replay attack [45]. For instance, the sender a transfers to the receiver b. If the receiver b replicates this transaction in the Blockchain networks, the transferred amount will be deducted twice from the sender a's account. To prevent this replay attack, the Ethereum includes a counter into each account, such that the transaction can be indexed corresponding to the state change of balances. This account based ledger also ensures the explicit contracting parties with the permanent addresses for guaranteeing the validity of the contracts.

In the Ethereum systems, the accounts can be categorised as the externally owned accounts and smart contract accounts [46]. The externally owned accounts are created by the public–private key pair as introduced in Sub-Section 3.2.1.2, by which the address of an account is mapped to the states of this account through using the key-value pair. The smart contracts accounts store the codes and executing states of the smart contracts. The smart contracts can be deployed by the externally owned accounts by calling the address of the smart contract accounts. However, the smart contracts accounts cannot initiate transactions by themselves, unless they are called by the externally owned accounts to execute transactions.

3.3.2 Data Structure

Ethereum is a decentralised application platform in which the states of every account, smart contract, and transaction are recorded. This requires an efficient structure to manage and store the data in the blocks. The data structure can map the address (key) of an account or a smart contract to its states (value) including the balances, transaction index, and operations of the smart contracts, which can be represented by a key-value pair [47].

Recall that in the Bitcoin system as introduced in Sub-Section 3.2.2.2, the transaction data is structured as the Merkle tree stored in the block. Each block contains certain amounts of transactions, e.g., 4,000 transactions per block in the Bitcoin system. Nonetheless, this data structure is not suitable for the Ethereum for the following reasons:

- Since the Ethereum system needs to store not only the transaction data, but also the account data, with the increasing number of new accounts, the size of the Merkel tree would correspondingly increase.

- The time for mining a new block is reduced to around 10 seconds in the Ethereum [48], which poses a challenge for generating a large-scale Merkle tree for each new block.

- In each time step, only some accounts change their states whereas other accounts remain unchanged. Updating all states of accounts would waste computational resources.

- Arbitrarily structuring the account data into the Merkle tree would result in the inconsistency for costumers and inefficiency for searching the account information.

To overcome these issues, the Ethereum uses modified Merkle Patricia tree to store the data [49]. To facilitate the illustration of the modified Merkle Patricia tree, we first introduce the concepts of the digital tree and Patricia tree as follows:

- *Digital tree* [50]: Digital tree structures a set of all the keys into a tree structure, so that a specific key can be efficiently searched from this set. The individual characters of a key are assigned to each node of the digital tree, and the nodes are lined by the order of characters of a key. The numbers of branches of the digital tree are determined by the value range of the characters. For instance, the value range of the Ethereum address is 160 bytes. Given certain inputs, the tree structure would remain consistent irrespective of the order for organising the keys.

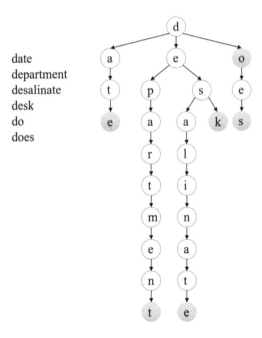

date
department
desalinate
desk
do
does

Figure 3.8 A set of six keys is structured as a digital tree. The keys are 'date', 'department', 'desalinate', 'desk', 'do', and 'does'. The individual characters of a key are assigned to each node. The nodes are linked by the order of characters of a key.

An example of a digital tree is presented in Figure 3.8, in which a set of six keys ('date', 'department', 'desalinate', 'desk', 'do', and 'does') is structured. A drawback of the digital tree is that it is not efficient to store the long keys, such as the 'department' and 'desalinate' in the example, with limited sharing node.

- *Patricia tree* [51]: To overcome the drawback of the digital tree, the Patricia tree uses the path compression to optimise the space. The nodes which are the only child are merged with their parent nodes.

  For the same example of the digital tree, the six keys can be optimally structured as the Patricia tree as presented in Figure 3.9. It is worth mentioning that when a new key is inserted into the Patricia tree, the compressed path may need to be decompressed.

Based on the concepts of the digital tree, Patricia tree, and Merkle tree as introduced in Sub-Section 3.2.2.2, the modified Merkle Patricia tree is a cryptographically authenticated Patricia tree, in which the ordinary pointers are replaced by the hash pointers as in the Merkle tree [52]. In the Ethereum system, the data of states of accounts and smart contracts, transactions, and receipts of transactions is structured by three modified Merkle Patricia trees, i.e., the state tree, transaction tree, and receipt tree, respectively [53]. These three modified Merkle Patricia trees are enclosed into a block by a miner, and the root hashes of these three modified Merkle Patricia trees are stored in the block header as shown in Figure 3.10.

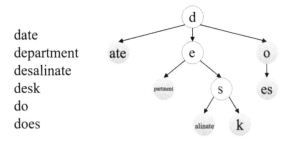

date
department
desalinate
desk
do
does

Figure 3.9 A set of six keys is structured as a Patricia tree. The keys are 'date', 'department', 'desalinate', 'desk', 'do', 'does'. The nodes which are the only child are merged with their parent nodes.

The functions for storing the root hash into the block header include:

- *Tampering-resistance*: Once a malicious node tampers with any data in a block, it would result in a different root hash.

- *Merkle-Proof*: The light nodes can request the Merkle proof provided by the full nodes as introduced in Sub-Section 3.2.3 to verify the transactions and balances.

- *Membership*: Whether an account is a member of the Ethereum network, it can be verified.

3.3.2.1 State Tree

The data related to all the externally owned accounts and smart contract accounts is structured in a modified Merkle Patricia tree, called the state tree [54]. The state

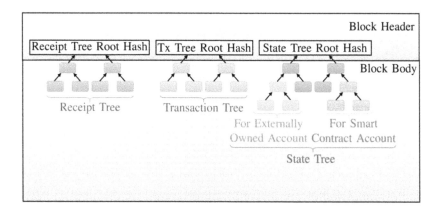

Figure 3.10 The data structure of a block in the format of the modified Merkle Patricia trees. The data of states of accounts and smart contracts, transactions, and receipts of transactions is structured by the state tree, transaction tree, and receipt tree, respectively.

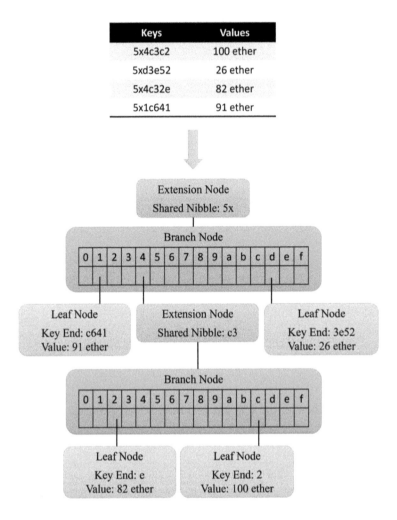

Figure 3.11 Four key-value pairs are structured into a state tree. For simplicity, only the account balances are considered as the states. The extension nodes contain the path-compressed characters as the shared nibbles. The branch nodes cover the value range of the characters. The leaf nodes contain the rest path-compressed characters of each key as the key end. The first extension node with the shared nibble '5x' is the root node, whose hash value is stored in the block header.

tree maps the address (key) of an account or a smart contract to its states (value) encoded by the recursive length prefix [55], forming a key-value pair.

As shown in Figure 3.11, four key-value pairs are structured into a state tree. For simplicity, only the account balances are considered as the states. The extension nodes contain the path-compressed characters as the shared nibbles. The branch nodes cover the value range of the characters. The leaf nodes contain the rest path-compressed characters of each key as the key end. The first extension node with the shared nibble '5x' is the root node, whose hash value is stored in the block header.

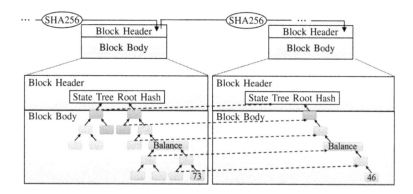

Figure 3.12 Schematic illustration for updating the state tree when a new block is generated. Only the nodes whose states are changed are updated by using a new branch to replace the previous branch, whereas the rest nodes remain unchanged.

When a new block is generated, only the nodes whose states are changed are updated by using a new branch to replace the previous branch, whereas the rest nodes remain unchanged as shown in Figure 3.12. This design not only improves the efficiency of updating the state information, but also enables the executing process of the smart contracts to be traceable. When the smart contracts need to revert to the previous state, it can simply trace the corresponding branch of the state tree in the previous block.

3.3.2.2 Transaction Tree

Analogous to the approach of enclosing the transactions in the Bitcoin system as introduced in Sub-Section 3.2.2.2, the transactions are stored in the leaf nodes and structured as the modified Merkle Patricia tree [56]. As presented in Figure 3.13, the leaf nodes are linked to their parent nodes through cryptographic hash function. The root hash of the transaction tree is stored in the block header. When a light node needs to verify whether a transaction is stored in the Blockchain networks, this light node can request the full node to provide the Merkle proof.

3.3.2.3 Receipt Tree

The receipt tree stores the outcomes of the transactions as receipts [57]. All the information related to a transaction, e.g., sender, receiver, and transaction amount, is recorded in a receipt. For this reason, every transaction can be matched to its corresponding receipt. Compared to the Bitcoin system, this additional receipt tree supports an efficient search of the executed outcomes by using the bloom filter [58], which is particularly useful when executing complex procedures of smart contracts.

The bloom filter is able to search an element from an impractically large scale set by producing the digest for this set using the cryptographic hash function. An example of producing a digest and searching elements of the set $\mathcal{A}$ by using the bloom filter is presented in Figure 3.14. The bloom filter maps the elements a, b, and c of

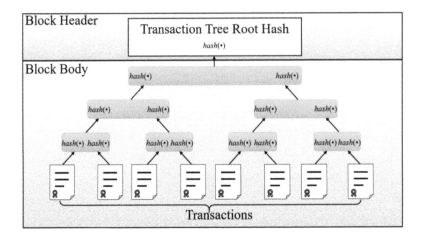

Figure 3.13 Data structure of the transaction tree. When a new block is generated, transactions are are stored in the leaf nodes and structured as the modified Merkle Patricia tree. The leaf nodes are linked to their parent nodes through using the cryptographic hash function.

the set $\mathcal{A}$ to a position in the digest by using the cryptographic hash function. The positions which have the hash values of corresponding elements are indicated as 1, whereas the positions which do not have the hash values of corresponding elements are indicated as 0. This design dramatically reduces the required memory for storing the set. If the membership of an element d/e needs to be checked, the hash value of this element would be calculated and located to a position in the digest, in which 1 indicates either the element e is a member of this set $\mathcal{A}$ or there is a hash collision, and 0 indicates the element d is definitely not a member of this set $\mathcal{A}$. This is the typical feature of the bloom filter, i.e., either possibly in set or definitely not in set [59]. In the Ethereum, the hash collision has been further overcome by including a set of multiple cryptographic hash functions.

It is worth mentioning that rather than including the information of all the accounts into the state tree, only the transactions and corresponding receipts within a new block are included into the transaction tree and receipt tree, respectively. Hence, the state tree shares certain branches between blocks, whereas the transaction tree and receipt tree are independent between blocks.

3.3.3 Smart Contracts

The creation of digital currency has decentralised the financial systems. The next field that can be decentralised is the contract management systems, which is realised by the creation of smart contracts. In the conventional contract management systems, contracting parties negotiate the content of contracts and sign the agreed contracts. The signed contracts are validated by a third party, e.g., the court. Violation of contracts would result in the legal arbitration afterwards. By contrast, the creation of smart contracts can prevent the violation of contracts beforehand. This

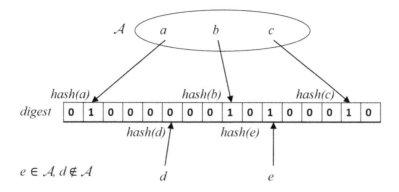

Figure 3.14 Producing a digest and searching elements of a set by using the bloom filter. The bloom filter maps the elements a, b, and c of the set $\mathcal{A}$ to a position in the digest by using the cryptographic hash function. The positions which have the hash values of corresponding elements are indicated as 1, whereas the positions which do not have the hash values of corresponding elements are indicated as 0. If the membership of an element d/e needs to be checked, the hash value of this element would be calculated and located to a position in the digest, in which 1 indicates either the element e is a member of this set $\mathcal{A}$ or there is a hash collision, and 0 indicates the element d is definitely not a member of this set $\mathcal{A}$.

is because smart contracts use programmable functions to standardise the content of contracts [60]. The standardised contracts are collectively validated by every node in the Blockchain networks and programmable functions are automatically self-enforced by the state update [61]. Every contracting party can follow this predefined functions to proceed transactions. If certain conditions are not met, the contracting parties are forbidden to enter the contracts from the very beginning.

One initiative of the smart contracts is the vending machine. A buyer chooses a product and pays the price of this product to the vending machine. By confirming that the paid price equals to the price of this product (condition), the product will be automatically transferred to this buyer in a self-enforcing manner. From this initiative, we can find that the innovation of smart contracts evolves the Blockchain from an application to a platform, under which the functions, e.g., the trading, negotiation, and auctions can be performed [2, 62]. Depending on Blockchain networks, all self-executing functions on this platform are collectively validated by all the nodes through reaching a consensus.

Smart contracts are a computerised transactional protocol which executes the predefined functions in a replicable, verifiable, secure, and immutable manner [62]. The primary features of smart contracts are summarised as follows:

- Self-enforcement: The programmable functions of smart contracts enable predefined actions or procedures to be automatically enacted.

- Low transactional cost: Due to the automation and self-enforcement, the smart contracts can reduce the costs of processing information of transactions and system states, which eases the burdens of information and communication infrastructures and reduces the transactional cost.

- Accessibility: The contents and executing states of smart contracts are open and accessible for all participants of Blockchain networks.

- Duplicability: The smart contracts can be duplicated by all participants of Blockchain networks, which means when a participant initiates a smart contract, this participant only needs to call the address of this smart contract stored in the Blockchain networks.

- Verifiability: The states of accounts and smart contracts as well as transactions can be verified by every participant in the Blockchain networks through verifying the root hash of the modified Merkle Patricia trees.

- Security: Smart contracts are secured by Blockchain networks through collectively maintaining the consensus by every participant.

In the following subsections, the languages, control structures, and deployment of smart contracts are introduced, with the Solidity [63] as an example.

3.3.3.1 Input and Output

The inputs of Solidity are certain number of parameters and outputs of Solidity are certain or arbitrary number of parameters. All the parameters in inputs and outputs need to be defined with the variable type. The default byte representation of a variable is set as zero. For example, the default values for *uint*, *int*, and *bool* are 0, 0, and *false*, respectively. For the statically sized arrays, e.g., *bytes1* to *bytes32*, the elements are initialised to the their own default values. For the dynamically sized arrays, e.g., *string* and *bytes*, the elements are initialised to empty array or string.

The input name needs to be defined whereas the output name can be omitted. The output can be defined after the *return* keyword. A function can return multiple outputs, in the form of *return (output 1, output 2, output 3, ..., output N)*. The default returns are initialised as zero unless defined otherwise. An instance of a public contract which accepts two integers from the external calls and returns two results with the product and sum of these two integers is as follows:

```
contract ContractExample {
        function calculation(uint _x, uint _y) public
            returns (uint _product, uint _sum){
                _product = _x * _y;
                _sum = _x + _y;
        }
}
```

As shown in this example, when defining the conditions of a function, the parentheses is necessary. For a single statement body, the Curly brackets could be omitted.

The type of tuple can also be used in the solidity as inputs or outputs. The tuple is in the form of a list which contains different data types with the constant length at the compiling time. Once the output returns a tuple with different types of values, these values would be assigned to either new variables or existing variables. As presented in the following instance, the output of the function *tuplefunction()* have multiple values inside the tuple, including *uint* and *bool*. These values will be assigned to the function *assignfunction()*.

```
contract Tuple {
        function tuplefunction() public returns (uint, uint, bool){
                return (5,6,true);
        }
        function assignfunction() public {
                (uint x, uint y, bool z) = tuplefunction();
                (x, y) = (y, x);
        }
}
```

3.3.3.2 Control Structures

Similar to the common semantics of scripting languages, such as JavaScript or C, the following basic control structures can be used:

- *if*;

- *else*;

- *while*;

- *do*;

- *for*;

- *break*;

- *continue*;

- *return.*

A smart contract can be called by the internal function call, external function call, and name call, with details explained as follows:

- *Internal function call:* In the internal function call, a function can be directly called by another function within the same smart contract. As shown in the following example, the *functionB* can be directly called by the *functionA* under the same contracts. Because the current memory is stored and passed to another internally called function, the internal function call uses simple jumps in the Ethereum virtual machine, which is an efficient way of calling functions.

```
contract InternalCall {
        function functionA(uint _x, uint _y) public returns (uint
            _output){
            return functionB();
        }
        function functionB() internal returns (uint _output){
            return functionA(2,5)*functionB();
        }
}
```

- *External function call:* In the external function call, a function is called externally from another contract by using a message call. The external function call cannot be converted into simple jumps as the internal function call and all function arguments therefore need to be copied to the memory. When a customer calls a function of the smart contract using the external function call, this customer can specify the value to be transferred and the gas to be paid to the miner using *.value()* and *.gas()*, respectively. As presented in the following instance, the *ContractA* can be externally called by the *ContractB*.

```
contract ContractA {
        function initiate() public payable returns (uint _output){
            return 56;
        }
}

contract ContractB {
        ContractA contractin;
        function init(address _addr) public {
            contractin = ContractA(_addr);
        }
        function contractcall() public {
            contractin.init.value(25).gas(8);
        }
}
```

In this example, if the *ContractA* is set to be able to accept the transfer, the modifier *payable* needs to be added. The term *ContractA(_addr)* indicates the type of the contract with the input address _addr known as *ContractA*, such that the constructor of this contract is not executed. The call would result in exceptions for the following cases:

1) If the address is invalid;

2) If the called contract returns an exception;

3) If the transaction cost exceeds the gas limit.

- *Name Call:* A function can be also called by names in an arbitrary order by including the call in the curly brackets. The called list of elements needs to match the elements of the function arguments. The following instance shows that the *functionB* calls the arguments of the *functionA* by names

```
contract NameCall {
        function functionA(uint _x, uint _y) public {
        }
        function functionB() public {
                func1({_y: 3, _x: 1});
        }
}
```

3.3.3.3 Creating New Contracts

A new contract with known full codes can be created by another contract by using the *new*. The requirement of known full codes can prevent the recursive creation. In the following example, *Contract1* is the new contract being created by *Contract2*. The function *CreateNew* only creates a new contract whereas the function *CreateNewPay* creates a new contract and specifies the transferred amount of currency into this contract by using *.value()*. It is noted that the cases of out of stack or the balances would result in exceptions.

```
contract Contract1 {
        uint _b;
        function Contract1(uint _a) public payable {
                uint _b = _a;
        }
}

contract Contract2 {
        Contract1 contract1 = new Contract1(8);
        function CreateNew (uint _a) public{
                Contract1 NewContract1 = new Contract1(8);
        }
        function CreateNewPay(uint _a, uint _amount) public payable {
                Contract1 NewContract1 = (new Contract1).value(_a)(
                        _amount);
        }
}
```

3.3.3.4 Conditions and Errors

The conditions of smart contracts can be set by using functions *assert* or *require*. If the defined conditions are not met, these functions would throw an exception. For the Solidity, the state reverting exceptions are used to undo all the state changes in the current call. The caller would receive a notice of the error. *Assert* function is used to test for internal errors and define invariant conditions. The function *require* is used for:

- Ensuring the defined conditions are valid. For instance, the inputs of a function need to meet certain conditions;

- Ensuring the state variables of smart contracts are met;

- Validating the output values of a call to external contracts.

In the following example, the function *require* is used to check the input conditions, i.e., the transferred value needs to be greater than or equal to 8. The function *assert* is used to check the internal errors, i.e., the balance after the transfer is updated correctly. A message in the type of string can be added to the *require* function, whereas the *assert* function cannot include any string message.

```
contract Condition {
        function conditioncheck(address _addr) public payable returns (
            uint _balance) {
                require(msg.value >=8, "minimum transfer requirement is
                    8");
                uint previousbalance = _addr.balance;
                _addr.transfer(msg.value);
                assert(_addr.balance == previousbalance − msg.value);
                return _addr.balance;
        }
}
```

The function *revert* can also be used to trigger exceptions and revert state changes in the current call. A string message can be provided to notify the callers with the detailed errors as shown in the following example:

```
contract Condition {
        function conditioncheck(address _addr) public payable returns (
            uint _balance) {
                if (msg.value < 3)
                revert("minimum transfer accepted is 3");
                return _addr.balance;
        }
}
```

3.3.3.5 *Deployment of Smart Contracts*

A programmed smart contract can be deployed on the Ethereum virtual machine [64] which provides a form of abstractions between the smart contracts programmes and the hardware. Ethereum virtual machine would be maintained by all the connected peers which are taken as clients, so as to keep the continuous, uninterrupted, and immutable operations under the protocols of Blockchain networks. The update of

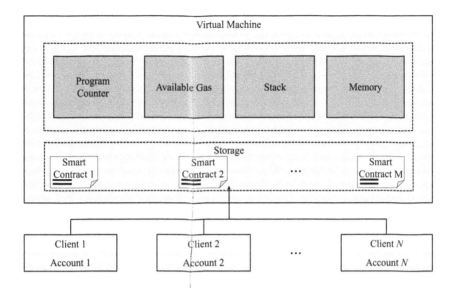

Figure 3.15 Example of the virtual machine for deploying smart contracts. All the participants in the Blockchain networks collectively maintain this virtual machine. The states of smart contracts and accounts changing from one block to another block are driven by the mechanisms defined by the virtual machine.

states of all accounts and smart contracts from one block to another block is driven by the mechanisms defined by this virtual machine. An example of virtual machine is presented in Figure 3.15.

The virtual machine imports the previous states and new enclosed transactions as inputs, and returns new states as deterministic outputs, through which the state transition can be driven [65]. This state transition is updated in the state tree and new transactions are included in the transaction tree. The execution of the virtual machine is in the format of a 1024-item depth stack machine, through which each item is in the form of a 256-bit word. The deployed smart contracts are executed as the virtual machine opcodes by executing the standardised stack operations.

An instance for deploying the Ethereum smart contracts written in the virtual machine by using the Remix IDE [66] is presented in Figure 3.16. Under this platform, a user can register an account with balance of 100 Ether. The gas limit, transferred value, and contract type can be defined by a user when calling the smart contracts or participating in a deployed smart contract. When a new smart contract is deployed, the initiator needs to ensure the input information meets the pre defined conditions of this contract.

3.4 CHAPTER SUMMARY

Fundamental theories of Blockchain technologies have been introduced in this chapter, including the first and second generations of Blockchain, i.e., Blockchain-based cryptocurrency and Blockchain-based smart contracts. These Blockchain technologies have the potential to support open and accessible local energy markets. The

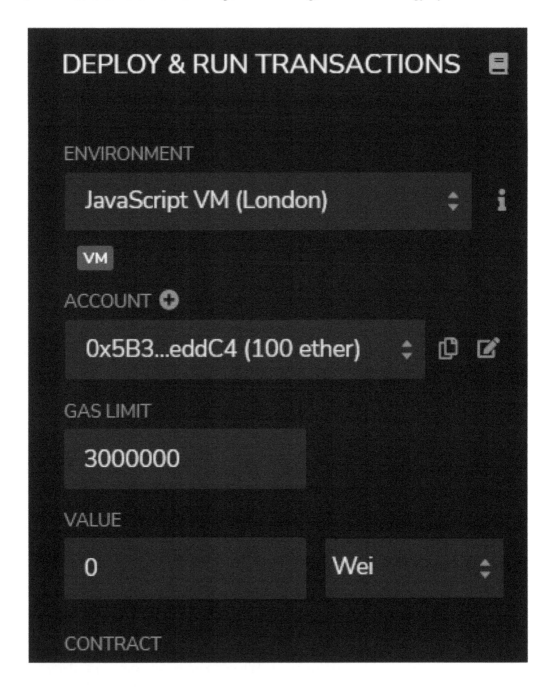

Figure 3.16 Deploying the Ethereum smart contracts written in the virtual machine by using the Remix IDE [66]. Under this platform, a user can register an account with balance of 100 Ether. The gas limit, transferred value, and contract type can be defined by a user when calling the smart contracts or participating in a deployed smart contract. When a new smart contract is deployed, the initiator needs to ensure the inputs information meets the pre defined conditions of a this contract.

cryptography theory, structures of Blockchain networks, and consensus ensure the security and trustworthiness of energy trading. The block structure and node types allow the accessibility of individual prosumers through using their local devices. Blockchain-based smart contracts turn the Blockchain to be a fundamental platform to support the self-enforcement, automation, and standardisation of energy trading procedures.

Bibliography

[1] G. E. Suh, D. Clarke, B. Gassend, M. Van Dijk, and S. Devadas, "Aegis: Architecture for tamper-evident and tamper-resistant processing," in *ACM International Conference on Supercomputing 25th Anniversary Volume*, 2003, pp. 357–368.

[2] W. Hua, F. Luo, L. Du, S. Chen, T. Kim, T. Morstyn, V. Robu, and Y. Zhou, "Blockchain technologies empowering peer-to-peer trading in multi-energy systems: From advanced technologies towards applications," *IET Smart Grid*, vol. 5, no. 4, pp. 219–222, 2022.

[3] Y. Zhou, A. N. Manea, W. Hua, J. Wu, W. Zhou, J. Yu, and S. Rahman, "Application of distributed ledger technology in distribution networks," *Proceedings of the IEEE*, 2022.

[4] D. Valdeolmillos, Y. Mezquita, A. González-Briones, J. Prieto, and J. M. Corchado, "Blockchain technology: a review of the current challenges of cryptocurrency," in *International Congress on Blockchain and Applications*. Springer, 2019, pp. 153–160.

[5] B. Preneel, "Cryptographic hash functions," *European Transactions on Telecommunications*, vol. 5, no. 4, pp. 431–448, 1994.

[6] D. R. Stinson and M. Paterson, *Cryptography: Theory and Practice*. CRC press, 2018.

[7] Q. Dang, *Recommendation for applications using approved hash algorithms*. US Department of Commerce, National Institute of Standards and Technology, 2008.

[8] M. L. Goldberger and K. M. Watson, *Collision Theory*. Courier Corporation, 2004.

[9] H. Gilbert and H. Handschuh, "Security analysis of sha-256 and sisters," in *International Workshop on Selected Areas in Cryptography*. Springer, 2003, pp. 175–193.

[10] O. Kitamura, "Comparative study on collision resistance of side structure," *Marine Technology and SNAME News*, vol. 34, no. 4, p. 293, 1997.

[11] G. Ateniese and B. de Medeiros, "Identity-based chameleon hash and applications," in *International Conference on Financial Cryptography*. Springer, 2004, pp. 164–180.

[12] D. MacKenzie, "Pick a nonce and try a hash," *London Review of Books*, vol. 41, no. 8, pp. 35–38, 2019.

[13] D. Puthal, N. Malik, S. P. Mohanty, E. Kougianos, and C. Yang, "The blockchain as a decentralized security framework [future directions]," *IEEE Consumer Electronics Magazine*, vol. 7, no. 2, pp. 18–21, 2018.

[14] J. Bonneau, A. Miller, J. Clark, A. Narayanan, J. A. Kroll, and E. W. Felten, "Sok: Research perspectives and challenges for bitcoin and cryptocurrencies," in *2015 IEEE Symposium on Security and Privacy*. IEEE, 2015, pp. 104–121.

[15] O. Pal, B. Alam, V. Thakur, and S. Singh, "Key management for blockchain technology," *ICT Express*, 2019.

[16] J. Patarin, "Asymmetric cryptography with a hidden monomial," in *Annual International Cryptology Conference*. Springer, 1996, pp. 45–60.

[17] R. L. Rivest, A. Shamir, and L. M. Adleman, "A method for obtaining digital signatures and public key cryptosystems," in *Secure Communications and Asymmetric Cryptosystems*. Routledge, 2019, pp. 217–239.

[18] N. Prusty, *Building Blockchain Projects*. Packt Publishing Ltd, 2017.

[19] J. Da Costa Cruz, A. S. Schröder, and G. von Wangenheim, "Chaining property to blocks–on the economic efficiency of blockchain-based property enforcement," in *International Conference on Business Information Systems*. Springer, 2018, pp. 313–324.

[20] C. Pop, M. Antal, T. Cioara, I. Anghel, D. Sera, I. Salomie, G. Raveduto, D. Ziu, V. Croce, and M. Bertoncini, "Blockchain-based scalable and tamper-evident solution for registering energy data," *Sensors*, vol. 19, no. 14, p. 3033, 2019.

[21] R. Beck, "Beyond bitcoin: The rise of blockchain world," *Computer*, vol. 51, no. 2, pp. 54–58, 2018.

[22] W. Hua, J. Jiang, H. Sun, F. Teng, and G. Strbac, "Consumer-centric decarbonization framework using stackelberg game and blockchain," *Applied Energy*, vol. 309, p. 118384, 2022.

[23] M. Yu, S. Sahraei, S. Li, S. Avestimehr, S. Kannan, and P. Viswanath, "Coded merkle tree: Solving data availability attacks in blockchains," in *International Conference on Financial Cryptography and Data Security*. Springer, 2020, pp. 114–134.

[24] C. Ge, Z. Liu, and L. Fang, "A blockchain based decentralized data security mechanism for the internet of things," *Journal of Parallel and Distributed Computing*, vol. 141, pp. 1–9, 2020.

[25] G. O. Karame, E. Androulaki, and S. Capkun, "Double-spending fast payments in bitcoin," in *Proceedings of the 2012 ACM Conference on Computer and Communications Security*, 2012, pp. 906–917.

[26] S. Delgado-Segura, C. Perez-Sola, G. Navarro-Arribas, and J. Herrera-Joancomarti, "Analysis of the bitcoin utxo set," in *International Conference on Financial Cryptography and Data Security*. Springer, 2018, pp. 78–91.

[27] A. Gervais, G. O. Karame, K. Wüst, V. Glykantzis, H. Ritzdorf, and S. Capkun, "On the security and performance of proof of work blockchains," in *Proceedings of the 2016 ACM SIGSAC Conference on Computer and Communications Security*, 2016, pp. 3–16.

[28] J. Garay, A. Kiayias, and N. Leonardos, "The bitcoin backbone protocol: Analysis and applications," in *Annual International Conference on the Theory and Applications of cryptographic techniques*. Springer, 2015, pp. 281–310.

[29] https://www.blockchain.com/explorer, Apr. 2021.

[30] https://bitcoin.stackexchange.com/questions/1863/why-was-the-target-block-time-chosen-to-be-10-minutes.

[31] I. Bashir, *Mastering Blockchain*. Packt Publishing Ltd, 2017.

[32] C. Decker and R. Wattenhofer, "Information propagation in the bitcoin network," in *IEEE P2P 2013 Proceedings*. IEEE, 2013, pp. 1–10.

[33] https://www.bitcoinmining.com/what-is-bitcoin-mining-difficulty/.

[34] A. Judmayer, N. Stifter, K. Krombholz, and E. Weippl, "Blocks and chains: introduction to bitcoin, cryptocurrencies, and their consensus mechanisms," *Synthesis Lectures on Information Security, Privacy, & Trust*, vol. 9, no. 1, pp. 1–123, 2017.

[35] M. Al-Bassam, A. Sonnino, and V. Buterin, "Fraud and data availability proofs: Maximising light client security and scaling blockchains with dishonest majorities," *arXiv preprint arXiv:1809.09044*, 2018.

[36] K. R. Ozyilmaz and A. Yurdakul, "Designing a blockchain-based IOT with ethereum, swarm, and lora: The software solution to create high availability with minimal security risks," *IEEE Consumer Electronics Magazine*, vol. 8, no. 2, pp. 28–34, 2019.

[37] E. K. Lua, J. Crowcroft, M. Pias, R. Sharma, and S. Lim, "A survey and comparison of peer-to-peer overlay network schemes," *IEEE Communications Surveys & Tutorials*, vol. 7, no. 2, pp. 72–93, 2005.

[38] J. Li, "Data transmission scheme considering node failure for blockchain," *Wireless Personal Communications*, vol. 103, no. 1, pp. 179–194, 2018.

[39] A. E. Gencer, S. Basu, I. Eyal, R. Van Renesse, and E. G. Sirer, "Decentralization in bitcoin and ethereum networks," in *International Conference on Financial Cryptography and Data Security*. Springer, 2018, pp. 439–457.

[40] M. Swan, *Blockchain: Blueprint for a new economy*. O'Reilly Media, Inc., 2015.

[41] V. Buterin *et al.*, "A next-generation smart contract and decentralized application platform," white paper, vol. 3, no. 37, 2014.

[42] J. Kehrli, "Blockchain 2.0-from bitcoin transactions to smart contract applications," *Niceideas, November. Available at: https://www. niceideas. ch/roller2/badtrash/entry/blockchain-2-0-frombitcoin (Accessed: 5 January 2018)*, 2016.

[43] N. Szabo, "Smart contracts," *Unpublished manuscript*, 1994.

[44] K. Christidis and M. Devetsikiotis, "Blockchains and smart contracts for the internet of things," *IEEE Access*, vol. 4, pp. 2292–2303, 2016.

[45] H. Chen, M. Pendleton, L. Njilla, and S. Xu, "A survey on ethereum systems security: Vulnerabilities, attacks, and defenses," *ACM Computing Surveys (CSUR)*, vol. 53, no. 3, pp. 1–43, 2020.

[46] A.-T. Panescu and V. Manta, "Smart contracts for research data rights management over the ethereum blockchain network," *Science & Technology Libraries*, vol. 37, no. 3, pp. 235–245, 2018.

[47] A. M. Antonopoulos and G. Wood, *Mastering Ethereum: Building Smart Contracts and Dapps*. O'reilly Media, 2018.

[48] H. J. Singh and A. S. Hafid, "Prediction of transaction confirmation time in ethereum blockchain using machine learning," in *International Congress on Blockchain and Applications*. Springer, 2019, pp. 126–133.

[49] V. Dhillon, D. Metcalf, and M. Hooper, "Unpacking ethereum," in *Blockchain Enabled Applications*. Springer, 2021, pp. 37–72.

[50] B. Pittel, "Paths in a random digital tree: limiting distributions," *Advances in Applied Probability*, pp. 139–155, 1986.

[51] P. Kirschenhofer, H. Prodinger, and W. Szpankowski, "On the balance property of patricia tries: external path length viewpoint," *Theoretical Computer Science*, vol. 68, no. 1, pp. 1–17, 1989.

[52] Z. Lu, Q. Wang, G. Qu, H. Zhang, and Z. Liu, "A blockchain-based privacy-preserving authentication scheme for vanets," *IEEE Transactions on Very Large Scale Integration (VLSI) Systems*, vol. 27, no. 12, pp. 2792–2801, 2019.

[53] S. Tikhomirov, "Ethereum: State of knowledge and research perspectives," in *International Symposium on Foundations and Practice of Security*. Springer, 2017, pp. 206–221.

[54] K. Iyer and C. Dannen, "The ethereum development environment," in *Building Games with Ethereum Smart Contracts*. Springer, 2018, pp. 19–36.

[55] A. Coglio, "Ethereum's recursive length prefix in acl2," *arXiv preprint arXiv:2009.13769*, 2020.

[56] H. Liu, X. Luo, H. Liu, and X. Xia, "Merkle tree: A fundamental component of blockchains," in *2021 International Conference on Electronic Information Engineering and Computer Science (EIECS)*. IEEE, 2021, pp. 556–561.

[57] Y. Huang, B. Wang, and Y. Wang, "Research and application of smart contract based on ethereum blockchain," *Journal of Physics: Conference Series*, vol. 1748, no. 4. IOP Publishing, 2021, p. 042016.

[58] J. Bonneau, "Ethiks: Using ethereum to audit a coniks key transparency log," in *International Conference on Financial Cryptography and Data Security*. Springer, 2016, pp. 95–105.

[59] R. Dautov and S. Distefano, "Targeted content delivery to IOT devices using bloom filters," in *International Conference on Ad-Hoc Networks and Wireless*. Springer, 2017, pp. 39–52.

[60] Y. Liu, Q. Lu, X. Xu, L. Zhu, and H. Yao, "Applying design patterns in smart contracts," in *International Conference on Blockchain*. Springer, 2018, pp. 92–106.

[61] C. Patsonakis, K. Samari, A. Kiayiasy, and M. Roussopoulos, "On the practicality of a smart contract pki," in *2019 IEEE International Conference on Decentralized Applications and Infrastructures (DAPPCON)*. IEEE, 2019, pp. 109–118.

[62] W. Hua, Y. Chen, M. Qadrdan, J. Jiang, H. Sun, and J. Wu, "Applications of blockchain and artificial intelligence technologies for enabling prosumers in smart grids: A review," *Renewable and Sustainable Energy Reviews*, vol. 161, p. 112308, 2022.

[63] R. Modi, *Solidity Programming Essentials: A Beginner's Guide to Build Smart Contracts for Ethereum and Blockchain*. Packt Publishing Ltd, 2018.

[64] C. Dannen, *Introducing Ethereum and Solidity*. Springer, 2017, vol. 318.

[65] Y. Hirai, "Defining the ethereum virtual machine for interactive theorem provers," in *International Conference on Financial Cryptography and Data Security*. Springer, 2017, pp. 520–535.

[66] http://remix.ethereum.org, Apr. 2021.

II

Applications in Smart Energy Systems

CHAPTER 4

Reforms in Energy Systems: Prosumers Era and Future Low-Carbon Energy Systems

Currently, there is a global shift in the way power systems operate. Traditionally, power systems involved one-way power flows generated by centralized power stations, transmitted through distribution networks, and delivered to passive consumers such as households, businesses, and industries. These consumers would only draw power from the grid and pay for their consumption through wholesale or retail electricity prices, acting as price takers. However, the power system is now transitioning towards a low-carbon, decentralized paradigm, where there are numerous distributed generation sources, energy storage systems, and electric vehicles looking to connect to the power grid, particularly on the demand-side. As a result, power now flows in both directions, from the supply side to the demand-side, or power exchange within the demand-side. At the same time, passive consumers are becoming active by generating energy locally to meet their energy needs or sharing energy with one another. This poses challenges as the power system was not initially designed for this type of operation, both in terms of physical infrastructure such as wires and transformers, and non-physical infrastructure such as information, communication, and control architectures. in addition, as energy systems transition towards a decentralized and localized paradigm, energy markets are also undergoing a similar shift.

This chapter introduces the transition of power systems and energy markets towards the net zero prosumers era. Section 4.1 introduces key stakeholders in current GB energy system. Section 4.2 describes the emerging role of prosumers, and Section 4.3 identifies potential architectures of information and control for prosumer networks, including peer-to-peer trading markets in Sub-Section 4.3.1, intermediary-based trading markets in Sub-Section 4.3.2, and microgrid-based trading markets in Sub-Section 4.3.3. In Section 4.4, the current and potential future regulatory

DOI: 10.1201/9781003170440-4

supports in facilitating such transition are reviewed, including barriers and principles for prosumers' engagement in Sub-Section 4.4.1, regulations for net zero power systems in Sub-Section 4.4.1 and balanced energy markets in Sub-Section 4.4.2. Section 4.5 illustrates technical challenges of future low-inertia power systems caused by increasing penetration of renewable energy sources, including the foundational concepts of inertia, frequency, and frequency response in Sub-Section 4.5.1, challenges caused by low-inertia power systems in Sub-Section 4.5.2, and solutions to those challenges in Sub-Section 4.5.3. Section 4.6 concludes this chapter.

4.1 KEY STAKEHOLDERS IN GB ENERGY SYSTEM

In this section, key stakeholders in GB energy systems are introduced, as a preliminary of energy system transitions. The stakeholders include the power system operator, transmission system operator, distribution network operator, energy suppliers, policy makers, regulators, and consumers, with details as follows.

4.1.1 Power System Operator

The power system operator in the GB energy system is the National Grid Electricity System Operator (ESO). ESO is responsible for managing the electricity system in GB, ensuring that supply and demand are balanced and that the grid remains stable and secure.

ESO is a separate business unit within the National Grid group, which owns and operates the high-voltage transmission network in GB. As the power system operator, the ESO is responsible for maintaining the frequency and voltage of the grid within tight limits, ensuring that there is enough generation to meet demand, and managing the flow of electricity across the transmission network.

ESO works closely with other market participants, including generators, suppliers, and network operators, to ensure the efficient and reliable operation of the GB electricity system. It also plays a key role in facilitating the transition to a low-carbon energy system by integrating increasing amounts of renewable generation and supporting the development of new technologies and market arrangements.

The income for the National Grid ESO comes from a number of sources, including charges for the use of the electricity transmission network and income from the balancing mechanism. These are detailed as follows:

- The ESO charges generators and suppliers for the use of the transmission network to transport electricity from power stations to homes and businesses. These charges are set by the energy regulator, Ofgem, and are designed to cover the costs of operating and maintaining the transmission network.

- The ESO also earns income from the balancing mechanism, which is used to balance the supply and demand of electricity on the transmission network in real time. The ESO pays generators and suppliers to increase or decrease their output to balance the system, and these payments are funded by charges to network users.

- In addition to these sources of income, the ESO also receives funding from the UK government to support the development and implementation of policies and initiatives to promote the transition to a low-carbon energy system.

Overall, the income for the National Grid ESO comes from charges for the use of the transmission network, income from the balancing mechanism, and funding from the UK government. The ESO uses this income to fund its operations, maintain the security and stability of the electricity system, and promote the transition to a low-carbon energy system.

4.1.2 Transmission System Operator

The transmission network operator is responsible for operating the high-voltage transmission network that connects the power stations to the distribution network operators. The transmission network operator in GB is the National Grid Electricity Transmission, which owns and operates the high-voltage transmission network across England and Wales, as well as the interconnectors that link GB to other countries. The transmission network operator is responsible for ensuring secure and efficient operation of the transmission network, maintaining the voltage and frequency of the grid, and managing the flow of electricity across the network.

National Grid ESO is responsible for managing the real-time operation of the electricity system and ensuring a balance between supply and demand, while National Grid Electricity Transmission is responsible for owning and operating the high-voltage transmission network that forms a key part of the electricity system. In the GB energy system, both the National Grid ESO and the National Grid Electricity Transmission are involved in running power flow analysis and frequency response.

The ESO is responsible for ensuring that the electricity system is balanced in real time, by managing the supply and demand of electricity on the transmission network. To do this, the ESO uses a range of tools and techniques, including power flow analysis and frequency response. Power flow analysis is used to model the flow of electricity across the transmission network and identify potential bottlenecks or constraints. Frequency response is used to maintain the stability of the grid by managing the frequency of the AC waveform.

National Grid Electricity Transmission is responsible for the operation and maintenance of the high-voltage electricity transmission network in England and Wales. As part of this role, it provides the infrastructure and equipment necessary to support the ESO in managing the electricity system, including power flow analysis and frequency response.

The income for National Grid Electricity Transmission comes from a number of sources, including charges for the use of the electricity transmission network, income from regulated activities, and other activities such as asset optimisation, with details as follows:

- National Grid Electricity Transmission also charges generators and suppliers for the use of the transmission network to transport electricity from power stations to homes and businesses.

- National Grid Electricity Transmission also earns income from regulated activities, such as providing connection services to new generators or modifying the transmission network to accommodate new energy sources or technologies. These activities are also subject to regulatory oversight by Ofgem, which sets the price controls and performance targets for National Grid Electricity Transmission.

- In addition to these sources of income, National Grid Electricity Transmission also engages in other activities, such as asset optimisation, where it seeks to maximise the value of its existing assets by exploring new business models or revenue streams.

Overall, the income for National Grid Electricity Transmission comes from charges for the use of the transmission network, income from regulated activities, and other activities such as asset optimisation. National Grid Electricity Transmission uses this income to fund its operations, maintain the security and stability of the electricity system, and invest in the infrastructure necessary to support the transition to a low-carbon energy system.

4.1.3 Distribution Network Operator

There are currently 14 distribution network operators in the GB energy system, each responsible for a particular geographic region. The distribution network operators are responsible for distributing electricity from the high-voltage transmission network to homes and businesses across their respective regions.

The 14 distribution network operators in GB are:

- Electricity North West Limited (ENWL)

- Northern Powergrid (NPg)

- Scottish and Southern Energy Power Distribution (SSEPD)

- ScottishPower Energy Networks (SPEN)

- Western Power Distribution (WPD)

- UK Power Networks (UKPN)

- Electricity North East Limited (ENEL)

- Northern Ireland Electricity Networks (NIE Networks)

- Northern Powergrid (Northeast) Limited (NPG-NE)

- Northern Powergrid (Yorkshire) plc (NPG-Y)

- Electricity Supply Board (ESB) Networks Limited

- WPD South West (WPD-SW)

- WPD South Wales (WPD-SWALEC)

- Scottish Hydro Electric Power Distribution (SHEPD)

These distribution network operators are regulated by the energy regulator, Ofgem, which sets price controls and performance targets to ensure that they operate efficiently and in the best interests of consumers.

4.1.4 Energy Suppliers

In the GB energy market, there are a large number of energy suppliers who provide electricity and gas to homes and businesses. These suppliers purchase energy from the wholesale markets and sell it on to customers, offering a range of tariffs and services to meet different needs.

Some of the largest energy suppliers in GB include:

- British Gas

- E.ON

- SSE

- EDF Energy

- Scottish Power

- npower

- Octopus Energy

- Bulb

- OVO Energy

- Shell Energy

In addition to these large suppliers, there are also a number of smaller, independent energy suppliers who compete in the market, offering innovative products and services to customers.

Energy suppliers in GB are subject to regulation by the energy regulator, Ofgem, which sets price caps and other regulations to ensure that customers are treated fairly and that the market operates efficiently. Customers are able to switch energy suppliers to find the best deal for them, and many suppliers offer incentives and rewards to encourage customers to switch.

4.1.5 Policy Maker

In the GB energy system, the policy maker is the UK government, which sets the policy and regulatory framework for the energy sector.

The Department for Business, Energy and Industrial Strategy (BEIS) is the UK government department responsible for developing and implementing policies relating to energy, including electricity, gas, and heat. BEIS works to ensure that the energy system is secure, affordable, and sustainable, while promoting innovation and growth in the energy sector.

BEIS works closely with other government departments, such as the Department for Environment, Food and Rural Affairs (DEFRA) and the Department for Transport, to develop cross-cutting policies that address the challenges of climate change, air quality, and sustainable development.

The UK government plays a crucial role in shaping the direction and priorities of the GB energy system, and in ensuring that the energy sector operates in the best interests of consumers and the wider society.

4.1.6 Regulators

In addition to setting policy, the UK government also regulates the energy sector through the energy regulator, Ofgem. Ofgem is responsible for promoting competition and protecting consumers in the energy market, and sets the price controls and performance targets for network operators and energy suppliers. Ofgem is a non-ministerial government department responsible for protecting the interests of gas and electricity consumers by promoting competition and regulating the industry.

As the regulator, Ofgem is responsible for setting price controls and performance targets for the energy networks and ensuring that they invest efficiently in maintaining and upgrading their infrastructure. It also sets price caps for the retail energy markets to protect consumers from being overcharged by energy suppliers.

In addition to regulating prices, Ofgem also promotes competition in the market and supports the development of new technologies and business models that can help to reduce carbon emissions and improve the efficiency of the energy system.

Ofgem plays a crucial role in ensuring that the energy system in GB operates efficiently, transparently and in the best interests of consumers.

4.1.7 Consumers

In the GB energy system, consumers refer to the individuals and businesses who use electricity and gas to power their homes and operations. These include households, small- and medium-sized enterprises, and large industrial users.

Consumers in GB are able to choose from a range of energy suppliers, each offering different tariffs and services to meet different needs. They are also able to switch suppliers and tariffs to find the best deal for them, as well as access support and advice from consumer groups and government agencies.

In addition to purchasing energy from suppliers, consumers in GB can also generate their own energy through the use of renewable energy technologies such as solar panels or wind turbines. This is known as distributed energy, and it is playing an increasingly important role in the energy system as more and more consumers seek to reduce their carbon footprint and take greater control over their energy usage.

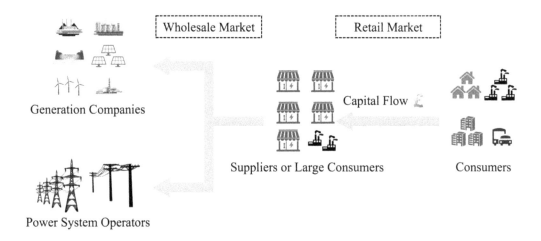

Figure 4.1   Schematic illustration of the capital flow in liberalised energy markets. The energy suppliers and large industrial consumers purchase the energy from the generation companies in the wholesale markets, and sell this energy to their consumers. Hence, there is a one-way capital flow from consumers to the suppliers, and subsequently the capital flow is distributed between the generation companies and the transmission or distribution system operators.

Consumers play an important role in the GB energy system, as they are the ultimate users of energy and the primary beneficiaries of the system's services. They are also important drivers of change, as they seek to reduce their energy consumption, generate their own energy, and play a more active role in the management of the energy system.

4.2   THE EMERGING ROLE OF PROSUMERS

In the liberalised energy markets, as shown in Figure 4.1, the energy suppliers and large industrial consumers purchase the energy from the generation companies in the wholesale markets and sell this energy to their consumers. Hence, there is a one-way capital flow from consumers to the suppliers, and subsequently the capital flow is distributed between the generation companies and the transmission or distribution system operators for their services of managing power grids. However, with the engagement of active consumers, this market design does not fit for the purpose any more. We need local energy markets to allow consumers to exchange energy and capital within their community, along with tailored auction and pricing mechanisms. As identified by the National Grid ESO [1], there are four different, credible pathways to decarbonise energy systems till 2050: steady progression, system transformation, consumer transformation, and leading the way. These four scenarios are evaluated on the dimensions of demand-side engagement, electrification of heat, energy efficiency, and supply-side flexibility. In any of these scenarios, there are increasing numbers of active consumers with their localised generation and storage assets, which enables consumers to produce, consume, and store energy through the distributed renewable sources [2], batteries [3], electric vehicles [4], and smart meters [5]. The conventional power systems are transitioning towards prosumers era.

The figure of prosumers was coined by Alvin Toffler in 1980 [6]. On the context of energy markets, prosumers are small-sized or medium-sized agents [7], e.g., residential, commercial and industrial users, who actively produce energy and feed surplus energy into a distribution network after self-consumption. When prosumers' demand cannot be met by self-generation, they import energy from main grids or other prosumers [7]. Accommodating increasing numbers of prosumers would require flexible market structures, efficient information infrastructures, and decentralised control architectures.

4.3 MARKET STRUCTURES FOR PROSUMER NETWORKS

A flexible structure of local energy markets towards the decentralised generation and consumption is crucial for the integration of emerging role of prosumers. The foundation to support this kind of flexible local energy markets is the decentralised information and control infrastructures. There are primarily three architectures commonly used in research and industrial practices, including the peer-to-peer trading markets, intermediary-based trading markets, and microgrid-based trading markets. The schematic illustration of these three types of energy markets design is presented in Figure 4.2. In the three sub-figures, the dots represent the control units and lines represent the information flows. The key features of these three architectures are summarised as follows, with details and implementations explained in the following subsections.

- *Peer-to-peer trading markets:* The first architecture on the left hand side is the fully decentralised structure. In this architecture, the energy sources will be directly controlled by prosumer themselves. These prosumers can decide their own offering or bidding prices. There behaviours will be driven by their own interests. The key challenge of this architecture is how to align individual prosumers' interests with the power system's benefits.

- *Intermediary-based trading markets:* The second architecture on the middle is the intermediary based structure. The control unit is the intermediary, such as aggregators or energy suppliers. Like the retail markets, the aggregators or suppliers determine the energy prices for its customers. All the resources within the community are pooled together to guarantee the community's benefits. The issue is it still relies on the third party, i.e., the intermediary.

- *Microgrid-based trading markets:* The third architecture on the right hand side is organised as microgrids. It can be either operated in an islanded mode independently, or connected to the utility grid. When microgrids are connected to the utility grid, prosumers can export their surplus power back to the grid. In this case, they are driven by the electricity prices. When microgrids are operated independently, prosumers can share energy within a microgrid and they are driven by the price of a microgrid.

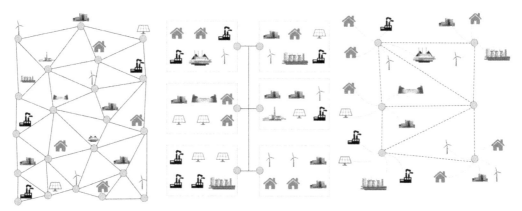

Peer-to-Peer Trading Markets   Intermediary-Based Trading Markets   Microgrid-Based Trading Markets

Figure 4.2 Schematic illustration of energy markets design towards the prosumers era. Under the peer-to-peer trading markets, prosumers interconnect with each other to trade electricity and other services; Under the intermediary-based trading markets, an ensemble of prosumers is organised by an intermediary to pool generation sources, flexible demand, and storage capacities; Under the microgrid-based trading markets, prosumers connect to microgrids and microgrids either connect to the main grid or operate at the islanded mode as indicated by the dashed lines. Dots indicate prosuming agents. Lines indicate the information exchange among agents.

4.3.1 Peer-to-Peer Trading Markets

Like other collaborative economy [8], such as Uber and Airbnb, the peer-to-peer trading allows the prosumers and consumers in the demand-side of power systems to exchange energy and other services, e.g., the demand-side management, energy storage [9], and carbon credits [10] [11], in a real-time, autonomous, and decentralised manner. The real-time would require some enabling technologies such as smart meters and advanced communication technologies with low latency. The autonomous means the distributed energy sources are directly controlled by their owners. The decentralisation would require some distributed ledger technologies, such as Blockchain. The role of the distribution system operator becomes a facilitator to facilitate prosumers' engagement and provide the distribution functions [12]. In comparison to other two structures, the peer-to-peer trading markets are the least structured markets. Instead of using central authorities, such as aggregators or energy suppliers, as control agents, individual prosumers become an independent agent to exchange energy and information with each other and perform control functions [13]. Hence, this framework enables a flexible market structure.

The question is what benefits the peer-to-peer energy trading would bring to the power systems, communities, and prosumers. Figure 4.3 categorises the benefits of the peer-to-peer energy trading from different temporal dimensions. The real-time benefits include balancing supply and demand locally, bill saving for consumers, provision of ancillary services, and flexibility provision. The mid-term benefits include

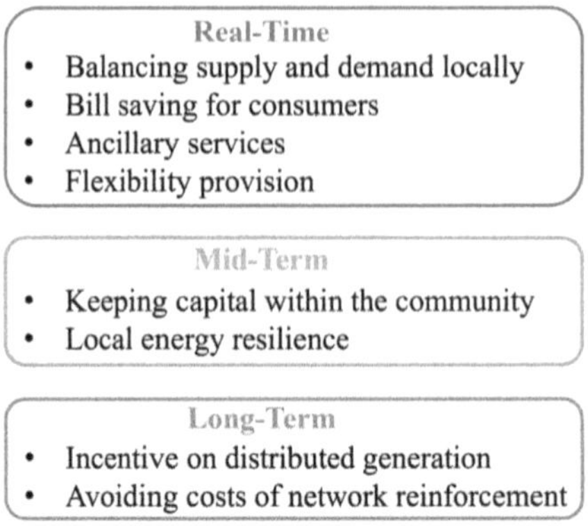

> **Real-Time**
> • Balancing supply and demand locally
> • Bill saving for consumers
> • Ancillary services
> • Flexibility provision

> **Mid-Term**
> • Keeping capital within the community
> • Local energy resilience

> **Long-Term**
> • Incentive on distributed generation
> • Avoiding costs of network reinforcement

Figure 4.3 Benefits of the peer-to-peer energy trading from different temporal dimensions. The real-time benefits include balancing supply and demand locally, bill saving for consumers, provision of ancillary services, and flexibility provision. The mid-term benefits include keeping capital within the community and enhancing local energy resilience. The long-term benefits include the incentive on distributed generation and avoiding huge costs of power network reinforcement.

keeping capital within the community and enhancing local energy resilience. The long-term benefits include the incentive on distributed generation and avoiding huge costs of power network reinforcement.

To help readers understand these benefits, an real world example is used: I have the highest solar output of my roof-top solar panel on a sunny day. After meeting my self-consumption, I still have extra power to share with my neighbours which would be otherwise wasted. My neighbour can buy this energy at a price lower than the retail electricity price. Hence, my neighbour could save the electricity bills. For myself, I could have extra revenue by selling power. For the community, the capital has been kept within the community. we can balance the supply and demand locally without going through the long-distance transmission and distribution. For the transmission system operators and distribution system operators, they can avoid or defer the massive investment on reinforcing the transmission and distribution networks. In addition, the peer-to-peer energy trading can help local energy resilience. During the situation of the blackout, if I have the battery storage, I can provide my stored energy as a service to the community. There are also other ancillary services, such as frequency response to maintain the frequency of power systems and flexibility provision. For example, if I want to reduce my demand at the peak demand period, I can also provide the demand reduction to the power system operators in a form of the flexibility service.

Nonetheless, the increasing burdens on control agents and information flows along with system states and control decisions made by agents amplify the volumes of information flows. This presents a challenge for the information infrastructures of power systems [14]. Another challenge of the peer-to-peer trading markets is how to ensure the system constraints and the security of supply without the intervention of power system operators.

In practice, the RWE [15] developed the peer-to-peer trading platforms integrating functions of the decentralised control, network management, communication, automation, and security. The Power Ledger [16] provides the peer-to-peer energy trading for 11,000 participants from residential and commercial consumers in Australia based on software solutions. This peer-to-peer trading market is supported by the Australia government, utilities, and distribution system operator.

4.3.2 Intermediary-Based Trading Markets

The intermediary-based trading markets are more structured than the peer-to-peer trading markets. Under the intermediary-based trading markets, an ensemble of prosumers is organised as a community or local organisation, e.g., smart buildings [17] and virtual power plants [18]. Each community is managed by an intermediary, e.g., aggregators [19] or retailers [20], as an agent to maintain the regional energy balance and provide energy services. All generation sources, flexible demand, and storage capacities within a community are pooled to collectively coordinate resources for local benefits. The intermediary can earn bonus from regulators or utilities for providing services to prosumers such as the efficiency update, demand response, and setup of renewable energy sources [21].

The concept of virtual power plant was originated from virtualisation and digitalisation of internet of things from 1990s. It is similar to the virtual technologies used in games, conferences, and businesses. The virtual reality is integrated into the operational management of power systems. According to Kraftwerke [22], virtual power plants are an internet based decentralised networks which can organise small or medium generation resources, e.g., solar panels and heat pumps, and flexibility resources, e.g., demand-side management and fast frequency response. All distributed resources are coordinated by a central controller, e.g., mainframe computer, while they remain independent operation and asset ownership.

An example of the intermediary is the Stem [23] which has designed a platform to provide the storage services and demand response for consumers in California through real-time optimisation and automated control. The utility in Oregon, Portland General Electric [24], linked up 525 households with solar-storage systems into a virtual power plant as grid resources, which assembled 4 MW batteries as a precursor to 200 MW of flexibility provision in distribution networks in terms of supply–demand balance. The company of Energy and Meteo Systems [25] in Germany has established a virtual power plant via the digital control centre with the services of the real-time data management, remote control of wind and solar generation, energy scheduling, demand-side management, and balancing group management. The data collection and controlling decisions are managed by the digital control centre without the need of new IT infrastructures. The Swell Energy [26] pooled the energy storage, solar panels, and battery management software to provide electricity to households in Los Angeles as parts of the grid capacity.

4.3.3   Microgrid-Based Trading Markets

The microgrid-based trading markets are the most structured framework, under which prosumers are connected to the microgrid and the microgrid can either connect to the main grid or operate at the islanded mode. When a microgrid connects to the main grid, prosumers can sell surplus generation to the main grid [7]. Prosumers would be incentivised to generate more energy for earning profits through exporting. When a microgrid operates at the islanded mode, the surplus generation can be stored within the microgrid or used for load shifting services [27]. Prosumers would be incentivised to strategically schedule their generation and consumption for local energy balance. The primary difference between the microgrid-based trading markets and intermediary-based trading markets is that there is no intermediary in microgrid-based trading markets to pool resources together. Individual resources of the generation and consumption can directly connect to the microgrid and then to the main grid. Rather than seeking for an intermediary's benefits, e.g., maximising the bonus, individual microgrids seek for their own benefits, e.g., maximising energy exports or achieving energy balance.

The microgrids interconnect to distribution networks and provide electricity to its nearby consumers through both software and control systems. The microgrids geographically serve for local consumers and therefore overcome the power losses of long-distance transmission in conventional electricity systems and internet based virtual power plants. In addition to the software-based centralised controller for asset management as that in virtual power plants, microgrids are equipped with hardware, such as inverters or switches, in assisting their operations.

As practical implementations, the LO3 Energy has developed the Brooklyn microgrid integrating buildings on site to support demand-side managements and improve communication infrastructures. The Asea Brown Boveri Ltd [28] provides microgrid solutions for customers to ensure the reliable, stable, and affordable power.

4.4   REGULATORY SUPPORTS

This section will first identify the regulatory barriers for the engagement of prosumers and list principles for a fair design of local energy markets. Second, policy supports for transitioning towards future net zero power systems in GB are introduced, including the carbon pricing schemes, contract for difference auction, and capacity auction. Third, regulations for balancing markets are presented.

4.4.1   Regulatory Barriers and Principles for Prosumers Engagement

Six primary barriers are firstly identified in this section as summarised in Figure 4.4, with details explained as follows:

- *Scale differences:* The scale of prosumers is small compared to large-scale generation companies or energy suppliers, but the number of them is large with the features of decentralisation and distribution.

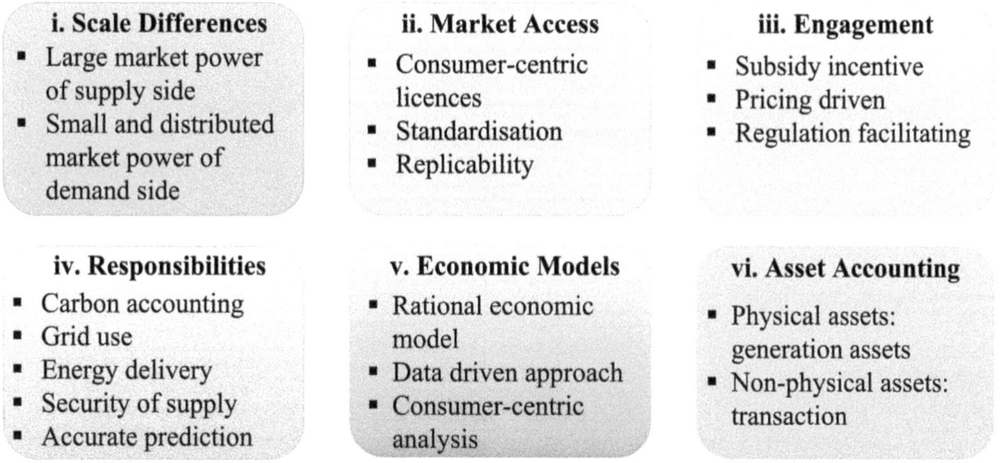

i. Scale Differences	ii. Market Access	iii. Engagement
■ Large market power of supply side ■ Small and distributed market power of demand side	■ Consumer-centric licences ■ Standardisation ■ Replicability	■ Subsidy incentive ■ Pricing driven ■ Regulation facilitating
iv. Responsibilities	v. Economic Models	vi. Asset Accounting
■ Carbon accounting ■ Grid use ■ Energy delivery ■ Security of supply ■ Accurate prediction	■ Rational economic model ■ Data driven approach ■ Consumer-centric analysis	■ Physical assets: generation assets ■ Non-physical assets: transaction

Figure 4.4   Key regulatory barriers for the engagement of prosumers.

- *Market access:* In energy markets, generation companies and energy suppliers would require licences to access the markets, but the number of prosumers is large. It is difficult to design the licence individually. The opportunity for policy makers is to design standardised licenses which are replicable for all prosumers.

- *Engagement:* The third barrier is how to incentivise the engagement of prosumers into the peer-to-peer energy trading, whether using energy pricing schemes naturally drive the behaviours of prosumers, providing extra monetary compensation, or making certain regulations.

- *Responsibilities:* The fourth barrier is how to account for the responsibilities of prosumers when they participate in the peer-to-peer energy trading. For example, how to account for the contribution of carbon reduction if prosumers install the air-source heat pump or roof-top solar panels. If prosumers exchange energy through the distribution networks, do they need to pay for the use of power grids. How to ensure the prosumers predict their generation accurately and how to ensure the security of supply.

- *Economic models:* The fifth barrier is how to analyse and model the behaviours of individual prosumers. We know that the household does not invest by a rational economic model. One opportunity is to use the data driven machine learning to learn their behaviours from historical data.

- *Asset accounting:* The last barrier is how to account for the assets of prosumers, including both physical assets, such as roof-top solar panels and air source heat pumps, and non-physical assets, such as the transaction information.

Based on these identified barriers, some principles for demand-side policy design are listed next. As presented in the scale of Figure 4.5, the current policies mainly focus on the supply side, with less attention on the demand-side. To make this scale

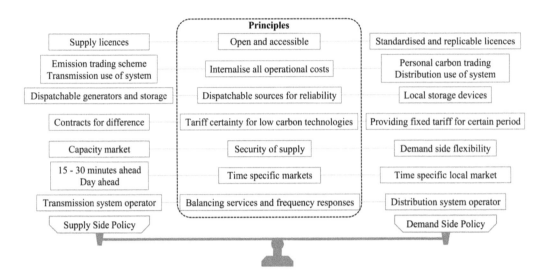

Figure 4.5 Principles for the demand-side policy design. The current policies mainly focus on the supply side, with less attention on the demand-side. To make this scale balance, demand-side policy measures need to be equally designed.

balance, we need to equally design demand-side policy measures. The detailed principles are explained as follows:

- *Scale differences:* To ensure an open and accessible energy markets, if the large-scale generators or suppliers need the licenses on the supply side, the same standardised and replicable licenses would be required on the demand-side.

- *Internalise all operational costs:* All the operational costs need to be internalised. For example, the personal carbon trading [29, 30] is needed in the demand-side similar to emission trading scheme in the supply side. The cost of grid use needs to be charged from prosumers when they participate in the peer-to-peer energy trading.

- *Dispatchable sources for reliability:* The third principle is to ensure the demand-side has the same dispatchable sources similar like the dispatchable generators and storage devices in the supply side to compensate active or reactive power, for ensuring the reliability and stability of power systems.

- *Tariff certainty for low-carbon technologies:* The fourth principle is to provide certainty for investing in low-carbon technologies. We have the contract for difference in the supply side to give generation companies some certainty on the low carbon investment. We also need the same certainty for prosumers.

- *Security of supply:* Similar to the capacity market, the demand-side flexibility needs to be aggregated for assisting power system operation and ensuring the security of supply.

- *Time-specific market:* The time specific local markets are necessary, which aligns to the wholesale and retail energy markets.

- *Balancing Services and Frequency Responses:* The demand-side should be incentivised to provide ancillary services and fast frequency response to power systems.

4.4.2 Policy Supports for Net Zero Transition

Policy makers aim to achieve the net-zero or net-negative electricity generation through determining the carbon price and providing the contract for difference auctions and capacity auctions, which is explained as follows.

4.4.2.1 Carbon Pricing Scheme

The carbon intensity describes the amount of carbon emissions from power systems per unit of electric energy generation, with a unit of gCO_2/kWh. Policy makers compare the actual carbon intensity with a emission target to see if the carbon emission reduction is on track.

The low carbon policy design has been focused by the international regulations and existing research. Market-based low carbon policies, also known as the carbon pricing, are an economic instrument to address the carbon emissions caused by the combustion of fossil fuels [31]. The carbon pricing enforces the pollutant emitters to compensate the environmental damage in a monetary manner. Therefore, the implementation of the carbon pricing increases the costs of using fossil fuels and subsequently stimulates the carbon mitigation [32]. Two primary forms of the carbon pricing are the carbon tax and emissions trading scheme. By the end of 2019, the carbon pricing schemes have been implemented in 46 countries, of which 25 countries adopt the carbon tax and the rest 21 countries adopt the emissions trading scheme [33]. The carbon pricing helps these countries achieve their low carbon targets by stimulating the energy conservation, improving the energy efficiency, and investing low carbon technologies. The details of these two forms of the carbon pricing are explained as follows:

- *Carbon tax:* The carbon tax levies a fixed rate on the carbon content of fossil fuels [34]. The rate of carbon tax is determined by the social cost of carbon which quantifies the marginal damage costs of carbon emissions to the society [35]. As a revenue of the policy maker, the carbon tax can be further redistributed for investing in low-carbon technologies or providing monetary compensation for the demand-side carbon mitigation, so as to achieve the carbon revenue neutrality. For related research please refer to [36, 37].

- *Emissions trading scheme:* The emissions trading scheme, also known as the cap-and-trade scheme, is an alternative policy to the carbon tax. Under the emissions trading scheme, the policy makers and regulators allocate a certain amount of carbon allowances for a given time period [38]. Emitters are obliged to have an enough amount of carbon allowances covering the amount of their carbon emissions. The surplus or scarcity of carbon allowances can be traded among emitters [39].

Nonetheless, an inappropriate carbon price determined by the emissions trading scheme would inefficiently incentivise the carbon emissions mitigation and fail to achieve the low carbon targets. The issue of the inappropriate carbon price presents a challenge for the emissions trading scheme in a majority of countries [40]. If the carbon price lies below the social cost of carbon or the rate at which the low-carbon targets can be achieved, it would insufficiently stimulate the mitigation of carbon emissions; If the carbon price in one region is higher than that in another region, the market competitiveness of carbon producers in the high-price region would be harmed. The carbon producers are prone to discharging carbon emissions in the low-price region, while the total amount of carbon emissions remains unchanged, which is defined as the carbon leakage issue [41]. In addition, the carbon producers will pass the cost of carbon allowance onto consumers in the form of higher prices on the products, e.g., higher electricity prices.

To overcome the issue of the inappropriate carbon price, the carbon price floor and ceiling are implemented in current international carbon markets by setting an additional price limits for the carbon emissions producers in certain regions [42]. For the case of the UK carbon market, because the carbon price of the EU emissions trading scheme is lower than the social cost of carbon in the UK, the carbon price has failed to incentivise the UK coal-to-gas transition before 2013 [40]. Afterwards, the UK has formulated the carbon price support for its own carbon producers as an additional carbon price floor to the EU emissions trading scheme. The US set a similar price floor and facilitated carbon auctions in 2009 [43]. By contrast, in New Zealand, a carbon price ceiling was enacted through the fixed price option to prevent high carbon prices and protect the market competitiveness of generators [44].

As two well-established policy instruments, the carbon tax and emissions trading scheme have the following aspects in common:

- Both the carbon tax and emissions trading scheme impose a price on carbon emissions for facilitating energy producers and consumers to internalise the social cost of carbon.

- Instead of the command-and-control based policy measures that specify actions for the carbon mitigation to be taken, the market based policy measures flexibly incentivise carbon producers to strategically respond to the prices.

- Market-based low carbon policies can generate public revenue through charging the carbon tax or selling carbon allowances.

The differences between the carbon tax and emissions trading scheme including the advantages and limitations of each policy design are as follows:

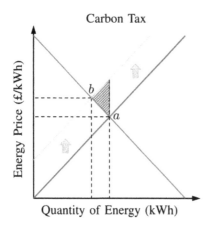
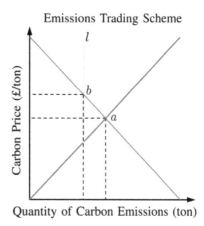

Figure 4.6 Comparison between carbon tax and emissions trading scheme from economics perspective. The implementation of the carbon tax would raise the energy price and reduce the energy demand. The emissions trading scheme would limit the total carbon emissions and raise carbon price.

- The carbon tax gives a certainty to the price of carbon emissions through the fixed tax rate, whereas the emissions trading scheme gives a certainty to the quantity of carbon emissions through the fixed carbon allowance [45].

- Carbon tax is easier to be implemented since it is based on the established tax systems. By contrast, the emissions trading scheme is more flexible since it can be extended with the financial innovations such as the peer-to-peer trading, options, banking, and borrowing.

- From the economics perspective as indicated in Figure 4.5, when the carbon tax is implemented, the energy price increases and the energy demand decreases from point a to point b. Consumers would find alternatives, e.g., the load shifting or load curtailment, electric vehicles, replacing the gas furnace with the heat pumps. By contrast, under the emissions trading scheme, when the total amount of carbon allowance is fixed according to the target of the carbon mitigation as indicated by the line l, the carbon price would increase from point a to point b. Facing the uncertainty of the carbon price, generators would find alternatives, e.g., improving the combustion efficiency, replacing the coal by the gas, and investing the renewable generation.

The comparison between the carbon tax and emissions trading scheme is summarised in Table 4.1.

4.4.2.2 Contract for Difference Auction

Policy makers aim to encourage the investment in renewable generation infrastructures through providing the contract for difference auction. The eligible generation technologies for participating in the contract for difference auction include renewable,

Table 4.1 Comparison between the carbon tax and emissions trading scheme

	Carbon tax	Emission trading scheme
Common	Impose pricing on environmental damage Market based policy instead of commend-and-control Create public revenue	
Difference	Certainty to the price of carbon emissions Easier to implement based on tax system	Certainty to the quantity of carbon emissions Flexible and extendable with financial tools

nuclear, and bioenergy with carbon capture and storage technologies, after receiving the required capacity issued by the policy maker. The detailed procedures of the contract for difference auction are described as follows:

- *Step 1:* Individual generation companies which are willing to participate in the contract for difference auction submit their bids with the information of technological type, bidding price, bidding capacity, and construction time. The bidding price of a generation company can be set as the costs of producing per unit of electricity.

- *Step 2:* Policy makers arrange the bidding prices in the ascending order, and rank the generation companies from the lowest bidding price to the highest bidding price.

- *Step 3:* Until the sum of bidding capacity reaches the capacity issued by policy makers, the auction ends. The counted generation companies win the contract for difference auction at the strike price. The final strike price of the contract for difference auction equals to the bidding price of the last counted company, and individual generation companies pay as their bids.

The flowchart of the contract for difference auction is presented in Figure 4.7.

4.4.2.3 Capacity Auction

Policy makers aim to ensure the security of supply through providing the capacity auction. The detailed procedures of the capacity auction are described as follows:

- *Step 1:* Policy makers evaluate the required capacity using the average cold spell (ACS) peak demand [46] and derated capacity as

$$p^{\mathrm{rc}} = d^{\mathrm{peak,ACS}} - p^{\mathrm{dc}}, \tag{4.1}$$

where

$$d^{\mathrm{peak,ACS}} = d^{\mathrm{peak}} \cdot \theta, \tag{4.2}$$

and

$$p^{\mathrm{dc}} = \sum_{g=1}^{\mathcal{G}} p_g^{\mathrm{cap}} \cdot \vartheta_g, \tag{4.3}$$

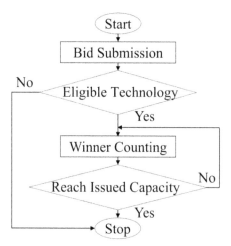

Figure 4.7 Flowchart of the contract for difference auction. Individual generation companies which are willing to participate in the contract for difference auction submit their bids. Policy makers arrange the bidding prices in the ascending order and rank the generation companies from the lowest bidding price to the highest bidding price. Until the sum of bidding capacity reaches the capacity issued by policy makers, the auction ends.

where p^{rc} is the total required capacity, $d^{\text{peak,ACS}}$ is the ACS peak demand, p^{dc} is the total derated capacity, d^{peak} is the total peak demand of all consumers, θ is the ACS coefficient, p_g^{cap} is the installed capacity of the power plant g, ϑ_g is the availability of the power plant g, and $\mathcal{G}$ is the index set of power plants.

- *Step 2:* If the required capacity is positive, i.e., $p^{\text{rc}} > 0$, policy makers initiate the capacity auction.

- *Step 3:* Individual generation companies which are willing to participate in the capacity auction submit their bids with the information of technological type, bidding price, bidding capacity, availability, and construction time. One potential strategy for submitting the bidding price of a generation company is that 1) if a generation company with its power plants is unprofitable, i.e., negative net present value (NPV), this generation company will submit the bidding price equals to its NPV, in achieving the zero NPV, and 2) if a generation company with its power plants is profitable, , i.e., positive NPV, this generation company will submit the bidding price as zero, which can be described as

$$\pi_m^{\text{cap}} = \begin{cases} -f_m^{\text{NPV}}, & \text{if } f_m^{\text{NPV}} < 0, \\ 0, & \text{if } f_m^{\text{NPV}} \geq 0, \end{cases} \tag{4.4}$$

where π_m^{cap} is the bidding price for the capacity auction of the generation company m, and f_m^{NPV} is the NPV of the generation company m.

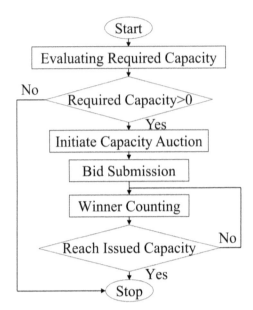

Figure 4.8 Flowchart of the capacity auction. Policy makers evaluate the required capacity. If the required capacity is positive, policy makers initiate the capacity auction. Individual generation companies which are willing to participate in the capacity auction submit their bids. Policy makers arrange the bidding prices in the ascending order, and count the generation companies from the lowest bidding price to the highest bidding price, until the sum of bidding capacity reaches the required capacity.

- *Step 4:* Policy makers arrange the bidding prices in the ascending order, and count the generation companies from the lowest bidding price to the highest bidding price.

- *Step 5:* Until the sum of bidding capacity reaches the required capacity, the auction ends. The final strike price of the capacity auction equals to the bidding price of the last counted company.

- *Step 6:* The counted generation companies win the capacity auction at the strike price.

The flowchart of the capacity auction is presented in Figure 4.8.

4.4.3 Regulation for Electricity Trading and Balance: A Case in the GB Electricity Market

The GB electricity market is transitioning to a more flexible form which allows individual customers to choose their electricity suppliers [47]. The suppliers can also buy electricity from generators of their own choice in order to meet the demand of their customers. It is noted that the electricity trading belongs to the non-physical trading [48], which means that the customers do not have the physical demand for any particular portion of electricity they want to buy, and suppliers do not have the

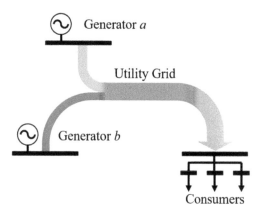

Figure 4.9 Schematic illustration of the non-physical trading of the electricity. The generators a and b feed their electricity generation to the utility grid, and the utility grid homogenises the electricity from different sources and then supply it to different consumers.

physical option for supplying their electricity to any particular customer. As shown in Figure 4.9, the generators a and b feed their generation to the utility grid, and the utility grid homogenises the electricity from different sources and supply to different consumers. The suppliers, as a non-physical trader, only account for the amounts and prices at which they purchase to the generators and charge from the customers. This results in the separation between the cash flows and power flows.

4.4.3.1 Settlement

Since the electricity is a non-physical product which cannot be stored (unless deploying the storage devices), the electricity needs to be generated, transmitted, distributed, and consumed continuously in a real-time manner. In the context of the GB electricity market, the electricity is considered to be generated, transmitted, distributed, and consumed in every half-hour interval. This half hour interval is defined as the settlement period [49] during which the amount and price of the electricity generation and consumption are determined.

During each half hour of the settlement period, the medium or large scale consumers, e.g., industries, and suppliers estimate their electricity demand and contract with generators at the estimated volume of demand [50]. The contracts can be submitted until the beginning of the settlement period. Afterwards, during the half hour of the settlement period, generators would be expected to generate the contracted volume of electricity and consumers would be expected to consume the contracted volume of electricity.

4.4.3.2 Imbalance Management

Ideally the supply and demand would be balanced in every half-hour settlement period. Nonetheless, the imbalance would be incurred for the following reasons:

- The suppliers inaccurately predict the electricity consumption of their consumers [51];

- The power output of a generator is less than the contracted amount of generation due to the part-loaded operation or technical issues [52];

- The technical issues are incurred in the transmission systems [53].

The issue of imbalance needs to be managed by an independent role in the power systems, i.e., the system operator. In the context of the GB energy market, the national grid acts as this system operator to ensure the supply-demand balance and address the issues incurred from the transmission and delivery.

For the generators, if they have the additional capacity which has not been contracted with any supplier or consumer during any half hour, they can make this additional capacity available to the system operator with an offering price which they would like to receive for this additional capacity. By contrast, if the generators would like to reduce the generation, they can provide a bidding price for reducing the generation.

For the suppliers and medium or large scale consumers, if they would like to curtail or shift their demand, they can make this additional volume of electricity available to the system operator with an offering price which they would like to receive for this additional volume. By contrast, if the suppliers and medium or large scale consumers would like to raise their demand, they can provide a bidding price for the increased demand.

Therefore, the *offer* describes the cases of which the generators raise their generation or the suppliers and medium or large scale consumers reduce their demand, whereas the *bid* describes the cases of which the generators reduce their generation or the suppliers and medium or large scale consumers raise their demand [54]. The system operator will accept these bids and offers in real-time, in order to balance the supply and demand.

After every settlement period, the actual generation and consumption in that half hour are metered and compared with the contracted generation and consumption considering the adjustments from bids and offers. The corresponding measures will be taken by the system operators to address the mismatch between the contracted volumes and actual volumes.

For the generators, if their actual generation is less than the contracted generation, they need to purchase the undelivered generation from the grid for meeting the contracted volume. If their actual generation is more than the contracted generation, they need to sell all the over-delivered generation to the grid.

For the suppliers and medium or large scale consumers, if their actual consumption is less than the contracted consumption, they need to sell all the additional electricity to the grid. If their actual consumption is more than their contracted consumption, they need to buy the additional electricity according to the actual volume of consumption.

These imbalance cases are subject to imbalance charges. The imbalance prices are categorised into the system buy price and system sell price [55]. The system buy price is paid by the participants which have a net deficit of the imbalance energy, and the system sell price is paid to the participants which have a net surplus of the imbalance energy. The fluctuation of these imbalance prices reflect the bids and offers of imbalance energy selected by the system operator, e.g., the national grid in the GB electricity market, in order to balance the energy flows in transmission systems and meet the reserve requirements.

The generators, suppliers, and medium or large scale consumers submit physical information through the balancing mechanism units, so that their bids and offers can be accepted by the system operator in advance of the market gate closure. The balancing mechanism unit is the unit which accounts for the energy inflows and out-flows of the total system, i.e., the transmission system and each distribution system combined. As the smallest unit, the balancing mechanism unit integrates a collection of generating units or consumption meters which can be independently monitored for the settlement. The settlement refers to a process of calculating the imbalance vol-umes and incurred prices. To ensure the accuracy of the settlement, the calculation of settlement is repeated four occasions covering 14 months once receiving a more accurate input data [56].

4.5 TECHNICAL CHALLENGES OF FUTURE LOW-INERTIA POWER SYSTEMS

Low-inertia power systems refer to power systems that have a low level of synchronous rotating mass, which is traditionally used to maintain system stability. These systems typically have a high penetration of renewable energy sources, such as wind and solar, which are characterised by their variable and intermittent nature.

The future of low-inertia power systems involves addressing the challenges asso-ciated with integrating large amounts of renewable energy into the grid. This will require the development of new technologies and strategies to maintain system sta-bility and ensure reliable power supply.

One approach is to use power electronics to emulate the behaviour of synchronous rotating mass, providing a synthetic inertia that can help stabilise the system. An-other approach is to develop advanced control systems that can quickly respond to changes in the system, allowing for more efficient use of existing resources.

The future of low-inertia power systems is likely to involve a combination of these and other technologies, as well as new approaches to grid planning and management. These efforts will be critical to ensure that renewable energy can be integrated into the grid in a way that maximises its benefits while maintaining the reliability and stability of the power system. This section will discuss technical challenges of future low-inertia power systems.

4.5.1 Frequency and Inertia

The inertia is the kinetic energy stored in the rotating masses of generators and loads, such as gas turbines or condensers, which are synchronously connected to a power system [57]. This kinetic energy can be exchanged with power systems whenever there

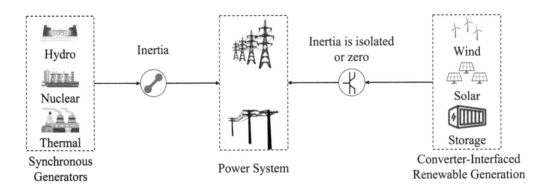

Figure 4.10 Transition towards future low-inertia power systems. On the left hand side, there are synchronous generators which can provide inertia to power systems, whereas on the right hand side, there are converter-interfaced generation, for which the inertia is either isolated, such as wind turbines, or zero, such as solar panels.

are instantaneous imbalances between generation and load, which is defined as the inertia response.

To achieve the future net zero energy system, there is increasing converter-interfaced renewable generation, combined with energy storage systems, replacing the conventional synchronous generators. As presented in Figure 4.10, on the left hand side, there are synchronous generators which can provide inertia to power systems, whereas on the right hand side, there are converter-interfaced generation, for which the inertia is either isolated, such as wind turbines, or zero, such as solar panels. The advantage is these converter-interfaced renewable energy sources have zero or near zero carbon intensities, which gives them the key role of decarbonising power systems. The disadvantage is low inertia would weaken the power system's ability to resist the disturbances caused by generation tripping or connection of new large load. The frequency is the key to low-inertia power systems. During the transient of power systems, if the load is greater than generation, for example, loss of a generator, the inertia stored in the rotating machines would be released to compensate this energy deficit. Therefore, the rotating machines slow down and frequency declines. By contrast, if the generation is greater than load, the excess energy would be converted to inertia to speed up rotating machines and increase the frequency.

Figure 4.11 shows an example of a generation contingency to illustrate the relationship between the inertia and frequency, and corresponding primary frequency response measures used in GB power systems. If there is a loss of large generation, the frequency would drop rapidly. The inertia stored in the rotating mass is released to slow the rate of change of frequency (RoCoF), through the inertial response. Next, the enhanced frequency response which is primarily provided by the energy storage system can release power to system in 1 s. When the monitoring and control devices of generators detect the decrease of frequency, they would increase their mechanical power to restore the frequency, which is called the primary frequency response. Both the enhanced frequency response and primary frequency response can bring

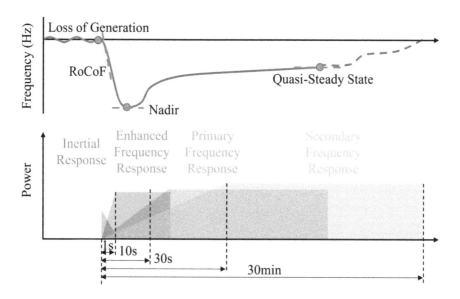

Figure 4.11 Relationship between the inertia and frequency, and corresponding primary frequency response measures. If there is a loss of large generation, the frequency would drop rapidly. The inertia stored in the rotating mass is released to slow the rate of change of frequency (RoCoF), through the inertial response. Next, the enhanced frequency response which is primarily provided by the energy storage system can release power to system in 1 s. When the monitoring and control devices of generators detect the decrease of frequency, they would increase their mechanical power to restore the frequency, which is called the primary frequency response. Both the enhanced frequency response and primary frequency response can bring the frequency back to a new equilibrium, called quasi-steady state frequency. The minimum point of frequency is called the frequency Nadir. The secondary frequency response then acts to restore the frequency to nominal frequency. At the same time of restoring the generation, the load decreases a bit, which is the load damping contributed by some flexible loads.

the frequency back to a new equilibrium, called quasi-steady state frequency. The minimum point of frequency is called the frequency Nadir. Later on, the secondary frequency response acts to restore the frequency to nominal frequency. At the same time of restoring the generation, the load decreases a bit, which is the load damping contributed by some flexible loads.

4.5.2 Challenges of Low-Inertia Power Systems

This subsection identifies primary challenges when transitioning towards low-inertia power systems. With the same example of the system disturbances as shown in Figure 4.12, the green line is the frequency response from a low-inertia power system. Figure 4.13 summarises the challenges caused by the transition from the high-inertia power system to low-inertia power system, in which the left hand side shows the primary

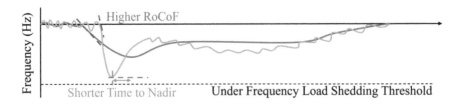

Figure 4.12 Comparison of frequency response between high-inertia power system and low-inertia power system. The green line is the frequency response from a low-inertia power system, and the red line is the frequency response from a high-inertia power system.

differences between the high-inertia and low-inertia power systems and the right hand side is the challenges caused by such differences, with details explained as follows:

- First, due to the intermittency of renewable energy sources and rapid change of demand with the connection of electric vehicles, the frequency becomes more volatile. This volatility brings the challenge for measuring the frequency and RoCoF. Accurate measurement of frequency and RoCoF are crucial for estimating the inertia of power systems in particular for the demand-side inertia which is expected to account for 30% of total inertia in future GB power systems [58].

- Second, the lower nadir, shorter time to the nadir, and higher RoCoF in the low-inertia power system would easier breach the threshold of under frequency load shedding or RoCoF threshold, and then trigger integrity protection schemes, such as loss of mains protection [59]. Consequently, there will be a cascading under and over frequency and system blackout.

- Third, it is challenging to contain the frequency in an effective and economic manner.

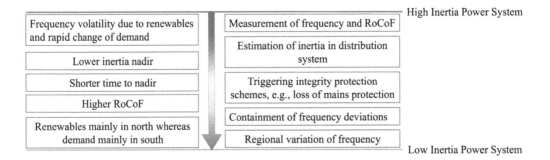

Figure 4.13 Challenges caused by the transition from the high-inertia power system to low-inertia power system. The left hand side shows the primary differences between the high-inertia and low-inertia power systems and the right hand side is the challenges caused by such differences.

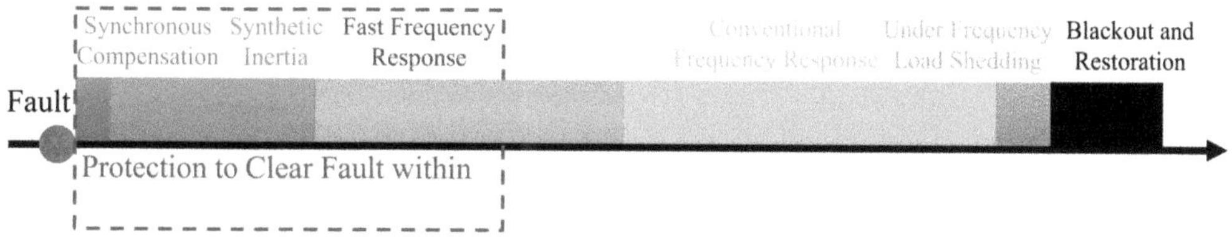

Figure 4.14 Time horizon for clearing the system fault after a disturbance.

- Last, for the GB power system, the low-inertia renewable energy sources are primarily generated at Scotland, whereas the demand primarily comes from south with limited transmission capacity. This would result in regional deviation of the frequency across power systems.

4.5.3 Solutions for Low-Inertia Power Systems

The solutions for overcoming challenges in low-inertia power systems include pre-fault solutions and in-fault solutions. The pre-fault solutions mainly consist of two parts: one is enhancing monitoring system to measure system dynamics in a real-time manner, and another is to assess the reliability of protection system. For the monitoring system, the wide area monitoring system was deployed in GB power systems with strategically located phasor measurement units [60]. For the conventional SCADA system, it could also use filters for the frequency measurement and window averaging for RoCoF calculation. The assessment of protection system includes examining validity of settings, either is over sensitive or less sensitive, assessing impacts of converter interfaced generation, and investigating mitigating actions.

During the system disturbance, the fault needs to be cleared by the protection system. So, for future low-inertia power system, it is key to act before the conventional primary frequency response. As presented in Figure 4.14, there are three key solutions:

- *Synchronous compensation [61]:* Synchronous compensation can provide instantaneous frequency support through transferring from motoring mode to generation mode.

- *Synthetic inertia [62]:* Synthetic inertia provided by converter-interfaced devices can act within 20 ms.

- *Fast frequency response [63]:* Fast frequency response can react in 2 s. For example, in GB power system, the fast frequency response through droop control is used as the enhanced frequency response.

4.6 CHAPTER SUMMARY

This chapter introduced the transition of power systems and energy markets towards the prosumers era by identifying potential information and control architecture of local energy markets. The current and potential future regulatory supports in facilitating such transition are reviewed. Future low-inertia power systems and incurred challenges are also discussed.

Bibliography

[1] https://www.nationalgrideso.com/future-energy/future-energy-scenarios/fes-2021.

[2] S. Grijalva and M. U. Tariq, "Prosumer-based smart grid architecture enables a flat, sustainable electricity industry," in *ISGT 2011*, 2011, pp. 1–6.

[3] H. Ibrahim, A. Ilinca, and J. Perron, "Energy storage systems—characteristics and comparisons," *Renewable and Sustainable Energy Reviews*, vol. 12, no. 5, pp. 1221–1250, 2008.

[4] J. Larminie and J. Lowry, *Electric Vehicle Technology Explained.* John Wiley & Sons, 2012.

[5] R. Zafar, A. Mahmood, S. Razzaq, W. Ali, U. Naeem, and K. Shehzad, "Prosumer based energy management and sharing in smart grid," *Renewable and Sustainable Energy Reviews*, vol. 82, pp. 1675–1684, 2018.

[6] A. Toffler, "The third wave: The classic study of tomorrow," *New York: Bantam*, 1984.

[7] Y. Parag and B. K. Sovacool, "Electricity market design for the prosumer era," *Nature Energy*, vol. 1, no. 4, pp. 1–6, 2016.

[8] J. Hamari, M. Sjoklint, and A. Ukkonen, "The sharing economy: Why people participate in collaborative consumption," *Journal of the Association for Information Science and Technology*, vol. 67, no. 9, pp. 2047–2059, 2016.

[9] A. Luth, J. M. Zepter, P. C. del Granado, and R. Egging, "Local electricity market designs for peer-to-peer trading: The role of battery flexibility," *Applied Energy*, vol. 229, pp. 1233–1243, 2018.

[10] E. Al Kawasmi, E. Arnautovic, and D. Svetinovic, "Bitcoin-based decentralized carbon emissions trading infrastructure model," *Systems Engineering*, vol. 18, no. 2, pp. 115–130, 2015.

[11] N. Liu, X. Yu, C. Wang, C. Li, L. Ma, and J. Lei, "Energy-sharing model with price-based demand response for microgrids of peer-to-peer prosumers," *IEEE Transactions on Power Systems*, vol. 32, no. 5, pp. 3569–3583, 2017.

[12] S.-W. Park, K.-S. Cho, and S.-Y. Son, "Voltage management method of distribution system in p2p energy transaction environment," *IFAC-PapersOnLine*, vol. 52, no. 4, pp. 324 – 329, 2019, iFAC Workshop on Control of Smart Grid and Renewable Energy Systems CSGRES 2019. [Online]. Available: http://www.sciencedirect.com/science/article/pii/S2405896319305671

[13] T. Morstyn, A. Teytelboym, and M. D. Mcculloch, "Bilateral contract networks for peer-to-peer energy trading," *IEEE Transactions on Smart Grid*, vol. 10, no. 2, pp. 2026–2035, 2019.

[14] L. Thomas, Y. Zhou, C. Long, J. Wu, and N. Jenkins, "A general form of smart contract for decentralized energy systems management," *Nature Energy*, vol. 4, no. 2, pp. 140–149, 2019.

[15] https://www.ofgem.gov.uk/data-portal, Mar. 2020.

[16] https://www.powerledger.io/.

[17] D. Minoli, K. Sohraby, and B. Occhiogrosso, "Iot considerations, requirements, and architectures for smart buildings—energy optimization and next-generation building management systems," *IEEE Internet of Things Journal*, vol. 4, no. 1, pp. 269–283, 2017.

[18] P. Asmus, "Microgrids, virtual power plants and our distributed energy future," *The Electricity Journal*, vol. 23, no. 10, pp. 72–82, 2010.

[19] L. Gkatzikis, I. Koutsopoulos, and T. Salonidis, "The role of aggregators in smart grid demand response markets," *IEEE Journal on Selected Areas in Communications*, vol. 31, no. 7, pp. 1247–1257, 2013.

[20] S. Nojavan, K. Zare, and B. Mohammadi-Ivatloo, "Optimal stochastic energy management of retailer based on selling price determination under smart grid environment in the presence of demand response program," *Applied Energy*, vol. 187, pp. 449–464, 2017.

[21] D. Li, W. Chiu, H. Sun, and H. V. Poor, "Multiobjective optimization for demand side management program in smart grid," *IEEE Transactions on Industrial Informatics*, vol. 14, no. 4, pp. 1482–1490, April 2018.

[22] Next Kraftwerke, "Virtual power plant: How to network distributed energy resources." May 2021.

[23] https://www.stem.com/.

[24] Portland General Electric, https://www.auto-grid.com/, May 2021.

[25] https://www.energymeteo.com/.

[26] Swell Energy, https://www.swellenergy.com/, May 2021.

[27] A. Di Giorgio and F. Liberati, "Near real time load shifting control for residential electricity prosumers under designed and market indexed pricing models," *Applied Energy*, vol. 128, pp. 119–132, 2014.

[28] https://global.abb/group/en.

[29] T. Fawcett, "Personal carbon trading: A policy ahead of its time?" *Energy Policy*, vol. 38, no. 11, pp. 6868–6876, 2010.

[30] W. Hua, J. Jiang, H. Sun, and J. Wu, "A blockchain based peer-to-peer trading framework integrating energy and carbon markets," *Applied Energy*, vol. 279, p. 115539, 2020.

[31] M. Fowlie, M. Reguant, and S. P. Ryan, "Market-based emissions regulation and industry dynamics," *Journal of Political Economy*, vol. 124, no. 1, pp. 249–302, 2016.

[32] W. Hua, Y. Chen, M. Qadrdan, J. Jiang, H. Sun, and J. Wu, "Applications of blockchain and artificial intelligence technologies for enabling prosumers in smart grids: A review," *Renewable and Sustainable Energy Reviews*, vol. 161, p. 112308, 2022.

[33] C. Ramstein, G. Dominioni, S. Ettehad, L. Lam, M. Quant, J. Zhang, L. Mark, S. Nierop, T. Berg, P. Leuschner *et al.*, *State and Trends of Carbon Pricing 2019*. The World Bank, 2019.

[34] G. E. Metcalf and D. Weisbach, "The design of a carbon tax," *Harv. Envtl. L. Rev.*, vol. 33, p. 499, 2009.

[35] W. D. Nordhaus, "Revisiting the social cost of carbon," *Proceedings of the National Academy of Sciences*, vol. 114, no. 7, pp. 1518–1523, 2017.

[36] W. Hua, D. Li, H. Sun, and P. Matthews, "Stackelberg game-theoretic model for low carbon energy market scheduling," *IET Smart Grid*, vol. 3, no. 1, pp. 31–41, 2020.

[37] W. Hua, J. Jiang, H. Sun, F. Teng, and G. Strbac, "Consumer-centric decarbonization framework using stackelberg game and blockchain," *Applied Energy*, vol. 309, p. 118384, 2022.

[38] A. M. Driga and A. S. Drigas, "Climate change 101: How everyday activities contribute to the ever-growing issue," *International Journal of Recent Contributions from Engineering, Science & IT (iJES)*, vol. 7, no. 1, pp. 22–31, 2019.

[39] R. N. Stavins, "Experience with market-based environmental policy instruments," in *Handbook of Environmental Economics*. Elsevier, 2003, vol. 1, pp. 355–435.

[40] D. Newbery, D. Reiner, and R. Ritz, "When is a carbon price floor desirable?" 2018.

[41] M. H. Babiker, "Climate change policy, market structure, and carbon leakage," *Journal of International Economics*, vol. 65, no. 2, pp. 421–445, 2005.

[42] B. Doda, "How to price carbon in good times... and bad!" *Wiley Interdisciplinary Reviews: Climate Change*, vol. 7, no. 1, pp. 135–144, 2016.

[43] L. H. Goulder and A. R. Schein, "Carbon taxes versus cap and trade: a critical review," *Climate Change Economics*, vol. 4, no. 03, p. 1350010, 2013.

[44] D. M. Newbery, D. M. Reiner, and R. A. Ritz, "The political economy of a carbon price floor for power generation," *The Energy Journal*, vol. 40, no. 1, 2019.

[45] M. Homam, "Economic efficiency of carbon tax versus carbon cap-and-trade," Homam Consulting and Business Solutions Inc., Tech. Rep., 2015.

[46] D. J. Brayshaw, C. Dent, and S. Zachary, "Wind generation's contribution to supporting peak electricity demand–meteorological insights," *Proceedings of the Institution of Mechanical Engineers, Part O: Journal of Risk and Reliability*, vol. 226, no. 1, pp. 44–50, 2012.

[47] A. C. Wright, "Reform of power system governance in the context of system change," *IET Smart Grid*, vol. 1, no. 1, pp. 19–23, 2018.

[48] D. W. Bunn and S. O. Kermer, "Statistical arbitrage and information flow in an electricity balancing market," *The Energy Journal*, vol. 42, no. 5, 2021.

[49] S. Littlechild, "Promoting competition and protecting customers? regulation of the gb retail energy market 2008–2016," *Journal of Regulatory Economics*, vol. 55, no. 2, pp. 107–139, 2019.

[50] W. Hua, H. Xiao, W. Pei, W.-Y. Chiu, J. Jiang, H. Sun, and P. Matthews, "Transactive energy and flexibility provision in multi-microgrids using stackelberg game," *CSEE Journal of Power and Energy Systems*, 2022.

[51] K. Lo and Y. Wu, "Risk assessment due to local demand forecast uncertainty in the competitive supply industry," *IEE Proceedings-Generation, Transmission and Distribution*, vol. 150, no. 5, pp. 573–581, 2003.

[52] S. Goodarzi, H. N. Perera, and D. Bunn, "The impact of renewable energy forecast errors on imbalance volumes and electricity spot prices," *Energy Policy*, vol. 134, p. 110827, 2019.

[53] J. P. Chaves-Ávila, R. A. van der Veen, and R. A. Hakvoort, "The interplay between imbalance pricing mechanisms and network congestions–analysis of the german electricity market," *Utilities Policy*, vol. 28, pp. 52–61, 2014.

[54] V. Guerrero-Mestre, A. A. S. de la Nieta, J. Contreras, and J. P. Catalao, "Optimal bidding of a group of wind farms in day-ahead markets through an external agent," *IEEE Transactions on Power Systems*, vol. 31, no. 4, pp. 2688–2700, 2015.

[55] https://www.elexon.co.uk/operations-settlement/balancing-and-settlement/imbalance-pricing/, Apr. 2021.

[56] https://www.elexon.co.uk/about/trading-electricty-market/, Apr. 2021.

[57] D. Duckwitz, "Power system inertia," Ph.D. dissertation, 2019.

[58] Y. Bian, H. Wyman-Pain, F. Li, R. Bhakar, S. Mishra, and N. P. Padhy, "Demand side contributions for system inertia in the gb power system," *IEEE Transactions on Power Systems*, vol. 33, no. 4, pp. 3521–3530, 2017.

[59] D. M. Laverty, R. J. Best, and D. J. Morrow, "Loss-of-mains protection system by application of phasor measurement unit technology with experimentally assessed threshold settings," *IET Generation, Transmission & Distribution*, vol. 9, no. 2, pp. 146–153, 2015.

[60] P. Ashton, G. Taylor, M. Irving, A. Carter, and M. Bradley, "Prospective wide area monitoring of the great britain transmission system using phasor measurement units," in *2012 IEEE Power and Energy Society General Meeting*. IEEE, 2012, pp. 1–8.

[61] B. Singh, R. Saha, A. Chandra, and K. Al-Haddad, "Static synchronous compensators (statcom): a review," *IET Power Electronics*, vol. 2, no. 4, pp. 297–324, 2009.

[62] I. Sami, N. Ullah, S. Muyeen, K. Techato, M. S. Chowdhury, and J.-S. Ro, "Control methods for standalone and grid connected micro-hydro power plants with synthetic inertia frequency support: A comprehensive review," *IEEE Access*, vol. 8, pp. 176 313–176 329, 2020.

[63] L. Meng, J. Zafar, S. K. Khadem, A. Collinson, K. C. Murchie, F. Coffele, and G. M. Burt, "Fast frequency response from energy storage systems—a review of grid standards, projects and technical issues," *IEEE Transactions on Smart Grid*, vol. 11, no. 2, pp. 1566–1581, 2019.

CHAPTER 5

Application of Artificial Intelligence for Energy Systems

In this chapter, the research and practices for implementing the artificial intelligence (AI) approaches into energy systems are introduced. Section 5.1 introduces how this would assist the operation, planning, and uncertainty prediction of energy systems. Sections 5.2, 5.3, 5.4, 5.5, and 5.6 review the studies for implementing the AI into energy systems through using the approaches of the optimisation, game theory, machine learning, stochastic approaches, and multi-agent system, respectively. From the perspectives of power system operators and market operators, Section 5.7 provides an example research for using the multi-agent system to model the operations and interactions for both power systems and energy markets. From the perspective of individual consumers, Section 5.8 provides an example research of how to extract energy patterns from individual consumers through using machine learning and map the extracted patterns to potential scheduling decisions. From the perspective of a community of consumers, Section 5.9 provides an example research of how to use reinforcement learning to control the operations of a multi-vector energy hub. To address the uncertainties caused by renewable energy sources and flexible demand, the example research in Section 5.10 describes how to use AI for accurate uncertainty predictions. Section 5.11 concludes this chapter.

5.1 INTRODUCTION

AI is capable of assisting the uncertainty forecasting, planning, and operational control for smart energy systems through exploiting massive volumes of data collected from smart meters or sensors. The extents for implementing the AI into smart energy systems are dependent on the development of the AI and digitalisations of energy systems, which can be categorised into the following four degrees:

- *Degree 1 (Responsiveness):* AI is able to enhance the conventional operation of energy systems on the situational awareness, black start, fault detection,

DOI: 10.1201/9781003170440-5

and contingency screening, with the improved computational accuracy and fast system response.

- *Degree 2 (Predictability):* AI can accurately predict the generation, consumption, system status, and other uncertain factors in a fine granularity, which allows the power system operators, i.e., transmission system operators and distribution system operators, to capture the system transition and potential risks.

- *Degree 3 (Prescription):* The functions of the responsiveness degree and predictability degree are systematically integrated to form an automatic management for entire energy systems, which maintains the outages and disturbances to the lowest level and strategically guarantees the security of supply and system resilience.

- *Degree 4 (Automation):* AI helps energy systems achieve full automation through the wide area control and network level optimisation. The energy system can keep self-healing without human interventions.

5.2 OPTIMISATION

The optimisation approaches include the programming techniques and heuristic algorithms. With respect to implementing the programming techniques, e.g., the linear programming, integer linear programming, mixed integer linear programming, and non-linear programming, to solve the optimisation problems of energy systems, Javaid et al. [1] proposed a linear programming model to assign power levels for controllable devices with the objective of costs minimisation, by which the power flows could be optimally controlled to accommodate power fluctuations. In ref. [2], a mixed integer non-linear bi-level programming was formulated to minimise the electricity bills of consumers under a marginal pricing scheme. To solve this problem, the original problem was converted as an equivalent single-level mixed integer linear programming based on the duality theory, integer algebra, and Karush–Kuhn–Tucker optimality conditions. Khushalani et al. [3] developed a service restoration algorithm for unbalanced distribution systems, by which the problem was formulated as a mixed integer non-linear programming.

With respect to implementing the heuristic algorithms, e.g., the particle swarm algorithm, genetic algorithm, artificial immune algorithm, and other heuristic algorithms to solve the optimisation problems of energy systems, Meng and Zeng [4] formulated a problem for maximising the profits of energy retailers by modelling the effects of real-time electricity prices on shiftable loads and curtailable loads. The problem was solved by the genetic algorithm. Olsen et al. [5] implemented the weighted sum bisection method to minimise carbon tax rate constrained by maintaining total carbon emissions from power systems below a prescribed target of carbon reduction. This research investigated the relationship between system investments and tax setting process and found that the carbon tax can encourage the investments on cleaner generation, transmission, and energy efficiency. Li et al. [6] proposed a hierarchical multi-objective scheduling model to integrate renewable energy sources and demand-side management. In this model, the utility seeks to minimise operating costs and the customers seek to maximise social welfare. The demand response aggregator as an intermediary seeks to maximise its net profits, which are the difference between bonus from utilities for providing demand-side management and the cost of offering compensation to customers. A selection criterion was designed to select the optimal solutions yielded by artificial immune algorithm without favouring any market participant. A user-centric multi-objective optimisation problem was further developed in [7] to achieve a trade-off between residential privacy and energy costs. This research developed a hybrid algorithm by combining a stochastic power scheduling with a deterministic battery control, which addressed the drawbacks of weighted-sum methods, i.e., combing objective functions with various scales, heuristically assigning weight coefficients, and misrepresentation of user preferences.

Nonetheless, the scalability and computational complexity limit the implementation of optimisation approaches on highly-complex problems of power systems scheduling. The scalability issue is caused when the scale of power system varies, since each scale requires predefined parameters and mathematical formulations. The computational complexity issue is caused when solving optimisation problems using heuristic algorithms, for which the optimal scheduling decisions are obtained by iteratively searching. At the instance of optimization solved by ι iterations, once it is combined with I types of generators and K types of loads, the computational complexity increases to $O\left(\iota^{I+K}\right)$ [8].

5.3   GAME THEORY

The game theory has been well documented in literature. The game-theoretic models, stakeholders, and solution approaches in the field of energy scheduling are summarised in Table 5.1. Belgana et al. [9] developed a multi-leader and multi-follower Stackelberg game-theoretic problem to find optimal strategies that could maximise the profits of utilities and minimise carbon emissions. The problem was solved by a hybrid multi-objective evolutionary algorithm. Meng and Zeng [10] proposed a 1-leader, N-follower Stackelberg game to maximise the profits of retailers at the leader level and minimise the electricity bills of consumers at the follower level considering the real-time pricing scheme. The genetic algorithm was used to solve the leader's optimisation problem and the linear programming was used to slove the follower's optimisation problem. Ghosh et al. [11] formulated a coupled constrained potential game to set the energy exchange prices for maximising the amount of energy exchange among prosumers and reducing the consumption from the utility grid. A distributed algorithm was proposed enabling individual prosumers to optimise their own payoffs. In ref. [12], an energy trading framework based on repeated non-cooperative game was designed enabling individual microgrids to optimise their own revenues. The reinforcement learning was exploited to estimate the payoff functions under incomplete information. The Cournot game was implemented in ref. [13] to model the competition between customers and utilities in distribution networks for satisfying the system reliability. Similarly, Zhang et al.[14] modelled local energy trading as a non-cooperative Cournot

Table 5.1 Comparison of game-theoretic models, stakeholders, and solution approaches in the field of energy scheduling

Literature	Game-Theoretic Model	Stakeholder	Solution Approach
Belgana et al. [9]	Stackelberg	Microproducers and consumers	Hybrid Multi-objective Evolutionary Algorithm
Meng and Zeng [10]	Stackelberg	Retailer and consumers	Genetic Algorithm and Linear Programming
Ghosh et al. [11]	Potential Game	Utility and prosumers	Distributed Algorithm
Wang et al. [12]	Stackelberg	Microgrid	Reinforcement Learning
Mohammadi et al. [13]	Cournot	Customers and utilities	Lagrangian function and KKT conditions
Zhang et al. [14]	Cournot	Energy providers	Optimal-Generation-Plan Algorithm

game to stimulate local energy balance and promote the penetration of renewable energy sources.

The game theory assumes all players are rational when they compete with each other. Nonetheless, during practical energy markets operation, individual players have various sensitivities to the incentive signals, which causes the individual decisions to deviate from theoretical rational decisions and thus reduces the model accuracy. For instance, when considering the small-scale consumers, e.g., residential users, the price-insensitive consumers normally use energy irrespective of pricing signals.

5.4 MACHINE LEARNING

To overcome the aforementioned issues of scalability and computational complexity by using optimisation approach, machine learning has been considered to assist or replace the step of solving optimisation problem by the intelligent heuristic algorithm, because it only requires historical data for extracting general features with the advantages of improved scalability and reduced computational complexity.

Using learning approaches for solving energy scheduling problems has been well studied in literature. The learning approaches can be categorised as supervised learning, unsupervised learning, and reinforcement learning. In supervised learning, the input is provided as a labelled dataset, such that the model can learn from the labels to improve the learning accuracy [15]. By contrast, in unsupervised learning, there is no labelled dataset, such that the model explores the hidden features and predicts the output in a self-organising manner [16]. In reinforcement learning, the model learns to react to the environment by self-adjusting through travelling from one state to another [17]. Zhang et al. [8] developed an online learning approach to replace heuristic algorithms for solving a cost minimisation problem under uncertain distributed renewable energy sources and load demand. Gasse et al. [18] proposed a learning model for extracting branch-and-bound variable selection policies to solve combinatorial optimisation, and testified that a series of computational complex problems could be efficiently solved. An energy management system was designed [19] to provide demand response services, by which the explicit model of consumers' dissatisfaction was replaced by the feature representations extracted through using reinforcement learning. Analogously, Ruelens et al. [20] combined heuristic algorithm with reinforcement learning to control a cluster of loads and storage devices, and Zhang et al. [21] integrated learning mechanism with optimisation techniques to obtain optimal demand response policies. The controller can help consumers reduce energy costs with improved computational efficiency.

Further research implemented deep neural networks as a regression algorithm into learning approaches. The convolutional neural network is a class of deep neural networks primarily used for analysing visual imagery, by which the network employs convolution for general matrix multiplication [22]. The convolutional operation imports low-level inputs, e.g. images, to learn general abstractions of a high-complexity problem without the use of manually predefined models [23]. Hence, the convolutional neural network is particularly suitable for the high-complexity problems. Owerko et al. [24] trained the convolutional neural network under imitation learning to approximate an optimal power flow solution. A well-trained convolutional neural network can scale to various power networks for accurately predicting optimal power flows. Du et al. [3] used the convolutional neural network to accelerate N-1 contingency screening of power systems, by which the convolutional neural network can generalise topological changes and uncertain renewable scenarios with improved computational efficiency. Claessens et al. [25] combined the convolutional neural network with reinforcement learning for high-complexity load control. The issue of partial observability was addressed through using the convolutional neural network to extract hidden state-time features. In ref. [26], the convolutional neural network was adopted as an online monitoring tool for predicting instabilities in power systems. This research demonstrated that a trained convolutional neural network was scalable in terms of varying load conditions, fault scenarios, topological structures, and generator parameters.

When the pattern recognition capability of the convolutional neural network is exploited, the approach of processing numerical data to the input of the convolutional neural network is the key for extracting hidden information. Choi et al. [27] processed time-series data of power systems from row vector to the matrix of greyscale image by restructuring the original datasets. Liao et al. [28] mapped different patches of bus matrix to various areas of power networks for voltage sag estimation. The variables representing power systems configuration were assigned as the dimension of depth from the input image.

Nonetheless, there are primary four issues for data-driven learning approaches:

- First, when the size of historical data is small, the overfitting issue would be caused by learning approaches. This would reduce the accuracy for predicting optimal scheduling decisions.

- Second, although the learning approaches can reduce the computational complexity and improve scalability from solving optimisation by heuristic intelligent algorithms, the predicted optimal decisions may deviate from the theoretical optimal decisions and result in the suboptimal solutions.

- Third, the predicted optimal decisions may not maintain the system constraints.

- Fourth, with respect to prosumer-centric energy scheduling, it could be useful to connect the intrinsic features of prosumers, e.g. pricing patterns, with potential scheduling strategies.

5.5 STOCHASTIC APPROACHES

Power system uncertainties caused by the intermittency of renewable energy sources and flexible demand present a challenge for accurately predicting generation and consumption. It is crucial for the reliability of power systems scheduling to consider the possible variations of these uncertainties. The probability approaches have been primarily focused in the literature for incorporating the analysis of system uncertainties into energy scheduling process.

Using a set of scenarios is a potential way to predict possible variations of uncertain variables, by which each variation is defined as a scenario [29]. The uncertain scenarios are generated from probabilistic distribution of historical data by using sampling approaches [30], such as the Monte Carlo simulation [31, 32], Latin hypercube sampling [33, 34, 35, 36] and stochastic analysis [37, 38]. Santos et al. [31] implemented the Monte Carlo simulation to generate renewable scenarios and carried scenarios optimisation by deterministic modelling. Similarly, Hemmati et al. [32] analysed the uncertainties of renewable energy resources and load deviation by the Monte Carlo simulation, and incorporated the uncertainty analysis into decision making process to maximise the profits of distributed generators in microgrids. Nonetheless, the Monte Carlo simulation through random sampling would cause the issues that are computationally intensive and inefficient. These issues can be further overcome by the Latin hypercube sampling which can reduce standard deviation of samples through space-filling. In ref. [33], the Latin hypercube sampling was used to generate uncertain scenarios for overcoming the computationally intensive and inefficient issues of Monte Carlo simulation and considered low probable conditions. Mavromatidis et al. [37] proposed a two-stage stochastic programming for the design of distributed energy systems considering the uncertainties of energy prices, emissions factors, heating demand, electricity demand, and solar radiation. In comparison to the deterministic methods, this study demonstrated that the stochastic method can yield a more accurate estimation of costs and carbon emissions. Huang et al. [38] designed an economic dispatch model for virtual power plants, by which the uncertainties caused by load prediction and power prediction were described by stochastic intervals. These intervals were subsequently integrated into a costs minimisation problem.

Further research efforts have been dedicated to improving the prediction accuracy and adaptability of scenarios. Liang et al. [34] proposed a non-parametric kernel density estimation method to yield the probability density distribution of uncertain variables. The scenarios were generated from the probability density distribution through using Latin hypercube sampling. In [35], a data-driven approach for scenarios generation was developed using generative adversarial networks. This approach can capture both temporal and spatial dimensions of uncertain variables, so as to improve scalability and diversity from probabilistic models. To select high-probable scenarios, Xiao et al. [36] proposed an approach to implement synchronous-back-to-generation-reduction for merging scenarios with a minimum probability distance. Nonetheless, when analysing system uncertainties, an approach for using real-time data to update uncertain scenarios needs to be studied to improve the prediction accuracy for uncertain scenarios.

5.6 AGENT-BASED SYSTEM

The agent-based system is a computational model for simulating actions and interactions of autonomous agents in order to analyse the behaviours of a system and drivers of its outcomes [39]. Applying the agent-based system to model the decision-making and interactions of stakeholders in power systems, including generation companies, consumers, policy makers, transmission system operators, and distribution system operators, has drawn increasing attentions. In ref. [40], the agent-based model was used for energy network modelling with the advantages of extendability, generalisation, and technological independence. Divényi and Dan [41] proposed a multi-agent model considering considering the technical combinations, wind-speed and temperature settings, heating constraints, fuel consumption, regulations, outages, and services. Researchers in ref. [42] developed an agent-based system for model predictive control of building energy systems to reduce the room energy consumption while maintaining indoor temperatures. Mittal et al. [43] designed an agent based approach to investigate solar adoption in residential sectors, in order to benefit all stakeholders.

5.7 RESEARCH EXAMPLE 1: MULTI-AGENT MODEL FOR ENERGY SYSTEM SCHEDULING

This example research introduces a multi-agent model for the energy system scheduling, in which a two-stage energy scheduling is performed, i.e., the day-ahead energy scheduling and real-time energy scheduling. The agent of the demand-side management controls the load shifting in the day-ahead market and load curtailment in the real-time market. The objectives of the energy scheduling includes reducing carbon emissions from energy systems, saving electricity bills for consumers, and improving operational profits for generators. This leads to a multi-objective optimisation problem which is solved by the multi-objective immune algorithm. The case studies have been conducted to demonstrate the effectiveness of the proposed model. It has been proved that the proposed multi-agent model for energy system scheduling contributes to the reductions of both carbon emissions and electricity bills. 0.11% (843.78 MW) of the demand-side management has been realised to guarantee the reliability of power networks during the real-time operation.

5.7.1 Introduction

The advantages of smart grids are enabled by advanced metering technologies, control methodologies, and communication technologies to be integrated into power systems [44]. These technologies contribute to the improvement of smart control and coordination for the purpose of efficient operation of power grids. Compared with the traditional power systems, smart grids become complex with the increasing penetration of renewable energy sources, flexible loads, large volumes of decisions made by multiple stakeholders, and complex interactions among agents. Conventional approaches for monitoring and managing the operation of energy systems including the supervisory control and data acquisition [45], system estimator [3], and contingency analyser [45] are incapable of guaranteeing the reliability and security of power supply, since they are not able to follow up the fast change of system states and offer a rapid response. As a consequence, it is necessary to implement the real-time control and management in smart grids. The advanced data acquisition and transmission technologies further facilitate the observability and controllability in achieving the active control of power networks. In addition, the multi-agent model is able to analyse the decision-making and interactions of actors in power systems.

With respect to the power generation scheduling, the day-ahead market financially arranges the bids of energy supply and demand one day in advance of the settlement, whereas the real-time market dynamically promotes the supply–demand balance in a real-time manner. In addition, a sophisticated day-ahead scheduling pursues the cost and carbon saving by optimising generation schedules in supply side. A stochastic environmental and economic dispatch of power systems was proposed in [46] for the generation scheduling, but the market operation was not considered in this research. Similarly, the research in [47] sought to strike a balance among maximum carbon reduction, minimum payment bills, and minimum costs through multi-objective optimisation. Nonetheless, there are still opportunities to incorporate the real-time control strategies into the generation scheduling by considering energy market operations in both day-ahead and real-time markets. To ensure the system coordination, the pricing signal is formulated by the market operators and the power allocation is performed by the power system operators, which means that the design of the multi-agent system involves both the design of energy markets and the design of power system operation. The market operators and power system operators are included into the grid management agent. Moreover, the demand-side management is capable of controlling the curtailable demand to reduce the peak demand and re-shape the load profile, which enhances to the security of supply and the reductions of overall costs and carbon emissions. The load shifting and load curtailment are two important techniques of the demand-side management implemented in the day-ahead and real-time markets, respectively.

Therefore, a dedicated study is proposed based on the multi-agent systems for the real-time control and management of power systems and energy markets, under the circumstance of simulated real-time operational conditions. Agents representing components of power systems and energy markets interact cooperatively to optimise operation of smart grids pursuing minimum payment bills for customers, and maximum profits and minimum carbon emissions for generators. Both the demand-side management and generation scheduling, as important scheduling functions, are demonstrated by the proposed research.

Compared to existing works, this paper has contributions as follows:

- This research develops the multi-agent system for the power scheduling and optimisation, integrated with the real-time simulation to balance the supply and demand, so that the negotiation rules can be delivered into the agent group for better inside coordination.

- This research considers the carbon emissions reduction, payment bills minimisation, and profits maximization into the market operation to achieve a fair low carbon smart grid scheduling.

The structure of this example research is organised as follows. Sub-Section 5.7.2 introduces the framework of the proposed multi-agent system including the agent design and coordination. Detailed mathematical formulations are given in Sub-Section 5.7.3. Sub-Section 5.7.4 presents the results of case studies for the daily power system scheduling. Finally, Sub-Section 5.7.5 draws the conclusion.

5.7.2 Framework of Multi-Agent System

This section introduces the framework of the proposed multi-agent model. The decision-making and coordination of stakeholders in both energy markets and power systems are modelled by multiple agents. Each intelligent thinking agent makes independent decisions in achieving its objectives, e.g. minimisation of electricity bills.

5.7.2.1 Agents Design

The multi-agent system consists of agents of the system operator, policy maker, demand-side management, generator, consumer, and supplier, with functions detailed as follows:

- *System operator agent*: The system operator agent monitors and manages the operation of power systems under the technical constraints.

- *Policy maker agent*: The policy maker agent formulates low carbon policies, e.g., the carbon pricing, to reduce carbon emissions from power systems.

- *Demand-side management agent*: The demand-side management agent provides services of the load shifting and load curtailment for consumers.

- *Generator agent*: The generator agent determines the power outputs of a single (or multiple) generating source(s) including coal, gas, nuclear, hydro, biomass, solar, onshore wind and offshore wind. The costs and profits of these sources are also evaluated by this agent.

- *Consumer agent*: The consumer agent, also called customer agent, controls and monitors the operation and status of loads and coordinates with the demand-side management agent.

- *Supplier agent*: The supplier agent, also called retailer agent, purchases the electricity from the generator agent in the wholesale market, and charges from the consumer agent in the retail market.

- *Market operator agent*: The market operator agent is responsible for market scheduling and matching bids and offers.

5.7.2.2 Agents Coordination

According to the architecture of energy markets and power systems, the power systems and energy markets are operated by those agents, as shown in Figure 5.1. For the operation of energy markets, generators generate electricity and submit bids to wholesale market. The retailers purchase the electricity by wholesale auction prices, before charging from their own customers. The market operator conducts the market scheduling. For the operation of power systems, the power system operator manages the supply–demand balance and allocates the power generation being subject to system constraints. The corresponding carbon allowances are assigned by the policy maker. The demand-side management agent performs the load shifting when receiving wholesale electricity prices, before the day-ahead energy scheduling is carried out, in order to pursue the minimum carbon emissions and maximum profits for generators, and minimum payment bills for consumers.

Interactions of agents and respective messages for the day-ahead generation scheduling and demand-side management are presented in Figure 5.2. The market operator matches offers and bids submitted by generators and suppliers, respectively, in the day-ahead energy markets. The demand-side management agent helps power system operator negotiate between the supplier and consumer on the load shifting. The power system operators also dispatch generators in order to maintain the supply–demand balance and reduce carbon emissions.

Interactions of agents and respective messages for the real-time generation scheduling and demand-side management are presented in Figure 5.3. The generation and consumption are dynamically matched in a real-time manner through the load curtailment and generation curtailment performed by the power system operators.

Apart from the multi-agent system, the computational intelligence techniques are also employed in the decision-making process of agents. The multi-objective immune algorithm [48] is adopted to solve the multi-objective optimisation problem. The multi-objective immune algorithm (see Algorithm 1) is a global searching algorithm with the robust computational capability. Through randomly generating potential solutions within the feasible range of decision variables and finding the global optimal solutions, the optimal decision-making and coordination problem can be effectively solved.

5.7.3 Problem Formulation

This section describes two key components of the designed multi-agent system, i.e., the problem formulation of demand-side management and generation dispatch.

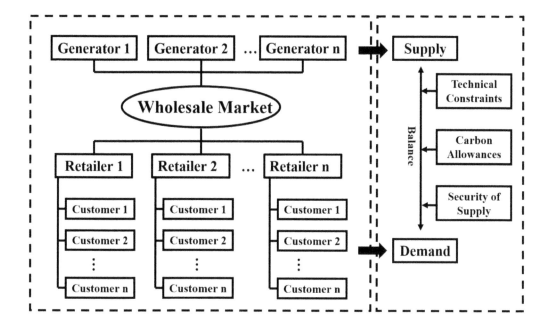

Figure 5.1 The operation of energy markets and power grids. For the operation of energy markets, generators generate electricity and submit bids to wholesale market. The retailers purchase the electricity by wholesale auction prices, before charging from their own customers. The market operator conducts the market scheduling. For the operation of power systems, the power system operator manages the supply–demand balance and allocates the power generation being subject to system constraints. The corresponding carbon allowances are assigned by the policy maker. The demand-side management agent performs the load shifting when receiving wholesale electricity prices, before the day-ahead energy scheduling is carried out, in order to pursue the minimum carbon emissions and maximum profits for generators, and minimum payment bills for consumers.

5.7.3.1 Demand-Side Management

The proposed scheme of the demand-side manage including the load shifting and load curtailment schedules connection moments of loads in the consumption side to realise objective demand curves. Load shifting and load curtailment are conducted in the day-ahead scheduling and real-time scheduling, respectively, through the demand-side management agent. The load shifting seeks to optimise the connection of shiftable loads one day in advance. By contrast, the load curtailment dynamically decreases the power consumption of curtailable loads during the real-time operations.

Correspondingly, the appliances are divided into the non-shiftable appliances, shiftable appliances and curtailable appliances for the purpose of applying the demand-side management [29]. Lights and refrigerators are examples of non-shiftable appliances for which the operations are not time-shiftable. By contrast, customers can shift the use of shiftable appliances such as dish washers, washing machines,

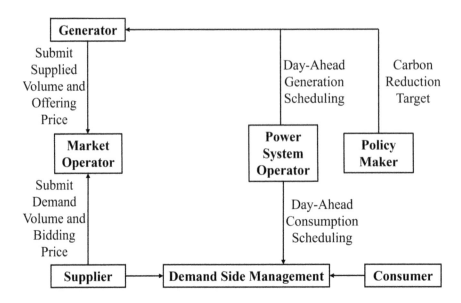

Figure 5.2 Interaction of agents in the day-ahead scheduling. The market operator matches offers and bids submitted by generators and suppliers, respectively, in the day-ahead energy markets. The demand-side management agent helps power system operator negotiate between the supplier and consumer on the load shifting. The power system operators also dispatch generators in order to maintain the supply–demand balance and reduce carbon emissions.

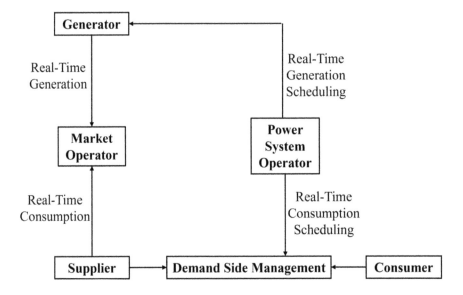

Figure 5.3 Interaction of agents in the real-time scheduling. The generation and consumption are dynamically matched in a real-time manner through the load curtailment and generation curtailment performed by the power system operators.

Algorithm 1

Require: objective functions, initial solution size n; maximum iteration: t_{max}.

1: generate a group of antibodies as initial population to represent the power dispatch over constraints.

2: remove dominated antibodies and remain non-dominated antibodies.

3: perform mutation operation over the remaining non-dominated antibodies to produce a set of antibodies.

4: **repeat**

5:    remove dominated antibodies.

6:    evaluate the remaining antibodies through satisfying the constraints and remove infeasible antibodies.

7:    **if** the population size is larger than the nominal size **then**

8:       update to normalise the antibodies

9:    **end if**

10: **until** the maximum iteration is reached.

Ensure: a solution which is able to maximise the minimum improvement in all dimensions is selected.

---

and water heaters from a higher electricity price period to a lower electricity price period. In addition, air conditioners and space heaters are examples of curtailable appliances. Although the operations of this type of appliances are not time shiftable, the consumption levels can be reduced. The details of problem formulations for the load shifting and load curtailment are given as follows:

- *Load shifting:*

  The load shifting technique of the demand-side management controls the time periods of appliances connection to adjust the total load consumption within objective load consumption. The change of load consumption at each time step is modelled as a linear function of the electricity prices.

  $$f_{\text{shift}}(t) = \alpha \cdot p_e(t) + \beta, \tag{5.1}$$

  where $f_{\text{shift}}(t)$ is the change of the load consumption at the time step t through the load shifting, $p_e(t)$ is the electricity price at the time step t, and α and β are coefficients of the load shifting relationship. During peak demand periods, $\alpha < 0$, because the energy consumption will be shifted away from this period with the increase of peak-time electricity prices. By contrast, during off-peak demand periods, $\alpha > 0$, because the energy consumption will be shifted to this period with the decrease of peak-time electricity prices. Hence, the total energy consumption can be shifted from peak demand periods to off-peak demand periods. Additionally, in order to keep the total consumption level of shiftable appliances the same during operational periods, we have

  $$\sum_{t=1}^{T} f_{\text{shift}}(t) = 0. \tag{5.2}$$

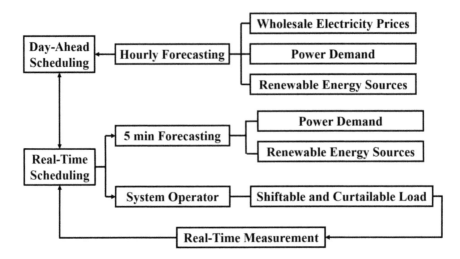

Figure 5.4 Architecture of the generation scheduling. This generation scheduling is carried out through the power system operator agent, in which the operational information is collected, before performing the scheduling decisions and allocating to each generator. The day-ahead scheduling aims for predictive generation dispatch and the real-time scheduling performs corrections through comparing the day-ahead schedules and real-time measurements.

- *Load curtailment:*

  The load curtailment is conducted during the real-time operation to reduce the total power consumption when it is necessary. The maximum level of the load curtailment is set considering the interests and acceptance levels of consumers as

  $$0 \leq f_{\text{curt}}(t) \leq f_{\text{curt}}^{\max}. \tag{5.3}$$

  where $f_{\text{curt}}(t)$ is the amount of curtailed load at the time step t, and $f_{\text{curt}}^{\max}$ is the maximum level of load curtailment.

5.7.3.2 Generation Scheduling

Given current energy market operations, two levels of generation scheduling, including the day-ahead scheduling and real-time scheduling, are considered in the proposed research. Figure 5.4 presents the architecture of the proposed generation scheduling. This generation scheduling is carried out by the power system operator agent, in which the operational information is collected, before performing the scheduling decisions and allocating to each generator. The day-ahead scheduling aims for predictive generation dispatch and the real-time scheduling performs corrections through comparing the day-ahead schedules and real-time measurements. The details of the day-ahead scheduling and real-time scheduling are given as follows:

- *Day-ahead scheduling:*

The proposed research formulates the day-ahead scheduling by considering the profits of generators and the reduction of carbon emissions, as well as saving the electricity bills for consumers, which leads to a multi-objective optimisation problem. Therefore, the objective functions in the day-ahead scheduling are to maximise profits, and minimise carbon emissions and electricity bills, being subject to system constraints including the demand level, power output limit, and ramp-up and ramp-down rates. Besides, in day-ahead energy markets, energy suppliers purchase electricity generated by generators from wholesale markets, before charging from consumers at retail electricity prices. For simplicity, the generators and suppliers are taken as a whole to consider the operational costs, and the retail electricity prices are taken the same as the wholesale electricity prices.

The objective of consumers is to minimise their electricity bills, which can be described as the bills of power consumption considering the load shifting. Thus, the payment bill minimisation problem for consumers can be modelled as follows:

Objective of payment bill minimization :

$$\min_{f_{\text{shift}}(t)} \sum_{t=1}^{T} [f_{\text{demand}}(t) - f_{\text{shift}}(t)] \cdot p_{\text{e}}(t) \qquad (5.4)$$

s.t.

$$\sum_{t=1}^{T} f_{\text{shift}}(t) = 0, \qquad (5.5)$$

where $f_{\text{demand}}(t)$ is the original power demand at the time step t, and $p_{\text{e}}(t)$ is the electricity price at the time step t.

By contrast, the objective of generators is to maximise their profits, which can be described as the revenues from consumers subtracting the operational costs. A cost function is defined to describe the cost of power generation by major sources, i.e., coal, nuclear, gas, wind, pumped storage, hydro, solar, and others, indexed as $u = 1, 2, ..., U$:

$$f_{\text{cost},u}(t) = \gamma_u \cdot g_u(t), \qquad (5.6)$$

where $f_{\text{cost},u}(t)$ is the cost of power generation by the source u at the time step t, $g_u(t)$ is the electricity generation by the source u at the time step t, and γ_u is the cost coefficient of the source u.

Considering the minimum power output $g_u^{\min}$ and maximum power output $g_u^{\max}$, we have

$$g_u^{\min} \leq g_u(t) \leq g_u^{\max} \qquad (5.7)$$

Therefore, the profits minimisation of generators can be modelled as follows:

Objective of profits maximization :

$$\max_{g_u(t), f_{\text{shift}}(t)} \sum_{t=1}^{T} \sum_{u=1}^{U} [f_{\text{demand}}(t) - f_{\text{shift}}(t)] \cdot p_{\text{e}}(t) - f_{\text{cost},u}(t), \qquad (5.8)$$

s.t.

$$g_u^{\min} \leq g_u(t) \leq g_u^{\max}. \tag{5.9}$$

Meanwhile, the reduction of carbon emissions is considered in the day-ahead scheduling. When generators seek to maximise their profits, they are also restrained by carbon emission allowances. If the carbon emissions exceed the allocated allowances, they have to afford the penalties formulated by policy makers. For the conventional sources including the coal, gas, pumped storage, and hydro, their carbon emissions are evaluated by carbon intensities which quantify the amount of carbon dioxide released per unit of energy produced [1]. By contrast, for the nuclear and renewable energy sources, due to the zero or near-zero carbon emissions during the operational process, the life-cycle carbon analysis is applied [1]. With the carbon intensities, an emission function $f_{\mathrm{carb},u}(t)$ is defined to describe the carbon emissions of power generation by major sources:

$$f_{\mathrm{carb},u}(t) = \delta_u \cdot g_u(t) \tag{5.10}$$

where δ_u is the emission coefficient of the source u.

Therefore, the carbon emissions minimisation of generators can be modelled as follows

Objective of carbon emissions reduction :

$$\min_{g_u(t)} \sum_{t=1}^{T} \sum_{u=1}^{U} f_{\mathrm{carb},u}(t). \tag{5.11}$$

Apart from this, there are two common constraints including the power balance constraint and ramp rate constraint during the operation of power systems as [1]

Power balance constraint :

$$[f_{\mathrm{demand}}(t) - f_{\mathrm{shift}}(t)] \leq \sum_{u=1}^{U} g_u(t). \tag{5.12}$$

Ramp rate constraint:

$$-r_u^{\mathrm{down}} \leq g_u(t) - g_u(t-1) \leq r_u^{\mathrm{up}} \tag{5.13}$$

where r_u^{down} and r_u^{up} denote the ramp-down and ramp-up rates of the generation source u.

- *Real-time scheduling:*

The proposed research formulates an operational strategy of the real-time scheduling, in which the supply–demand balance is matched through adopting feedback of real-time simulation in five minute intervals and forecasting data in the next five minutes. Hence, the continuous matching between the supply and demand is monitored and controlled. When the real-time matching exceeds the forecasting demand, the adjustments of power outputs are allocated to each generator in proportion to their maximum generation capacities. The amount of allocation power generation $f_{\text{alloc},u}(t)$ for the source u is given as follows:

$$f_{\text{alloc},u}(t) = \frac{g_u^{\text{max}} - g_u(t)}{\sum_{u=1}^{U} g_u^{\text{max}} - g_u(t)} \cdot \sum_{u=1}^{U} \left[g_u(t) - g_u^{\text{real}}(t) \right], \qquad (5.14)$$

where g_u^{max} is the maximum power output of the source u, and $g_u^{\text{real}}(t)$ is the real-time actual power output of the source u.

During the peak demand period, considering the capacity limitations of power generation sources, the demand-side management agent starts to perform the load curtailment to reduce the total consumption level of curtailable loads.

5.7.4 Case Studies

In order to demonstrate the performance of the proposed multi-agent system based model for the scheduling of power system generation and consumption, case studies have been conducted. Gridwatch provides the UK power outputs of all forms of generation feeding to the grid in 5-minute interval. One year of such data in 2016 is employed to forecast the daily electricity generation by major sources through the autoregressive method [49]. The hourly wholesale electricity price data and corresponding demand data [50] are also employed to forecast the electricity price and demand for each hour. We assume that the maximum of 5% of the load curtailment during the peak demand period from 16 to 22 h would be possible through the pricing incentive. The cost coefficients are adopted from the UK levelised cost of electricity generation [50].

The simulation results obtained from the load shifting by the demand-side management agent in the day-ahead market are presented in Figure 5.5. It is clear that with the incentive of hourly real-time wholesale power prices, the peak demand period has been shifted to the off-peak demand period.

The comparison of scheduling objectives for the day is presented in Table 5.2. It is clear that both payment bills and carbon emissions have been reduced through the day-ahead scheduling. After the scheduling of the multi-objective optimisation, there are about 1.90 % of reduction in the electricity bills and 29.39 % of reduction in the carbon emissions by the load shifting. This further proofs that the day-ahead scheduling scheme is capable of lowering or shifting the peak loads, so that the payment bills and carbon emissions can be reduced. By contrast, the profits of generators keep almost unchanged, which means that with the increase of revenues from consumers, the cost also increases to involve the generation and consumption scheduling.

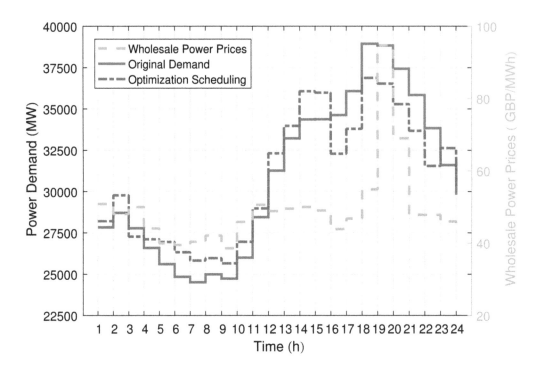

Figure 5.5 Load profile of the day-ahead hourly scheduling. The x-axis is the time step of the energy scheduling. The left y-axis is the power demand with the unit of MW, and the right y-axis is the wholesale power prices with the unit of GBP/MWh.

Table 5.2 The comparison of scheduling objectives including the electricity bills, carbon emissions, and profits

	Electricity bills (10^6 GBP)	Carbon emissions (10^6 ton)	Profits (10^6 GBP)
Without scheduling	63.2749	0.2559	26.8039
With scheduling	62.0741	0.1807	26.8061

Furthermore, Figure 5.6 presents the real-time scheduling for the day, which compares the original demand generated from forecasting results with optimisation scheduling and real-time measurements. It can be seen that the demand-side management contributes to bringing the real-time consumption curve to the optimisation scheduling curve as close as possible through power allocations. Additionally, 0.11% (843.78 MW) load curtailment has been realised to guarantee the reliability of power networks. This real-time scheduling corrects the supply-demand match in every 5-minute interval, so that the real-time measurement curve fluctuates around the scheduling curve to guarantee the scheduling accuracy and reliability.

Therefore, the case studies for the day-ahead and real-time scheduling proves that the proposed model strikes a balance between the supply and demand and helps

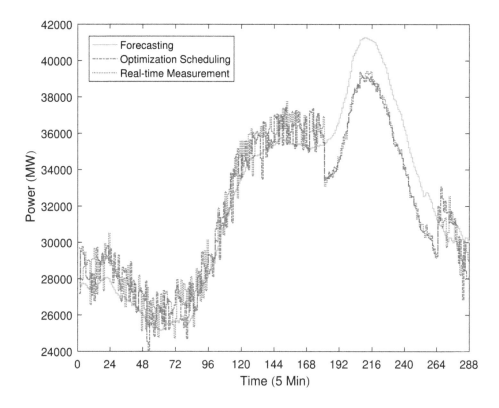

Figure 5.6 Load profile of the real-time scheduling. The x-axis is the time step of the energy scheduling. The y-axis is the power with the unit of MW.

realise the load curtailment and load shifting. The agents coordination allows the practical market operations to be applied into the scheduling process.

5.7.5 Research Summary

This example research proposes a multi-agent system design for low-carbon smart grids in both the generation scheduling and demand-side management for the day-ahead and real-time operations. The results of case studies demonstrate the effectiveness of designed control and management model and the possibility to employ smart grid technologies into the multi-agent system. Through multi-objective optimisation scheduling, the minimum payment bills of consumers, minimum carbon emissions, and maximum profits for generators have been realised through fairly dispatching. With the demand-side management including the load shifting and load curtailment, the scheduling brings the real-time operation to the objective demand curve.

5.8 RESEARCH EXAMPLE 2: ARTIFICIAL INTELLIGENCE FOR PRICING PATTERNS RECOGNITION

The deployment of smart meters facilitates increasing number of consumers to produce or store electricity at home, forming the role of prosumers [51]. However, the economic potential of individual prosumer's energy scheduling is limited, whereas multiple types of energy sources, loads, and energy behaviours require high-dimensional optimisation and accurate modelling. This research example proposes a prosumer-centric low carbon pricing patterns recognition through using the convolutional neural networks to scale to high-dimensional systems and generalise unseen inputs through extracting features from images. The deep network architecture with multiple layers of representations is compatible with various system conditions to abstract inherent dynamic price elasticities of generation, consumption, and carbon emissions from prosumers. The uncertainties caused by distributed renewable energy sources and flexible demand are considered through using the proposed scenarios generation algorithm. Case studies demonstrate that the proposed approach is capable of capturing the underlying hidden features of prosumer-centric pricing patterns and providing an accurate evaluation of scheduling potentials.

5.8.1 Introduction

Increasing energy demand has driven global energy-related carbon emissions rising to a historic high. The power sector accounts for nearly two-thirds of emissions growth [52]. Facing this environmental challenge, low carbon policy targets on phasing out fossil fuel based power generation through facilitating distributed renewable energy sources and charging carbon taxes from high-emission sources. In the EU, 200 million smart meters have been deployed by 2020 to support the integration of distributed renewable energy sources [53], so that consumers can produce or store energy at home via solar panels [54], electric vehicles [55], and batteries [56]. The role of consumers is transforming to prosumers when consumers actively schedule their own power generation and consumption.

Nonetheless, with the involvement of prosumers' role, there are several challenges for energy markets:

- First, although the carbon tax is set on generators to cut into their profits, the generators will pass some of these carbon costs on as the increase of electricity price for consumers. This is notably essential for prosumers who take carbon responsibility as both generators and consumers.

- Second, the increasing scale of prosumers will amplify uncertainties in distribution networks due to the intermittency of distributed renewable energy sources and flexible consumption patterns.

- Third, the conventional scheduling tools are transformed from previous generators' domain to individual prosumer's domain, which means that these tool needs to adapt with various scales of prosumers.

For the first challenge, since the carbon tax is a part of operating costs and affects electricity prices, it is worth investigating the price elasticity of carbon emissions which is analogous as the price elasticities of demand and supply, as an intrinsic characteristic of prosumers to extract their pricing patterns. There is an intensive literature on static price elasticity to analyse the price responsiveness [57, 4]. A scheduling algorithm was proposed to shift the elastic loads to high renewable generation periods in [57]. Consumers' price elasticity was analysed using the linear regression and categorised as the price sensitive, price insensitive, and price optimiser for strategic demand-side management in [4]. The static elasticity is an average and fixed value based on long-term observations. It is an efficient tool to analyse overall energy markets' trends, whereas it may degrade for capturing prosumer-centric pricing patterns, because individual behaviour does not always confront with statistical features and it may change over time in a strategic manner responding to the change of prices. With the help of smart grids and smart meters, real-time data including electricity price, generation, consumption, and carbon emissions can be bidirectionally transmitted [44]. This enables the dynamic price elasticity to be analysed [58]. It is instrumental to incorporate dynamic price elasticity into the scheduling model as a prerequisite and involve carbon emission behaviours responding to electricity prices.

For the second challenge, accounting uncertainties into prosumers' energy scheduling contributes to reliability of scheduling tools. Statistical method is a potential idea to involve uncertainties. Motivation behind the statistical method is that distribution of uncertain variables can be extracted from historical data, and randomness can be introduced based on distribution analysis. Hence, potential scenarios which describe all possible distributions of uncertain variables can be generated. Specifically, in prosumer-centric uncertainty analysis, these scenarios correspond to the possible generation and storage strategies as well as consumption patterns. Monte Carlo simulation [59] and Latin hypercube sampling [60] are two popular methods for scenarios generation. Compared to the Monte Carlo simulation, the Latin hypercube sampling yields a better performance due to the fast convergence and reduction of over-concentration by space-filling. However, the multidimensionality of uncertain variables and stochastic distribution will cause high computational burdens and it is difficult to find an optimal scheduling decision over large-scale scenarios. There are still opportunities for the scenarios reduction to keep the most probable and dissimilar scenarios.

With respect to the prosumer-centric energy scheduling, the model-based optimisation has been focused in the current literature, by which the scheduling objective is designed according to certain criteria and subjects to system constraints. Liang et al. [61] proposed a game theory strategy for prosumers' bidding in retail market through solving bi-level optimisation problem. Analogously, a Stackelberg game-theoretic model was implemented in energy sharing provider to facilitate energy sharing of prosumers in [62]. An optimal scheduling which minimises prosumers' profits in the day-ahead market was proposed in [63]. Nonetheless, in practical operation, the optimisation model needs to be implemented in a specific prosumer considering its operating conditions with predefined parameters and predictions, which causes computational and economic burdens due to the increasing scale of prosumers. Additionally, when the scheduling becomes a multi-objective optimisation problem, the objective functions, such as costs and carbon emissions, are not always under the same scale. Hence, the optimal solution is not suitable in practical operations. Unlike optimisation model for energy market scheduling, the learning method, as a model-free method, only requires historical data accumulation. This increases the system scalability from the perspectives of operation and implementation. The scale difference can be captured from historical data by learning method.

This research uses the convolutional neural network [64] as a deep learning architecture to automatically extract underlying features of prosumer-centric pricing patterns. The convolutional neural network is particularly suitable for our research due to the reduced dimensionality and computational burden by importing pixel, which enables low level sensor inputs to learn multiple abstractions without manually predefined features. In the literature [65, 66], the convolutional neural network was heavily used as a forecasting tool in power system operations, by which the historical data is converted from single-dimension time series to multi-dimension pixel through equally dividing the original dataset to fed into the convolutional neural network. However, the fidelity of original data is affected during such conversion. In our research, the dynamic price elasticity over the entire scheduling horizon is processed to keep the integrability of data and the interconnection between generation, consumption, and carbon emissions is coupled by the colour overlay.

Therefore, this research approaches low-carbon prosumer-centric scheduling to solve aforementioned issues considering several gaps in existing studies and has contributions as follows:

- In contrast to the static elasticity in existing works, this research properly captures the temporal dependency of generation, consumption, and carbon emissions through using convolutional neural network to extract dynamic elasticities of prosumers for pricing patterns recognition without affecting the fidelity of original data.

- A scenarios analysis algorithm is designed including scenarios generation and reduction to involve uncertainties of distributed renewable energy sources and flexible demand in a statistical manner.

- This research properly takes the advantage of convolutional neural network for capturing unseen features of pricing patterns and exploits pricing patterns for scheduling with a goal of minimising costs and carbon emissions, which allows our scheduling tool to be more scalable than the model-based optimisation with minimum assumptions on model structures.

The remainder of this work is summarized as follows. In Sub-Section 5.8.2, the system model and implementation are introduced including scenarios analysis algorithm and scheduling strategy. The neural network architecture for price patterns recognition based on price elasticity is described in Sub-Section 5.8.3. Sub-Section 5.8.4 conducts case studies to demonstrate the proposed model. Sub-Section 5.8.5 draws the conclusion.

5.8.2 Problem Formulation

The proposed system model and its implementation are described in this sub-section. The scenarios analysis algorithm and scheduling strategy are also introduced as a preliminary to the designed approach of recognising pricing patterns.

The proposed model is designed for prosumers in the day-ahead market to schedule their power generation, consumption, and incurred carbon emissions by extracting their intrinsic pricing patterns. In the context of this research, the prosumers refer to households with distributed energy sources, e.g., roof-top solar panel, air-source heat pump, diesel generator, storage system, electric vehicle, and hot water cylinder. The half-hour scheduling interval is used according to the settlement period in GB energy market. A schematic illustration of the proposed model and its implementation is presented in Figure 5.7. Once receiving the information of generation, consumption, and incurred carbon emissions of an individual prosumer for the current day, the database is updated and returns scheduling decisions to the prosumer. First, based on the historical data of a prosumer, generation and consumption scenarios are generated through using the developed scenarios generation algorithm. The generated scenarios represent potential variations of uncertain variables including distributed energy sources and flexible loads. Similar scenarios are subsequently merged and low probable scenarios are dropped to further enhance the accuracy of the energy scheduling model. Second, with the information of real-time pricing for the following day, the dynamic price elasticities of supply, demand, and carbon emissions are calculated and processed as images. Third, the reduced scenarios and elasticity images are imported into the neural networks to extract intrinsic features of prosumers. The outputs of neural networks are optimal scheduling decisions, under the strategies of saving the costs of using grid electricity and reducing carbon emissions. The training labels are obtained from solving a multi-objective optimisation problem, which indicates how far from predicted decisions of neural networks to optimal decisions from solving the optimisation problem.

5.8.2.1 Scenarios Analysis

The goal of scenarios analysis is to accurately evaluate possible variations of uncertain variables based on statistical distribution of historical data. Each variation is represented by a scenario. These uncertain variables include generation from distributed energy sources and consumption from flexible loads. Let the vector $\mathbf{p}_t^e$ denote the electricity generation by the source e at the scheduling time t, containing $|\mathcal{N}|$ scenarios as $\mathbf{p}_t^e = p_{t,1}^e, ..., p_{t,n}^e, ..., p_{t,|\mathcal{N}|}^e$. Let vector $\mathbf{d}_t$ denote the electricity consumption at the scheduling time t, containing $|\mathcal{N}|$ scenarios as $\mathbf{d}_t = d_{t,1}, ..., d_{t,n}, ..., d_{t,|\mathcal{N}|}$. Through scenarios analysis, the most probable scenarios will be selected and corresponding occurrence probabilities will be obtained.

Accurately evaluating the distribution of uncertain variables is a prerequisite for the scenarios generation. Compared with the approaches of using parametric estimation to formulate the distribution of uncertain variables, such as Weibull distribution [67] and normal distribution [68], using the non-parametric estimation is more suitable for capturing stochastic features of distributed energy sources and consumption from flexible loads, since it is primarily dependent on the historical observations without any assumption of parameters. Thus, the kernel density estimation [69] is used in our research to estimate the probability density function of uncertain variables. Furthermore, for the purpose of involving adequate randomness of multi-dimensional

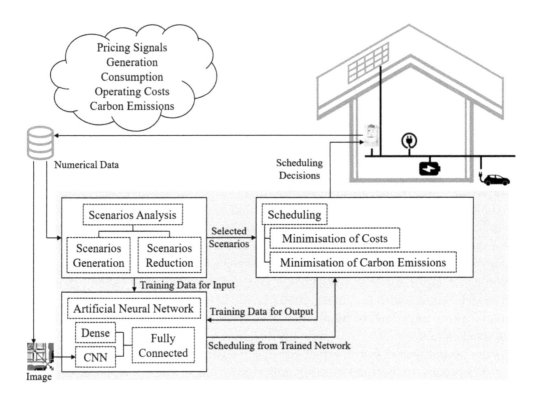

Figure 5.7 Schematic illustration of the proposed model and its implementation. Once receiving the information of generation, consumption, and incurred carbon emissions of an individual prosumer for the current day, the database is updated and returns scheduling decisions to the prosumer. First, generation and consumption scenarios are generated and then reduced through using the developed scenarios analysis algorithm. Second, the dynamic price elasticities of supply, demand, and carbon emissions are calculated and processed as images. Third, the reduced scenarios and elasticity images are imported into the neural networks to extract intrinsic features of prosumers. The outputs of neural networks are optimal scheduling decisions.

uncertain variables, the approach of Latin hypercube sampling [70] is used to generate scenarios based on the statistic distribution of these variables. Compared to approaches of random sampling, such as Monte Carlo simulation [59], Latin hypercube sampling is able to avoid over-concentration through space-filling, which means that samples are generated over the entire feasible range of historical observations. Monte Carlo simulation also requires longer computing time due to its slow convergence. In addition, the time-series data structure is inefficient to capture the most current features of dataset, because the early received data may cause deviation from key features due to dated information. The queues structure is designed to store data. Each queue is subject to the first-in first-out principle [71]. When new data is received, the first data added to the queue will be firstly dropped. The multiple uncertain variables are saved in corresponding queues with the most current characteristics.

For simplicity, using x to represent an uncertain variable, i.e., p_t^e and d_t, and x^m to represent the sample m from historical observations. The unknown density function of an uncertain variable is represented by $f(x)$. The kernel density estimation can be formulated as

$$\hat{f}(x) = \frac{1}{|\mathcal{M}| \cdot h} \sum_{m=1}^{|\mathcal{M}|} K\left(\frac{x - x^m}{h}\right), \tag{5.15}$$

where $\hat{f}(x)$ is the estimated kernel density function, $|\mathcal{M}|$ is the total number of samples, h is the bandwidth smoothing parameter, and $K(\bullet)$ is the kernel density function. Gaussian kernel [22] is used in this research due to its high efficiency and simple mathematical expression. A kernel is placed around every sample m, so that the estimated kernel density function is obtained by the sum of $|\mathcal{M}|$ kernels.

Based on the obtained kernel density function, scenarios can be generated to reflect possible variations of uncertain variables. To generate desired $|\mathcal{N}|$ scenarios, firstly, the cumulative density function of the uncertain variable x, denoted as $y = F(x)$, is equally divided into $|\mathcal{N}|$ intervals. Every interval corresponds to a scenario x_n with the occurrence probability as

$$\Pr(x_n) = \frac{1}{|\mathcal{N}|}. \tag{5.16}$$

Secondly, a point is randomly selected from each interval to calculate the value of this uncertain variable by using its inverse function as

$$x_n = F^{-1}(y_n). \tag{5.17}$$

We have

$$y_n = \left(\frac{1}{|\mathcal{N}|}\right) r_n + \frac{n-1}{|\mathcal{N}|}, \tag{5.18}$$

where r_n is a random variable being subject to the uniform distribution. Therefore, $|\mathcal{N}|$ initial scenarios are generated.

Although a large amount of scenarios is able to cover more possible variations of uncertain variables, it would increase computational burdens. It is necessary to merge similar scenarios and omit scenarios with low occurrence probabilities, by which primary characteristics of uncertain variables can be maintained. The scenario reduction approach is designed as shown in Algorithm 2.

5.8.2.2 Scheduling Strategy

In our research, the objectives of the energy scheduling of an individual prosumer include the minimisation of its operating costs and carbon reduction. Considering price elasticities of the generation, consumption, and incurred carbon emissions of a prosumer, denoted as ξ_{p^e}, ξ_d, and ξ_e, respectively, the scheduling decisions are electricity generation $\mathbf{p}_t^e$ and consumption $\mathbf{d}_t$. Therefore, the minimum costs and carbon emissions c_t^* e_t^* are taken as the training labels of neural networks. The selected scenarios and price elasticities are imported as inputs of the neural networks. The relationship

---

Algorithm 2 Scenario Reduction Algorithm

Require: Initial scenarios set $\mathcal{X} = x_1, ..., x_{|\mathcal{N}|}$, desired size of scenario set $|\mathcal{N}_d|$, removed scenarios set $\mathcal{X}_r$.

1: Calculate the probability distance between the scenario i and the scenario j, denoted as $d(i, j)$, $i, j \in \mathcal{N}$.

2: **repeat**

3:    For each scenario n, find the closest scenario k as $d(n, k) = \min d(n, q), q \neq n$, $n, k, q \in \mathcal{N}$.

4:    Given the probability of scenario n, i.e., $\Pr(x_n)$, calculate the probability of the scenario k by $\Pr(x_k) = \Pr(x_n) \cdot d(n, k)$.

5:    Delete the scenario r with the minimum probability, i.e., $\Pr(x_r) = \min \Pr(x_k)$, $r \in \mathcal{N}$.

6:    Update the scenario set $\mathcal{X}$, removed scenarios set $\mathcal{X}_r$, and corresponding occurrence probability by $\mathcal{X} = \mathcal{X} - \{x_r\}$, $\mathcal{X}_r = \mathcal{X}_r + \{x_r\}$, $\Pr(x_n)=\Pr(x_n)+\Pr(x_r)$

7: **until** $|\mathcal{N}|=|\mathcal{N}_d|$

Ensure: Selected scenarios x_n, and corresponding occurrence probabilities $\Pr(x_n)$.

---

between the training labels and inputs of neural networks can be described as

$$\{c_t^*, e_t^*\} = f_{nn}(\mathbf{p}_t^e, \mathbf{d}_t, \xi_{p^e}, \xi_d, \xi_e) \tag{5.19}$$

where $f_{nn}(\bullet)$ is the relationship function parametrised by training the neural networks.

5.8.3 Pricing Pattern Recognition

This subsection introduces the concept of price elasticity and describes the architecture of neural networks used for pricing pattern recognition.

To realise the goal of pricing pattern recognition, price elasticities of generation, consumption, and carbon emissions need to be analysed as intrinsic features of an prosumer. These features are subsequently processed to images as an input of the convolutional neural network. We will first introduce the conception of price elasticity, before describing the data processing approach. For simplicity, only the concept of price elasticity of carbon emissions is detailed, since the price elasticities of generation and consumption have the analogous formulations. The price elasticity of carbon emissions measures the responsiveness of carbon emission behaviours to a change in the retail electricity price. When the electricity price changes by $\Delta\pi$ from π_0, resulting in the change of carbon emissions by Δe from e_0, the price elasticity of carbon emissions is defined as:

$$\xi_e = \left(\frac{\Delta e}{e_0}\right) \cdot \left(\frac{\pi_0}{\Delta\pi}\right). \tag{5.20}$$

Based on the concept of price elasticity, the cross elasticity for every two different scheduling time can be calculated over the entire scheduling horizon $|\mathcal{T}|$. Those three types of price elasticities, i.e., price elasticities of generation, consumption, and carbon emissions, form a three-dimensional array, i.e., $\xi_e, \xi_{p^e}, \xi_d \in \mathbb{R}^{|\mathcal{T}| \times |\mathcal{T}| \times 3}$, by which each elasticity corresponds to one colour channel of 'R G B' in an image. Unlike the input data processing approaches which transform the time-series vector to a matrix by reshaping it, our designed processing approach keeps the fidelity and integrity of entire scheduling horizon. In addition, the colour overlay can reflect the interrelationship of generation, consumption and carbon emissions, and the colour gradients can reflect the time variations of energy behaviours, which can be automatically recognised by the convolutional neural network.

The relationship between the optimal values of scheduling objectives and decision variables is parametrised by training neural networks. The neural networks consist of a convolutional neural network to recognise intrinsic features of a prosumer and a dense hierarchical perception to import numerical data of generation and consumption over the entire scheduling horizon. The outputs from these two layers are subsequently merged by fully connected layers. An overview of the architecture of neural networks is presented in Figure 5.8.

The inputs of the convolutional neural network are the $|\mathcal{T}| \times |\mathcal{T}| \times 3$ elasticity image containing ξ_e, ξ_{p^e}, and ξ_d, and the inputs of dense hierarchical perception are $\mathbf{p}_t^e$ and $\mathbf{d}_t$ containing $|\mathcal{N}_d|$ selected scenarios. The neural networks return the minimum costs c_t^* and carbon emissions e_t^* as outputs. The convolutional neural network convolves the three-dimensional elasticity image with multiple filters and optimises the learning weights. The local elasticity feature over consecutive scheduling intervals within the filter size and temporal transient feature of elasticity over several scheduling intervals can be detected. In addition, multiple filters can detect the relationship of generation, consumption, and carbon emissions from the colour overlay. All the detected features are further processed by higher layers and ultimately stacked as a feature map. Assume a convolutional layer contains k filters, the feature map L_k is expressed as

$$L_k = \alpha \cdot (\mathbf{W}_k * \xi) + b_k, \qquad (5.21)$$

where α is the activation function, $\mathbf{W}_k$ is the weight matrix, $*$ is the convolutional operation, ξ is the image input, and b_k is the bias term. Pooling layer is followed to downsample the convolutional outputs by increasing translation invariance of neural networks. After multiple convolutional and pooling layers, all feature maps are combined by the fully connected layer as outputs of the convolutional neural network.

Meanwhile, $\mathbf{p}_t^e$ and $\mathbf{d}_t$ are fed into a dense hierarchical perception with an intermediate hidden layer and multiple nodes to process hidden representations. Finally, the outputs of convolutional neural network and hierarchical perception are merged into fully connected layers to extract combined hidden features. The final output layer maps the processed hidden features to optimum costs c_t^* and carbon emissions e_t^*. This 2-stream network structure has been demonstrated to yield a good performance in [72].

5.8.4 Case Studies

The performance of our proposed model is demonstrated by case studies. The consumption data of prosumers is obtained from residents in England. The generation

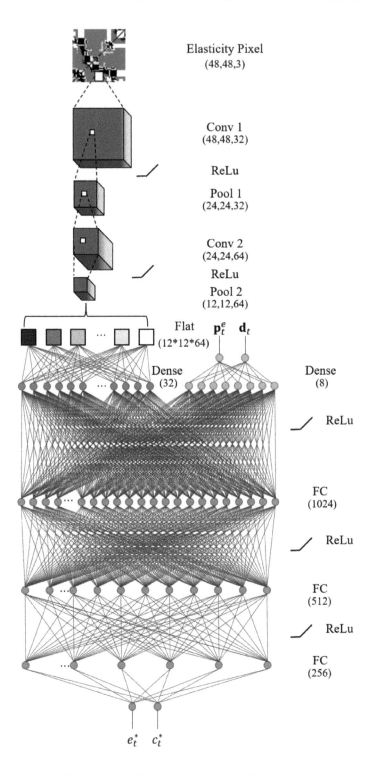

Figure 5.8 Schematic illustration of neural networks for pricing pattern recognition. The neural networks consist of a convolutional neural network to recognise intrinsic features of a prosumer and a dense hierarchical perception to import numerical data of generation and consumption over the entire scheduling horizon. The outputs from these two layers are subsequently merged by fully connected layers.

data of distributed energy sources is obtained from ref. [73]. The scenario analysis algorithm firstly generates 500 scenarios based on 365-day historical observations, before reduced to 10 selected scenarios.

To evaluate the performance of designed neural networks, the optimisation is implemented as a benchmark. The optimal solution of the energy scheduling problem is obtained by solving a multi-objective optimisation. Based on the scenarios analysis, the objective functions are minimising the expectation of costs and carbon emissions with respect to occurrence probabilities $\Pr(\mathbf{d}_t)$ and $\Pr(\mathbf{p}_t^e)$ as

Objective I: min. costs

$$\min_{\mathbf{p}_t^e, \mathbf{d}_t} \mathbb{E} \left\{ \sum_{t=1}^{T} [(\mathbf{d}_t \cdot \Pr(\mathbf{d}_t) - \mathbf{p}_t^e \cdot \Pr(\mathbf{p}_t^e)) \cdot \pi_t + \mathbf{p}_t^e \cdot \Pr(\mathbf{p}_t^e) \cdot \gamma^e] \cdot \Delta t \right\}, \qquad (5.22)$$

Objective II: min. carbon emissions

$$\min_{\mathbf{p}_t^e, \mathbf{d}_t} \mathbb{E} \left\{ \sum_{t=1}^{T} \mathbf{p}_t^e \cdot \Pr(\mathbf{p}_t^e) \cdot \delta^e \cdot \Delta t \right\}, \qquad (5.23)$$

where γ^e is the cost coefficient of the energy source e, δ^e is the carbon intensity of the energy source e, and Δt is the scheduling interval. The first term in the Equation (5.22) corresponds to the electricity bills for using grid electricity, and the second term corresponds to the generating costs. The multi-objective optimisation problem is solved by using the MATLAB® optimisation toolbox. This benchmark yields a theoretical minimum carbon emissions and costs. The goal of our designed neural networks is to obtain scheduling decisions that enable the predicted objective values to be close to the theoretical minimum values.

The learning approach is developed by Python using PyTorch [74]. The selected parameters of the convolutional neural network are shown in Table 5.3. The inputs of the convolutional neural network are elasticity images in the shape of a $48 \times 48 \times 3$ array. The first convolutional layer has thirty-two 5×5 filters. To keep the invariance of the input image, the stride $(1,1)$ is selected for the convolutional layer, before using the pooling layer to reduce spatial dimensions with 2×2 pool size and $(2,2)$ stride. Next, the output of the pooling layer is fed into the second convolutional layer which has sixty-four 2×2 filters with the stride of $(1,1)$, followed by a pooling layer with 2×2 pool size and $(2,2)$ stride. The outputs of convolutional layers are flatten to form feature maps, before processed by a hidden layer with 32 nodes. Meanwhile, the vectors of generation and consumption are taken as inputs to a feedforward network with one 8-node hidden layer. The combined outputs of convolutional neural network and feedforward network are merged by hierarchical perceptions with 3 hidden layers and 1024, 512, and 256 nodes. Final outputs of neural networks are minimum carbon emissions e_t^* and costs c_t^* in each scheduling interval. The rectifier nonlinearity (ReLu) [75] is used as the activation function, and the stochastic gradient descent [76] is used as an optimiser to train neural networks with 500 epochs and the learning rate of 0.001.

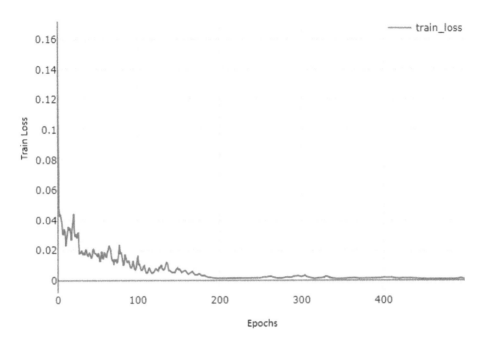

Figure 5.9 Variations of the train loss with the increase of epochs.

Table 5.3 Selected parameters of the convolutional neural network

Layer	Output Size	Filter Size	Stride
Input	$48 \times 48 \times 3$	-	-
Convolution1	$48 \times 48 \times 32$	5×5 @32	(1,1)
Pooling1	$24 \times 24 \times 32$	2×2	(2,2)
Convolution2	$24 \times 24 \times 64$	2×2 @64	(1,1)
Pooling2	$12 \times 12 \times 64$	2×2	(2,2)

We firstly import the data of whole 365 days to test the convergence performance of the training process. The relationship between the training losses and epochs is presented in Figure 5.9. It can be seen that the model converges after 200 epochs. Next, the effects of the data size on the learning accuracy are evaluated. With the increase of the data size from 50 days, 100 days to 200 days, the learning accuracy of optimal decisions of energy consumption is presented in Figure 5.10. It can be seen that with increase of the data size, the predicted scheduling decisions are close to theoretical optimal decisions.

The dispatched generation of a prosumer and corresponding carbon intensity are presented in Figure 5.11. Our proposed scheduling realises both carbon reduction and daily total costs saving through strategically responding to pricing signals considering the price elasticity of prosumers.

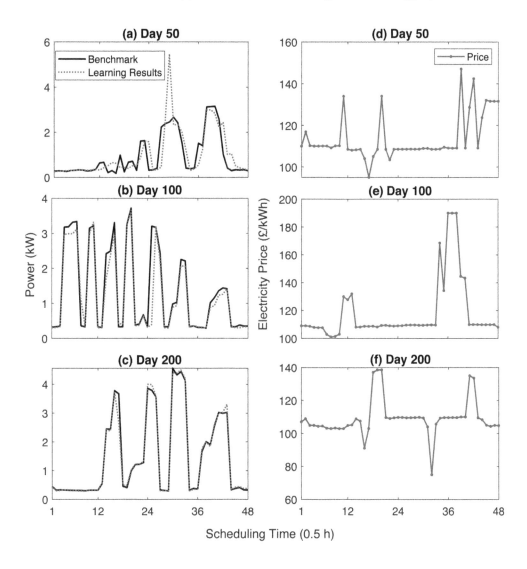

Figure 5.10 Learning process for electricity consumption at day 50, day 100, and day 200 in (a), (b), and (c), respectively, with corresponding electricity prices in (d), (e), and (f), respectively.

5.8.5 Research Summary

To extract intrinsic low-carbon pricing patterns of prosumers and involve uncertainties caused by distributed energy sources and flexible loads, a learning approach based on neural networks was designed for the energy scheduling of a prosumer, which improves the scalability and reduces the computational burden. The proposed scenarios analysis algorithm statistically evaluates the variations of uncertain variables. The convolutional neural network is capable of capturing temporal transient features of elasticity and relationship of generation, consumption, and carbon emissions. Simulation results demonstrate that the proposed learning approach can converge after 200

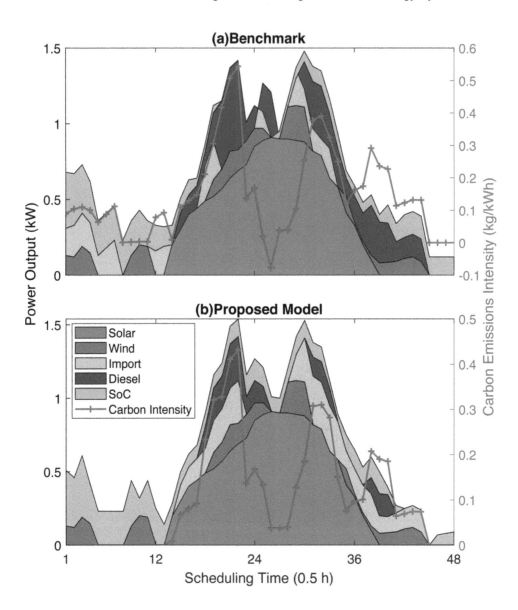

Figure 5.11 Comparison of the power output and carbon intensity of the benchmark in (a) and proposed model in (b).

epochs with 365-day data and accurately yield optimal scheduling decisions. Both carbon reduction and cost saving are achieved.

5.9 EXAMPLE RESEARCH 3: REINFORCEMENT LEARNING FOR LOW-CARBON ENERGY HUB SCHEDULING

Energy hub scheduling becomes the essence for optimally incorporating electricity, heat, and renewable energy sources. A scalable scheduling model which is suitable for flexible energy sources and operating conditions holds the key to modelling performances. This example research investigates the price elasticity as the intrinsic characteristics of energy hub operators and analyses the state action dependency and action transient dependency based on the conditional random field. With these characteristics, a reinforcement learning model is developed to schedule the exchange of the power and natural gas and dispatch the energy sources within the energy hub with reduced assumptions of model parameters and improved model scalability. Case studies are conducted on the real-time digital simulator through interacting between scheduling decisions and monitored operating conditions. It is demonstrated that the conditional random field-based reinforcement learning can approach the theoretical optimal solutions after 50-day training. Scheduling decisions are particularly affected by received pricing information during peak-demand periods. The developed method can save 9.76% of daily operating costs and mitigate 1.388 ton of carbon emissions from simulation results.

5.9.1 Introduction

Scheduling of integrated multiple energy vectors coupling electrical and thermal networks has received considerable attentions in recent literature [77, 78, 79, 80]. Motivation behind the scheduling is that the electricity, natural gas, and distributed renewable energy sources can be systematically optimised to improve operating performances including carbon mitigation [77], costs reduction, system resilience [78], and security of supply [79]. The scheduling problems are normally solved by the model-based optimisation, through which the scheduling objectives are designed by certain targets and subject to operational constraints [80]. The model based optimisation gives a theoretical optimal solution for energy hub scheduling, whereas some challenges raise from such approach: 1) The scale of energy hub varies by topological structures and technical combinations, which requires accurate predefined parameters for each scale of energy hub; 2) The uncertainties of renewable generation and flexible demand need to be accurately predicted; and 3) How the operators of the energy hub would react to the pricing signals, i.e., their price elasticity needs to be captured.

For the challenges of scalability and uncertainties, the model-free approach, e.g., reinforcement learning, provides a potential solution [64]. The reinforcement learning dynamically updates actions under a control policy through interacting with environments [64]. For the operation of energy hub, the reinforcement learning outperforms the model based optimisation from two aspects: 1) The reinforcement learning only requires historical data without predefined parameters and formulations which allows designed models to be scalable and compatible for scales of systems. This can also save the computational and economic burdens compared to the optimisation approach; 2) For multi-objective optimisation problems, to which the values of objective functions are under different scales or units, e.g., minimising carbon emissions in the unit of ton and minimising operating costs in the unit of GBP, the scale difference can be avoided by learning from historical data. In studies [81, 82], the reinforcement learning has been applied for power systems to assist or replace model-based optimisation approaches. However, existing works are based on historical data in an offline manner. combining the reinforcement learning with real-time simulation of energy hub enables the scheduling model to dynamically adapt with operational conditions and dispatching decisions.

In order to shape the patterns of supply and demand, the time-of-use pricing scheme has been implemented in retail energy markets. Incentivised by the time-of-use pricing scheme, energy hub operators strategically determine the imports of gas and electricity and energy dispatch, in order to minimise the operating costs. The price responding strategies and price elasticity have been intensively investigated in existing works, e.g., [83, 84, 85]. Based on existing works, there is a potential extension to include statistical model for capturing both decision transient feature and price elasticity of decision-making from energy hub operators.

In this example research, the conditional random field [86] is exploited as a regression approach to extract the temporal transitions of scheduling behaviours incentivised by the prices of gas and electricity. The conditional random field can statistically support the reinforcement learning for optimal decision-making and adapting system dynamics. Compared with the naive Bayes model [41], the logistic regression of the linear conditional random field can model the discrete decisions of energy imports and dispatch and express the Q-function of reinforcement learning as an expectation of operating costs for a range of scenarios. Therefore, this paper approaches the energy hub scheduling through solving limitations in existing research as:

- A scalable model which can adapt different sizes and energy sources needs to be designed.

- The dependency of scheduling decisions on the prices of gas and electricity, and the time-transition of decisions are not captured by existing models.

- The price elasticity needs to be considered into scheduling decisions.

Through addressing these research gaps, this example research offers the following contributions as:

- The conditional random field was exploited to capture dynamic price elasticity of various energy sources, as well as the dependency of scheduling decisions on the prices of gas and electricity, and the time-transition of decisions.

- The algorithm of reinforcement learning was developed to improve the scalability of scheduling models, such that the model can be dynamically updated through interacting with the real-time digital simulator.

- The model can reduce carbon emissions caused by transmission losses and imports from electricity and gas.

- Simulation results demonstrate that the proposed model can capture the dependency of scheduling decisions on energy prices and the dependency of transient action through real-time recursive weight updates. Both operational costs and carbon emissions were saved through the proposed scheduling model.

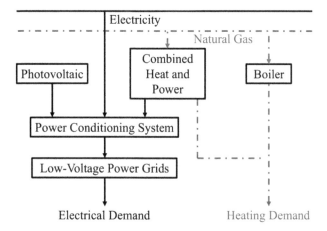

Figure 5.12 Structure of the energy hub with electricity and natural gas networks. The energy hub contains the power conditioning system, solar photovoltaic, combined heat and power, boiler, and low-voltage power grids. The black solid lines are the electrical power flow and the red dashed lines are the thermal flow.

The remainder of this work is summarised as follows. In Sub-Section 5.9.2, the model of energy hub is introduced considering the carbon mitigation and cost saving. The conditional random field-based reinforcement learning for price elasticity modelling is shown in Sub-Section 5.9.3. Section 5.9.4 performs case studies to demonstrate the proposed model. Section 5.9.5 draws the conclusion.

5.9.2 System Model

In this subsection, the model of the energy hub is mathematically described and the technical constraints during the operation of the energy hub are defined.

5.9.2.1 Energy Hub Components

The structure of the energy hub integrating electricity and gas networks is shown in Figure 5.12. It contains the power conditioning system, solar photovoltaic, combined heat and power, boiler, and low-voltage power grids.

The functions of each components are briefly introduced as follows:

- *Power conditioning system:* The power conditioning system is installed at the solar photovoltaic and combined heat and power as a voltage source inverter to improve the stability of grid voltage [87].

- *Combined heat and power:* The combined heat and power converts natural gas into electricity and recovers the generated heat for supplying heating loads [88].

- *Boiler:* The boiler converts natural gas into heat through heating contained fluid. The efficiency of boiler is higher than the combined heat and power [89].

- *Solar photovoltaic:* The solar photovoltaic is a renewable energy source commonly used in consumers' domain to convert solar energy into direct current electricity.

- *Low-voltage power grids:* The low-voltage power grids deliver the electricity to demand-side of the energy hub.

5.9.2.2 Technical Constraints

The technical constraints during the operation of the energy hub are mathematically formulated as follows:

- The power conditioning system injects or ejects reactive power as a voltage source inverter, which is constrained by the real electrical power and apparent electrical power capacity as

$$(p_e^{\text{pcs}})^2 + (q_e^{\text{pcs}})^2 \leq (s_e^{\text{pcs,max}})^2, \tag{5.24}$$

  where p_e^{pcs} is the active power output of the power conditioning system, q_e^{pcs} is the reactive power output of the power conditioning system, and $s_e^{\text{pcs,max}}$ is the maximum capacity of the power conditioning system.

- The electrical power output of the combined heat and power is constrained by its efficiency as

$$p_e^{\text{chp}} = \eta_e^{\text{chp}} \cdot p_g^{\text{chp}}, \tag{5.25}$$

  where p_e^{chp} is the active electrical power output of the combined heat and power, η_e^{chp} is the efficiency for converting the gas to electricity through using the combined heat and power, and p_g^{chp} is the gas consumption of the combined heat and power.

- The thermal power output of the combined heat and power is constrained by its efficiency as

$$p_h^{\text{chp}} = \eta_h^{\text{chp}} \cdot p_g^{\text{chp}}, \tag{5.26}$$

  where p_h^{chp} is the thermal power output of the combined heat and power, and η_h^{chp} is the efficiency for converting the gas to heat through using the combined heat and power.

- The active electrical power of the combined heat and power is constrained by its apparent electrical power capacity as

$$0 \leq p_e^{\text{chp}} \leq s_e^{\text{chp,max}}, \tag{5.27}$$

  where $s_e^{\text{chp,max}}$ is the apparent electrical power capacity of the combined heat and power.

- The reactive electrical power of the combined heat and power is constrained by its apparent electrical power capacity and maximum electrical active power output as

$$-\sqrt{(s_e^{\text{chp,max}})^2 - (p_e^{\text{chp,max}})^2} \le q_e^{\text{chp}} \le \sqrt{(s_e^{\text{chp,max}})^2 - (p_e^{\text{chp,max}})^2}, \quad (5.28)$$

where $p_e^{\text{chp,max}}$ is the maximum electrical active power output of the combined heat and power, and q_e^{chp} is the reactive electrical power output of the combined heat and power.

- The heat output of the combined heat and power is constrained by its apparent thermal power capacity as

$$0 \le p_h^{\text{chp}} \le s_h^{\text{chp,max}}, \quad (5.29)$$

where p_h^{chp} is the thermal power output of the combined heat and power, and $s_h^{\text{chp,max}}$ is the apparent thermal power capacity of the combined heat and power.

- The heat output of the boiler is constrained by its efficiency as

$$p_h^{\text{boiler}} = \eta_h^{\text{boiler}} \cdot p_g^{\text{boiler}}, \quad (5.30)$$

where p_h^{boiler} is the thermal power output of the gas boiler, η_h^{boiler} is the thermal efficiency of the gas boiler, and p_g^{boiler} is the gas consumption of the gas boiler.

- The heat output of the boiler is also constrained by its capacity as

$$0 \le p_h^{\text{boiler}} \le s_h^{\text{boiler,max}}, \quad (5.31)$$

where $s_h^{\text{boiler,max}}$ is the maximum thermal capacity of the gas boiler.

- The electrical power output of the solar photovoltaic is constrained by its efficiency as

$$p_e^{\text{pv}} = \eta_e^{\text{pv}} \cdot p_r, \quad (5.32)$$

where p_e^{pv} is the electrical power output of the solar photovoltaic, η_e^{pv} is the efficiency of the solar photovoltaic, and p_r is the solar irradiance.

- The reactive electrical power of the solar photovoltaic is constrained by its apparent electrical power capacity and maximum electrical power output as

$$-\sqrt{(s_e^{\text{pv,max}})^2 - (p_e^{\text{pv,max}})^2} \le q_e^{\text{pv}} \le \sqrt{(s_e^{\text{pv,max}})^2 - (p_e^{\text{pv,max}})^2}, \quad (5.33)$$

where $s_e^{\text{pv,max}}$ is the maximum electrical power capacity of the solar photovoltaic, $p_e^{\text{pv,max}}$ is the maximum electrical power output of the solar photovoltaic, and q_e^{pv} is the electrical reactive power output of the solar photovoltaic.

5.9.2.3 *Carbon Emissions Tracing*

The operation of an energy hub generates carbon emissions, primarily through the exchanges of electricity and natural gas with utility energy networks and transmission line losses within the energy hub. As part of their operational costs, energy hub operators are required to pay a carbon tax. To mitigate these emissions, a proposed scheduling model takes into account the carbon emissions generated by the operation of the energy hub. A general way to quantify carbon emissions is by using carbon intensity, which is the amount of produced carbon emissions per unit of the exchange of power or natural gas as

$$i = \frac{r}{p}, \tag{5.34}$$

where i is the carbon intensity with the unit of kg/kWh, p is the power flow or natural gas flow, and r is the quantity of carbon emissions per unit of the exchange of the power or natural gas per unit time with the unit of kg/h.

The model tracks two portions of carbon emissions. The first portion of carbon emissions is caused by transmission line loss within the energy hub. The topology structure of power networks must be considered to address this type of carbon emissions. Since power flow distribution is based on proportional sharing principles [90], the carbon emission distribution follows the same principle. Specifically, the carbon emission rate of an outflow branch is expressed as the sum of carbon emissions of the inflow branch and the bus-connected source as

$$r_j = \sum_{i \in z} p_{i,j} e_i + \sum_{s \in z} p_s e_s, \tag{5.35}$$

where e_i and e_s are carbon intensities in branch i and bus-connected source s, respectively, and $p_{i,j}$ is the share of power flow in jth branch coming from ith branch p_i.

According the to proportional sharing principle [90], we have

$$\frac{p_{i,j}}{p_j} = \frac{p_i}{\sum_{i \in z} p_{i,j} + \sum_{s \in z} p_s}. \tag{5.36}$$

The second portion of carbon emissions comes from the exchanges of electricity and natural gas with the utility energy networks. The proposed model considers the total carbon emission rate from natural gas r_g, electricity r_e, and transmission loss r_j during the scheduling horizon T, which should be less than the carbon emission limit. This carbon emission limit can be defined based on regulations or policies.

$$\sum_{t=1}^{T} r_g \cdot t + r_e \cdot t + r_j \cdot t \leq e^{\max}, \tag{5.37}$$

where R_e and R_g are carbon emission rates for electricity and natural gas, respectively, and $e^{\max}$ is the carbon emission limit defined based on regulations or policies.

5.9.3 Proposed Algorithm

This subsection presents the proposed algorithm for modelling price elasticity using the conditional random field-based reinforcement learning in the context of energy hub scheduling. To further elaborate on the proposed algorithm for price elasticity modelling, it is worth noting that the dynamic price elasticity of energy sources in the energy hub refers to the responsiveness of energy demand to changes in energy prices. By modelling price elasticity, the proposed algorithm allows for more effective management of energy flows in the energy hub, resulting in improved energy efficiency and reduced operational costs.

The conditional random field-based reinforcement learning approach is a powerful tool for modelling price elasticity in the context of energy hub scheduling. By leveraging the principles of reinforcement learning, the proposed algorithm is able to analyse historical data on energy prices and energy hub scheduling decisions to learn from past experiences and optimise future energy flows. This approach allows for the dynamic adjustment of energy hub scheduling decisions based on changing market conditions and energy prices.

In addition to improving energy efficiency and reducing operational costs, the proposed algorithm has important environmental implications. By enabling more efficient use of energy resources, the algorithm can help to reduce greenhouse gas emissions and other negative environmental impacts associated with energy production and consumption. Overall, the proposed algorithm represents a significant advance in the field of energy hub management and has the potential to drive significant improvements in the sustainability and efficiency of energy systems.

The action space A for energy hub scheduling is defined as follows: decisions for electricity exchange from the main grid A_e, decisions for natural gas exchange from the gas network A_g, decisions for dispatching gas to combined heat and power for producing electricity $A_{e,\text{CHP}}$ and heat $A_{h,\text{CHP}}$, and decisions for dispatching gas to the boiler A_{Boiler}. The control actions for energy hub scheduling are binary variables that represent the corresponding components to be switched on or off:

$$a_e, a_g, a_{e,\text{chp}}, a_{h,\text{chp}}, a_{\text{boiler}} \in \{0, 1\}. \tag{5.38}$$

The action vector $\mathbf{a}$ is subsequently defined as:

$$\mathbf{a} = (a_e, a_g, a_{e,\text{chp}}, a_{h,\text{chp}}, a_{\text{boiler}}). \tag{5.39}$$

The state space S of energy hub scheduling consists of the price of electricity S_e and the price of natural gas S_g. The state vector is subsequently defined as the electricity price π_e and natural gas price π_g, resulting in

$$\mathbf{s} = (\pi_e, \pi_g). \tag{5.40}$$

At the beginning of each scheduling interval t, the market operator announces the electricity and gas prices $\mathbf{s}^t = (\pi_e^t, \pi_g^t)$ to the energy hub operator. The energy hub operator then decides and dispatches the scheduling results $\mathbf{a}^t = (a_e^t, a_g^t, a_{e,\text{chp}}^t, a_{h,\text{chp}}^t, a_{\text{boiler}}^t)$ at the end of scheduling interval t. The proposed algorithm aims to introduce the dynamic price elasticity of energy sources in the energy hub. Given observed states $\{\mathbf{s}^1, ..., \mathbf{s}^t\}$ and past actions $\{\mathbf{a}^1, ..., \mathbf{a}^{t-1}\}$, the algorithm analyzes the probability of the energy hub operator's decisions for energy exchanges from main networks and inside dispatch $\mathbf{a}^t$ at t. Using this analysis, the reinforcement learning algorithm is performed to obtain an optimal control policy for energy hub scheduling.

5.9.3.1 Conditional Random Field for Elasticity Modelling

The proposed algorithm for modelling price elasticity of energy sources and action transient dependency is based on the linear conditional random field approach [91], which differs from the Hidden Markov Model approach in that it accounts for the dependence of action at the current time on all observed states $\{\mathbf{s}^1, ..., \mathbf{s}^t\}$. This is important for practical scheduling problems, where the energy hub operator may make global decisions in response to price signal variations to minimise total daily operating costs.

The linear conditional random field is subject to the Markov property [92], which means that conditioned on $\mathbf{s}^t$, action $\mathbf{a}^t$ at time t is independent of action $\mathbf{a}^k$ at time $k, (k \neq t)$, given $\mathbf{a}^{t+1}$ and $\mathbf{a}^{t-1}$, as

$$p(\mathbf{a}^t \mid \mathbf{a}^1, ...\mathbf{a}^t, \mathbf{s}^1, ...\mathbf{s}^t) = p(\mathbf{a}^t \mid \mathbf{a}^{t-1}, \mathbf{a}^{t+1}, \mathbf{s}^1, ..., \mathbf{s}^t). \tag{5.41}$$

The Markov property of the linear conditional random field allows for the modelling of the conditional probability of energy hub operator's decisions for energy exchanges at a given time based on the observed state. Specifically, the conditional probability is modelled as a product of state and transient feature functions, with weighting factors describing the strength of the dependencies. The normalization factor is used to ensure that the probabilities add up to one.

The conditional probability $p(\mathbf{a}^t|\mathbf{s}^t)$ is modelled as

$$p(\mathbf{a}^t|\mathbf{s}^t) = \frac{1}{Z(\mathbf{s}^t)} \prod_t \exp\left(\mu^t \Phi^t(\mathbf{a}^t, \mathbf{s}^t)\right) \prod_{t-1} \exp\left(\lambda^{t,t-1} \Psi^{t,t-1}(\mathbf{a}^t, \mathbf{a}^{t-1})\right), \tag{5.42}$$

where

$$\Phi^t(\mathbf{a}^t, \mathbf{s}^t) := \mathbf{a}^t \mathbf{s}^t \tag{5.43}$$

is the state feature function to describe the dependency of action $\mathbf{a}^t$ on state $\mathbf{s}^t$ at time t;

$$\Psi^{t,t-1}(\mathbf{a}^t, \mathbf{a}^{t-1}) = \mathbf{a}^t \mathbf{a}^{t-1} \tag{5.44}$$

is the transient feature function to describe the dependency of action $\mathbf{a}^t$ at time t on action $\mathbf{a}^{t-1}$ at time $t-1$, μ^t and $\lambda^{t,t-1}$ are weighting factors to describe the strength of these dependencies, and

$$Z(\mathbf{s}^t) = \sum_{\mathbf{a}^t} \prod_t \exp\left(\mu^t \Phi^t(\mathbf{a}^t, \mathbf{s}^t)\right) \prod_{t-1} \exp\left(\lambda^{t,t-1} \Psi^{t,t-1}(\mathbf{a}^t, \mathbf{a}^{t-1})\right) \tag{5.45}$$

is a normalization factor.

Motivated by the state feature function and transient feature function, the weighting factors μ^t and $\lambda^{t,t-1}$ can be defined as

$$\mu^t := \frac{f_{\mathbf{s}^t}(\mathbf{a}^t)}{t}, \tag{5.46}$$

$$\lambda^{t,t-1} := \frac{f_{\mathbf{a}^{t-1}}(\mathbf{a}^t)}{t}, \tag{5.47}$$

where $f_{\mathbf{s}^t}(\mathbf{a}^t)$ is the total amount of time in which the action $\mathbf{a}^t$ is performed as on $(\mathbf{a}^t=1)$ given state $\mathbf{s}^t$ and $f_{\mathbf{a}^{t-1}}(\mathbf{a}^t)$ is the total amount of time in which the action $\mathbf{a}^t$ is performed as on $(\mathbf{a}^t=1)$ given that action $\mathbf{a}^{t-1}$ is performed as on $(\mathbf{a}^{t-1}=1)$.

μ^t and $\lambda^{t,t-1}$ can be updated at each time step when receiving new pieces of information $\gamma_{\mathbf{s}^t}(\mathbf{a}^t)$ and $\gamma_{\mathbf{a}^{t-1}}(\mathbf{a}^t)$ as

$$f_{\mathbf{s}^t}(\mathbf{a}^t) = f_{\mathbf{s}^{t-1}}(\mathbf{a}^{t-1}) + \gamma_{\mathbf{s}^t}(\mathbf{a}^t), \tag{5.48}$$

and

$$f_{\mathbf{a}^{t-1}}(\mathbf{a}^t) = f_{\mathbf{a}^{t-2}}(\mathbf{a}^{t-1}) + \gamma_{\mathbf{a}^{t-1}}(\mathbf{a}^t). \tag{5.49}$$

Therefore, weighting factors μ^t and $\lambda^{t,t-1}$ can be updated recursively as

$$\mu^t = \frac{f_{\mathbf{s}^{t-1}}(\mathbf{a}^{t-1}) + \gamma_{\mathbf{s}^t}(\mathbf{a}^t)}{t-1} \cdot \frac{t-1}{t} = \mu^{t-1} + \frac{1}{t}[\gamma_{\mathbf{s}^t}(\mathbf{a}^t) - \mu^{t-1}], \tag{5.50}$$

$$\lambda^{t,t-1} = \lambda^{t-1,t-2} + \frac{1}{t}[\gamma_{\mathbf{a}^{t-1}}(\mathbf{a}^t) - \lambda^{t-1,t-2}]. \tag{5.51}$$

5.9.3.2 Reinforcement Learning

The proposed algorithm for energy hub scheduling is based on a linear conditional random field model that considers the price elasticity of energy sources and action transient dependency. This approach allows for more strategic decision-making by the energy hub operator, who can respond to changes in pricing signals to minimise operating costs. Our approach also incorporates the revenue from providing electricity and heat services within the energy hub as a means of compensating for costs.

The minimal cost of the current state is defined as the expectation of the probability of energy imports from the main grid becoming minimum, i.e., $p(\mathbf{a}^t = 0 \mid \mathbf{s}^t)$ and the probability of inside dispatch of energy hub becoming maximum, i.e., $p(\mathbf{a}^t = 1 \mid \mathbf{s}^t)$ given the current state as

$$\begin{aligned}
c(\mathbf{s}^t) =& \mathbb{E}\{p_e \cdot a_e^t \cdot t \cdot \pi_e + p_g \cdot a_g^t \cdot t \cdot \pi_g - [(p_e^{\text{pcs}} + p_e^{\text{pv}} + \\
& a_{e,\text{chp}}^t \cdot p_e^{\text{chp}}) \cdot t \cdot \pi_e + (a_{h,\text{chp}}^t \cdot p_h^{\text{chp}} + a_{\text{boiler}}^t \cdot p_h^{\text{boiler}}) \cdot t \cdot \pi_g] \mid \mathbf{s}^t\} \\
=& (p_e \cdot t \cdot \pi_e + p_g \cdot t \cdot \pi_g) \cdot p(\mathbf{a}^t = 0 \mid \mathbf{s}^t) \\
& - [p_e^{\text{chp}} \cdot t \cdot \pi_e + (p_h^{\text{chp}} + P_h^{\text{boiler}}) \cdot t \cdot \pi_g] \cdot p(\mathbf{a}^t = 1 \mid \mathbf{s}^t) \\
& - (p_e^{\text{pcs}} + p_e^{\text{pv}}) \cdot t \cdot \pi_e.
\end{aligned} \tag{5.52}$$

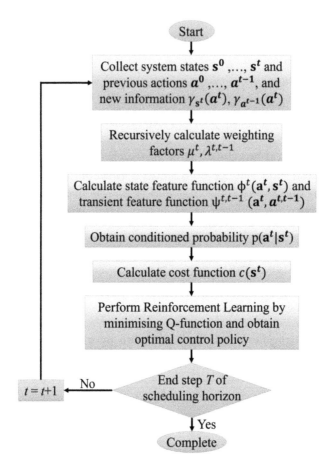

Figure 5.13 Flowchart of the proposed conditional random field-based reinforcement learning algorithm.

The objective of the reinforcement learning approach is to find an optimal control policy $(h : S \to A)$ that minimises the total cost of energy hub scheduling from the initial state $\mathbf{s}^1$ to the current state $\mathbf{s}^t$.

To achieve this, the Q-function is used to represent the total discounted cumulative reward following a given policy as

$$Q^h(\mathbf{a}^t, \mathbf{s}^t) = \mathbb{E}\{c(\mathbf{s}^0) + \xi c(\mathbf{s}^1) + \xi^2 c(\mathbf{s}^2) + \dots + \xi^t c(\mathbf{s}^t)\}, \tag{5.53}$$

where ξ is the discounting factor.

The optimal policy h^* is then obtained by minimising the Q-function subject to the given constraints of Equations (5.9.2.2) - (5.33) as

$$h^* \in \arg \min_{\mathbf{a}^t, p_e, p_g, q_e^{\mathrm{CHP}}, q_e^{\mathrm{PV}}} Q^*(\mathbf{a}^t, \mathbf{s}^t) \tag{5.54}$$

The flowchart in Figure 5.13 visually represents the proposed algorithm for conditional random field-based reinforcement learning for energy hub scheduling.

Overall, this approach provides a more comprehensive and dynamic solution for energy hub scheduling that takes into account the complexities of pricing and revenue generation, as well as the strategic decision-making required to optimise energy hub operations.

5.9.4 Numerical Results

In order to evaluate the effectiveness of the proposed model, a numerical simulation is conducted using a modified four-bus medium voltage distribution system, which is illustrated in Figure 5.14. This system features a combined heat and power unit and a boiler installed at bus 3, where the electricity and gas networks are coupled. Additionally, a PV system is installed at bus 2.

For comparison purposes, we set the same parameter values as given in ref. [79] and compare the performance of the proposed model in terms of operating cost and carbon emissions. By conducting this simulation, we can demonstrate the effectiveness of the proposed model and provide insight into the potential benefits of using the model in a practical application.

5.9.4.1 Simulator

The proposed energy hub scheduling model is implemented in real-time using the Real-Time Digital Simulator as shown in Figure 5.15. The advantages of the Real-Time Digital Simulator are summarised as follows:

- *Real-time simulation:* Real-Time Digital Simulator operates continuously and in real-time, providing a real-time environment for energy hub components. This enables the components to be connected through medium voltage interfaces and interact with the real power system components.

- *Accurate modelling:* Real-Time Digital Simulator has the ability to model complex systems accurately, including real-time modelling of smart meters and controllers using the Data Acquisition and Actuator module.

- *Hardware-in-the-loop simulation:* Real-Time Digital Simulator allows for hardware-in-the-loop simulation, which means that real components can be connected to the simulation. This provides a more accurate representation of the system than software-only simulations.

- *Fast and efficient testing:* Real-Time Digital Simulator enables fast and efficient testing of energy hub scheduling algorithms and strategies, allowing for quick optimisation of system performance.

- *Risk-free testing:* Since the simulation is performed in a controlled environment, testing can be performed without risking damage to real components or power systems. This makes it a safe and cost-effective way to test new energy hub scheduling strategies.

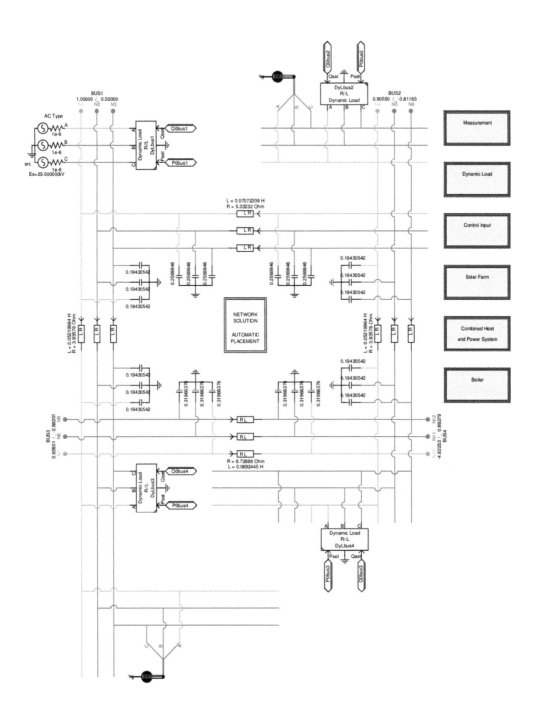

Figure 5.14 Schematic illustration of a four-bus medium voltage distribution system. This system features a combined heat and power unit and a boiler installed at bus 3, where the electricity and gas networks are coupled. Additionally, a PV system is installed at bus 2.

Figure 5.15 Interactions between MATLAB®, Real-Time Digital Simulator, and Data Acquisition and Actuator module. After performing the conditional random field-based reinforcement learning using MATLAB, the scheduling results are transmitted back to Real-Time Digital Simulator using the Giga-Transceiver Analogue Input Card. Meanwhile, real-time operating signals are transmitted from Real-Time Digital Simulator to MATLAB using the Giga-Transceiver Analogue Output Card.

After performing the conditional random field-based reinforcement learning using MATLAB, the scheduling results are transmitted back to Real-Time Digital Simulator using the Giga-Transceiver Analogue Input Card. Meanwhile, real-time operating signals are transmitted from Real-Time Digital Simulator to MATLAB using the Giga-Transceiver Analogue Output Card. This provides a real-time scheduling and performance evaluation of the energy hub model. The implementation on Real-Time Digital Simulator ensures that the scheduling decisions are more accurate and reliable, as it simulates the real-time operation of the energy hub model. By incorporating the conditional random field-based reinforcement learning approach, the proposed model can adapt to changing energy prices and make optimal scheduling decisions, leading to reduced carbon emissions and operating costs.

5.9.4.2 Evaluation of Model Performance

To evaluate the effectiveness of our proposed reinforcement learning model, we designed a benchmark optimisation problem, as follows, with the same decision variables and constraints as our proposed model:

$$\min \sum_{t=1}^{T} p_e \cdot a_e^t \cdot t \cdot \pi_e + p_g \cdot a_g^t \cdot t \cdot \pi_g - [(p_e^{\text{pcs}} \cdot t + p_e^{\text{pv}} \cdot t$$

$$+ a_{\text{chp}} \cdot p_e^{\text{chp}} \cdot t)\pi_e + (a_{\text{chp}}^t \cdot p_h^{\text{chp}} \cdot t + a_{\text{boiler}}^t \cdot p_h^{\text{boiler}} \cdot t) \cdot \pi_g]. \tag{5.55}$$

The objective of the benchmark optimisation is to minimise the daily overall costs, which yields a theoretically optimal scheduling solution with predefined cost coefficients and carbon intensities. In contrast, the goal of our proposed algorithm is to obtain a learning policy that approximates the optimal solution.

The simulation was performed using 50 days of historical hourly data. At each new time step, the weighting factors μ^t and $\lambda^{t,t-1}$ were updated recursively using Equations (5.50) and (5.51), respectively, to describe the strength of state-action and state-transient dependencies. The probability distribution of the weighting factors for electricity and gas prices is shown in Figure 5.16, where each column represents the distribution of the dependency for each control action on various price levels from low to high, and each row represents the difference of the dependency for various control actions responding to the same price level.

From Figure 5.16, it can be observed that the probability of μ^t is higher during the peak price period corresponding to peak demand, which indicates that the action $\mathbf{a}^t$ is more dependent on received price information during this period. In contrast, the probability of $\lambda^{t,t-1}$ is relatively independent of the price fluctuation and thus presents a homogeneous distribution because it is only relevant to the transient between states. These results demonstrate that the proposed model successfully captures the dependency features and can adapt to the changing energy prices.

Figure 5.17 illustrates the performance of reinforcement learning over 10, 30, and 50 days with respect to various electricity prices. The selected outputs of the combined heat and power, power exchange, and corresponding real-time electricity prices are used as examples to compare the learning results with the benchmark optimisation. The figure clearly shows that as the learning progresses with the accumulation of historical data, the learning results gradually approach the optimal solution regardless of the electricity price.

The results indicate that the proposed reinforcement learning algorithm is effective in adapting to changing energy prices and learning optimal scheduling decisions in real time. The algorithm uses the historical data to continuously update the weighting factors and improve the scheduling performance, leading to reduced carbon emissions and operating costs. The comparison of the learning results with the benchmark optimisation highlights the benefits of using the proposed algorithm, which takes into account the temporal variations of scheduling behaviours influenced by energy prices.

Overall, our proposed reinforcement learning model offers several advantages, including the ability to learn from real-time data, adapt to changing operating conditions, and improve the scheduling performance compared to benchmark optimisation. The use of Real-Time Digital Simulator for real-time simulation further enhances the accuracy and effectiveness of our proposed model.

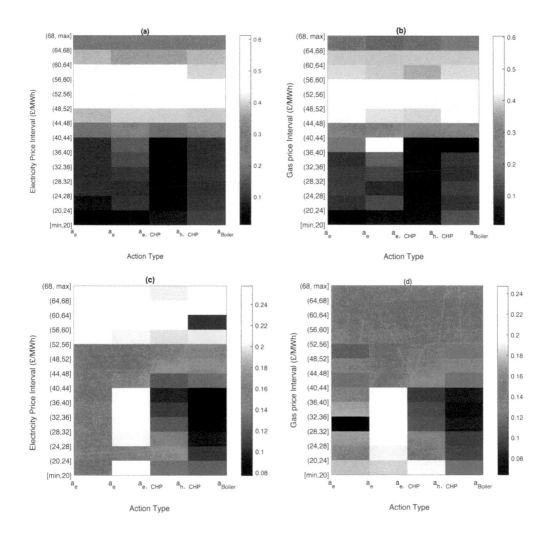

Figure 5.16 Probability distribution of weighting factors of μ^t and $\lambda^{t,t-1}$ at each interval for (a) μ^t with π_e, (b) μ^t with π_g, (c) $\lambda^{t,t-1}$ with π_e, and (d) $\lambda^{t,t-1}$ with π_g. The x axes represent the action type and the y axes represent the price interval.

5.9.4.3 Evaluation of Cost and Carbon Reduction

We compare our proposed algorithm with cost minimisation problem as

$$\min_{p_e, p_g, q_e^{\text{chp}}, q_e^{\text{pv}}} c_e(p_e) + c_g(p_g), \tag{5.56}$$

where p_e and p_g are electricity and natural gas importing from main energy networks, respectively, and c_e and c_g are corresponding costs. The cost minimisation problem is subject to the same constraints as our model.

Figure 5.18 shows the scheduling outputs and the corresponding average carbon intensity. The results demonstrate that our proposed model can effectively reduce daily carbon emissions by 1.388 tons, from 6.956 tons to 5.568 tons, primarily by reducing the peak-time electricity imported from the main grid and increasing the

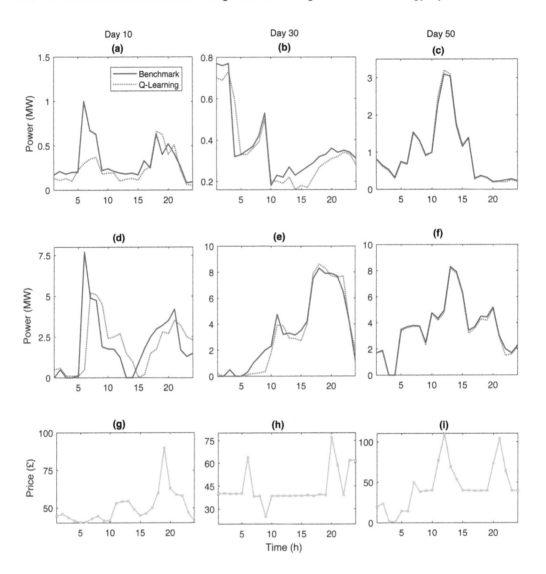

Figure 5.17 Illustration of learning process for electricity generated by combined heat and power in (a), (b), and (c), power exchange in (d), (e), and (f) with electricity prices in (g), (h), and (i). The day number is indicated at the top of each column.

proportion of gas. This is due to the fact that the energy hub operator is more sensitive to the peak-time price with the consideration of price elasticity. Moreover, the proposed model can also reduce the daily operating costs by 9.76% from £ 3012 to £ 2718, taking into account the revenue from internal energy supply. These results suggest that our proposed model can achieve significant environmental and economic benefits, while ensuring the energy hub operates efficiently and reliably. Furthermore, the simulation results demonstrate that the reinforcement learning approach can learn from historical data and adapt to changing energy prices, and thus obtain near-optimal scheduling solutions without the need for extensive mathematical modelling and predefined assumptions.

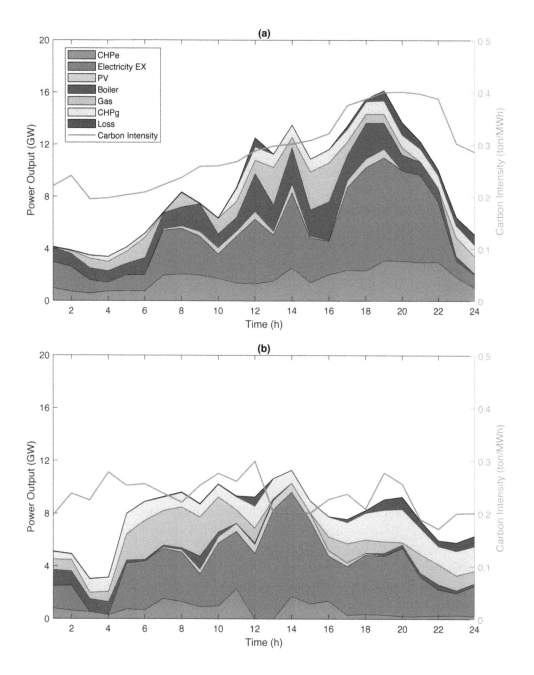

Figure 5.18 Comparison of energy scheduling and carbon intensity for (a) Scheduling from (22) and (b) proposed algorithm.

5.9.5 Research Summary

In this example research, a novel approach to energy hub scheduling is proposed, which aims to improve scalability and incorporate intrinsic price elasticity analysis. Unlike current model-based optimisation approaches that rely on predefined parameters and assumptions, the proposed approach is based on a conditional random field-based reinforcement learning model. This model enables the evaluation of state action feature and action transient feature, which allows for the incorporation of temporal variations in scheduling behaviours influenced by energy prices.

The proposed scheduling strategy is developed based on real-time digital simulation, which enables the Q-function to dynamically adapt to the system operating conditions. Simulation results show that the weighting factors successfully describe the dependency features of scheduling decisions on pricing signals during peak demand periods. The reinforcement learning algorithm can approximate the theoretical optimal scheduling after only 50 days.

In addition to improving the scalability and incorporating price elasticity analysis, the proposed algorithm is also effective in reducing both carbon emissions and operating costs. This demonstrates the practical usefulness of the proposed approach and its potential for wide application in real-world energy hub systems. Further research can explore the optimisation of the Q-function for more complex energy hub systems with multiple inputs and outputs.

5.10 EXAMPLE RESEARCH 4: ARTIFICIAL INTELLIGENCE FOR ENERGY SYSTEMS SCHEDULING UNDER UNCERTAINTIES

Besides supplying increasing energy demands, modern energy systems are playing a major role in reducing carbon emissions and tackling climate change challenges [93]. One key task for future energy systems is to further improve the energy efficiency in supplying diverse energy demand forms, e.g., electricity loads, heating loads, and cooling loads. However, some load forms, such as heating loads, are still relying on a single energy source, which could be carbon-intensive. For example, in the UK 80% homes are relying on natural gas for heating, which accounts for 30% of the overall natural gas consumption in the UK market [94].

Among the potential solutions, an integrated multi-vector energy system has shown its advantage in meeting the various load demands with hybrid energy sources [95][96]. With versatile energy converters, the multi-vector energy system operator can meet the load demands in a more efficient and economical manner [79]. For example, the heat load can be supplied by natural gas boilers, the recovered heat from combined heat and power (CHP) system, or their combinations [97][98].

5.10.1 Introduction

The energy markets can offer more options to help drive down the operating cost of meeting load demands. Since the energy flows between multiple energy inputs and outputs can be controlled via the energy converters, multi-vector system operators don't need to rely on a single energy source. Instead, they can access multiple wholesale energy markets, according to its own operating situation.

Moreover, renewable energy sources are envisioned to be a significant part of the total energy capacity, where in the UK it is forecasted to boost from 34% of the installed capacity to 60% in 2030 [99]. Renewable energy sources not only provide clean and low-carbon energy, but also have the potentials in reducing the system's operating cost. Meantime, mature data services are widely available for the operators to make a better energy scheduling. For example, in the UK, the National Grid provides day-ahead forecast services to the electricity load demand, and wind and solar energy generations[100].

Making optimal decisions on the day-ahead market is a very challenging task [101]. This is because when making energy scheduling on the day-ahead market, the operators should rely on day-ahead forecasts to make a decision, where the actual demands are not known. The mismatch between forecasts and actual values introduces uncertainties to the system operations, where renewable energy sources and load demands are all major uncertainty sources leading to significant impacts on the operating costs.

For a real-world system, the fundamental objective of the multi-vector energy system is to convert the energy from inputs to meet the load demands. However, the change of required energy input amount in real-time energy dispatch will also have impacts on the safety of the whole energy grid operations, which requires services like balancing services to deal with such short-term energy fluctuation in the grid. The balancing service will introduce extra costs to the system operation, which is also depending on the uncertainties of the system. In order to minimize the total operating costs on both the day-ahead market and the balancing market, the multi-vector energy system is challenged to make optimal energy schedules on the day-ahead market.

Moreover, the uncertainties will result in a mismatch between energy scheduling and the actual required dispatching. The change of real-time energy dispatching will also have impacts on the safety of the whole energy system operations, which requires services like balancing services to deal with such short-term energy fluctuations. The balancing service will introduce extra costs, which is also depending on the uncertainties of the system. In order to minimize the total operating costs in a multi-vector energy system, it is challenging to make an optimal energy scheduling on the day-ahead market [102].

In the existing literature, the modelled economic reserve scheduling problem was studied for a multi-vector energy system with both electricity and natural gas in [103], where renewable energy forecasting errors were accounted by an interval with fixed proportion. In [104], the day-ahead solar power forecast and load demands were modelled with a discrete probability distribution, where the two-stage stochastic mixed-integer linear programming method was used to address the day-ahead energy scheduling problem. A multi-objective optimization-based method was proposed for the day-ahead scheduling of a multi-vector energy system with thermal, wind and solar energies [105], where renewable energy was described by a deterministic probability distribution model.

Addressing the impacts from various uncertainty sources has been a key research and practical challenge in the modern energy system, where artificial intelligence (AI) has shown a great advantage over traditional methods. For example, compared to the traditional deterministic model based load forecasting methods, existing works have proposed various AI-based methods with better accuracy performance, including different architectures like artificial neural networks (ANN) [49], recurrent neural networks (RNN) [106], and deep neural networks (DNN) [107]. With the development of renewable energy, the research of more accurate forecasts on renewable energy productions is also becoming a hot topic.

Methods are already available on common renewable energy forms including solar power [108] and wind power [109]. There have been several successful attempts in addressing the day-ahead scheduling problem. In ref. [110], an extreme learning machine was integrated into the day-ahead scheduling and real-time dispatching schemes, which was trained with a historical data to make accurate forecasts. In [111], the long short-term memory neural networks were trained to make probabilistic forecasts on load demands and wind energy generations, which was integrated into the look-ahead dispatching schemes.

Note that the accuracy of the forecasts depends on the dataset, the performance index and the specific scenarios. The local forecasts with only historical dataset might not benefit the day-ahead scheduling compared to the sophisticated forecasting service with more considerations, such as weather forecasts.

The day-ahead scheduling problem for an electricity system with solar energy as the single uncertainty source was studied in [112], where the kernel method was used. The training method was later improved using neural networks in [113], which incorporates linear programming into the training procedure. However, these existing works didn't address the problem caused by multiple uncertainty sources, which is the main focus of this section. In addition, the existing linear programming-based training procedure is time-consuming and computationally exhaustive. These issues will be addressed by a two-stage deep learning training method in this section.

Forecasting errors in renewable energy sources and load demands will cause the system to make a deviated decision on the day-ahead market, which results in the punishment cost as modelled in (5.62). It is difficult to avoid forecasting errors in a real-world multi-vector energy system, as real forecasting errors could be very hard to characterize. As illustrated in Fig. 5.19 and Fig. 5.20 using British electricity load demands and on-shore wind energy in 2017 [100], it can be seen that the electricity load demands' forecasting errors show a different distribution compared with the wind energy generations' forecasting errors.

The actual forecasting errors in extreme conditions can even exceed 600% in the wind energy production, which clearly indicate a different distribution pattern compared with the electricity load forecasting error. The simultaneous consideration of multiple uncertainty sources will make things more complicated, which will be addressed by the proposed deep learning-based methods detailed in the following section.

It can be seen that the electricity load demands' forecasting errors show a much smaller variance range than that of the wind energy generations' forecasting errors. The actual forecasting errors in extreme conditions can even exceed 600% in the wind energy production, which clearly indicates a different distribution pattern compared with the electricity load forecasting error. Unlike the ideal assumptions in theoretical research, the actual distribution of forecasting errors is much more complex, which makes it very challenging to be approximated by a small set of finite discrete probability distributions. The simultaneous consideration of multiple uncertainty sources

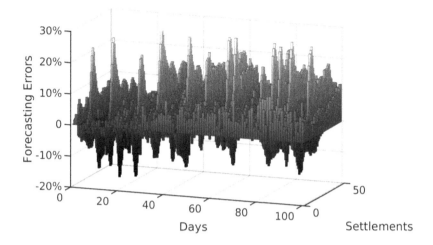

Figure 5.19 Historical electricity load demand's day-ahead forecasting errors in the U.K., an example of 100 days in 2017 [100].

will make things more complicated, which will be addressed by the proposed deep learning-based methods detailed in the following section.

The state-of-the-art method in dealing with the uncertainties is to use discrete probability to approximate the distributions of these forecasting errors. However, these methods are subject to several problems as follows:

- The real distributions of forecasting errors are continuous in nature, where discrete distributions cannot provide a very accurate approximation and lead to bad decisions thereafter.

- The discrete probability distribution is manually derived, which is based on experience and on a case-by-case method, which is very hard to generalize and the performance is therefore subject to the trial-and-error procedure.

- In the case of several uncertainty components, the joint discrete probability distribution of all the uncertainty variables needs to be generated based on the individual discrete probability distribution of a single uncertainty variable. The inaccuracy from each discrete probability distribution will be mixed with each other. It may even lead to an inaccurate joint discrete probability distribution, thereupon deteriorate the energy dispatching performance and increase the total costs.

- During the calculation, the system needs to exhaustively search the solution space to find the energy dispatch solution, which is very time-consuming. The computation burden and time consumption will get worse as the uncertainty components increase, which is due to the fact that the uncertainty variables have enlarged the solution dimensions.

Remarks: From the view of the energy system, another challenge of the distribution approximation based methods is that the forecasting errors cannot be directly

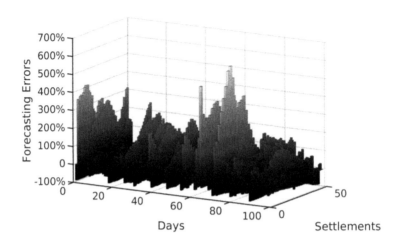

Figure 5.20 Historical wind generation's day-ahead forecasting errors in the U.K., an example of 100 days in 2017 [100].

measured. This is because the forecasting errors only describe how accurate the forecasting can be, and mostly in a statistically manner (e.g., the average error, or the variance of the error). However, is the better forecasting accuracy definitely means an increase of the energy system performance that relying on it? And also its reverse question, is the worse forecasting accuracy definitely means a decrease? This is actually a very practical problem, for example if the energy system operator is considering two bids on the forecasting service provider, should it choose the one with average error of 4.5% or the other one with average error of 5.5%? With the unknown relation between the forecasting errors and the end performance in the energy system, it is impossible to determine which service is better.

To overcome this challenge, it requires a reconsideration on the forecasting accuracy in the energy system. In ideal cases, the accuracy should be defined based on its impacts on the end performances in the energy system, i.e., the accuracy is measured by means of its end usage. This is a challenging issue, because it requires a collaboration from multiple fields instead of the data service for forecasting alone. In the next section, the energy system impacts due to the forecasting accuracy are to be addressed via the AI methods, which would shine a light upon this challenging issue.

5.10.2   Data-Driven Approach in Addressing Uncertainties

A multi-vector energy system could have various energy converters, which can supply different forms of load demands by controlling conversions between multiple energy vectors. An example of a multi-vector energy system is illustrated in Fig. 5.21. Let $\mathbf{S} = \{S_1, \ldots, S_N\}$ denote the energy source vector, whose elements are energy sources, e.g., wind energy, solar energy, electricity and natural gas. Let $\mathbf{L} = \{L_1, \ldots, L_M\}$ denote the load demand vector, whose elements are the load demands, e.g., electricity, hot water and heat.

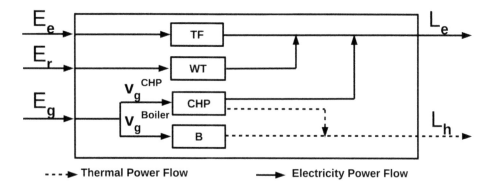

Figure 5.21 A multi-vector energy system consists of a Transformer (TF), a Wind Turbine (WT), a Combined Heat & Power (CHP) system and a Boiler (B).

Suppose there are K energy converter components in the multi-vector energy system, e.g., transformers (high voltage electricity to medium or low voltage electricity), natural gas boilers (natural gas to heat) and CHP systems (natural gas to electricity and heat). Correspondingly, their energy conversion efficiency can be represented by $\boldsymbol{\eta} = \{\eta_1, \ldots, \eta_K\}$. During the operations, the multi-vector energy system needs to make dispatching decisions on the energy sources to different energy converter components, which can be represented by the non-negative dispatching matrix $\mathbf{v}$ with the dimension of $N \times K$.

For the energy converter component k, its corresponding dispatched energy input is $v_{kn}S_n$. The objective is to make decisions on the energy sources amount $\mathbf{S}$ and the dispatching matrix $\mathbf{v}$ to meet with the load demands $\mathbf{L}$ as follows:

$$\mathbf{L} = H_{\boldsymbol{\eta}}(\mathbf{S}, \mathbf{v}), \qquad (5.57)$$

where $H_{\boldsymbol{\eta}}(\cdot)$ denotes the energy dispatching operation via the converters.

5.10.2.1 Ideal Energy Dispatching

Here we consider a system with energy sources in the form of electricity S_e, renewable energy S_r and natural gas S_g. Meantime, the load demands are in the forms of electricity and heat, denoted by $\mathbf{L} = \{L_e, L_h\}$, respectively. The modelling method studied in this section can be directly extended to the more complex scenarios, e.g., more energy sources and load demands forms.

During normal operations, the multi-vector energy system inputs are the forecasted load demands $\mathbf{L}^f = \{L_e^f, L_h^f\}$ and the forecasted renewable energy generation S_r^f. The system has to decide the amount of electricity and gas $\mathbf{S}^f = \{S_e^f, S_g^f\}$ to be purchased from the day-ahead market, as well as the energy dispatching matrix $\mathbf{v}^f$. This decision is usually made based on the day-ahead market price for each energy carrier, i.e., C_e^0 for electricity price and C_g^0 for gas price in this case.

Suppose renewable energy is free to the multi-vector energy system, the total cost on the day-ahead market is then the sum of the cost for purchasing electricity and natural gas. The objective of making energy dispatching decision is to minimize the total cost given as follows,

$$\min_{S_e^f, S_g^f, \mathbf{v}^f} \qquad C_e^0 S_e^f + C_g^0 S_g^f \qquad (5.58\text{a})$$

$$\text{s.t.:} \qquad \mathbf{L}^f - H_{\boldsymbol{\eta}}(\mathbf{S}^f, \mathbf{v}^f) = 0, \qquad (5.58\text{b})$$

$$\sum_{S_n^f \in \mathbf{S}^f} v_{kn}^f S_n^f \leq P_k^{\max}, \forall k, \qquad (5.58\text{c})$$

where dispatched energy for meeting the load demands is characterized by (5.58b). For each energy converter, it has to be operated within its rated power as characterized in (5.58c).

By solving (5.58a)–(5.58c), the multi-vector energy system makes energy dispatching decisions $\{S_e^f, S_g^f, \mathbf{v}^f\}$ to achieve the least operating cost. Therefore (5.58a) – (5.58c) can be equivalently written as an energy dispatching function $f_{\boldsymbol{\eta}}(\cdot)$ as follows,

$$\{S_e^f, S_g^f, \mathbf{v}^f\} = f_{\boldsymbol{\eta}}(L_e^f, L_h^f, S_r^f). \qquad (5.59)$$

However, the real-world load demands and renewable energy are very complex and subject to many physical and uncertain factors. For example, the electricity load varies with the day of the week, time of the day and individual consumers' behaviours. Meanwhile, the renewable energy such as wind energy is subject to many physical factors including weather and temperature.

5.10.2.2 Practical Energy Dispatching

The real-world forecasts are with uncertainties, which are usually different from their corresponding actual values. In this section, the uncertainties are characterized by forecasting errors including electricity load demands δ_e, heat load demands δ_h and renewable energy productions δ_r defined as follows,

$$\delta_e = \frac{L_e^a - L_e^f}{L_e^f}, \delta_h = \frac{L_h^a - L_h^f}{L_h^f}, \delta_r = \frac{S_r^a - S_r^f}{S_r^f}. \qquad (5.60)$$

If these forecasting errors are known in advance, then the optimal energy dispatching should be given as follows,

$$\{S_e^a, S_g^a, \mathbf{v}^a\} = f_{\boldsymbol{\eta}}((1 + \delta_e)L_e^f, (1 + \delta_h)L_h^f, (1 + \delta_r)S_r^f). \qquad (5.61)$$

Due to the uncertainties in load demands and renewable energy generations, the day-ahead energy scheduling cannot always meet the actual optimal system load demands in real time.

This will have an impact on the safety and stability of the energy network [100]. For instance, the electricity system operators (ESO) need to rely on balancing services to balance the electricity supply and demand in real-time with certain costs, which are denoted as C_e^+ and C_e^-, respectively.

Similarly, the transmission system operators (TSO) [114] provide such balancing services to meet the short-term fluctuations in natural gas demand or supply, whose

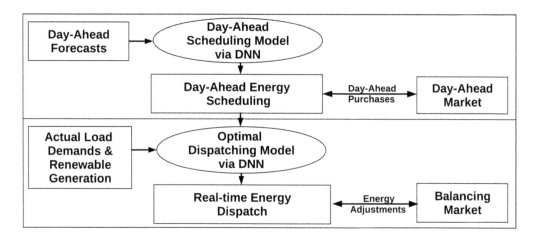

Figure 5.22 The proposed deep learning embedded multi-vector energy system scheduling and dispatching scheme.

costs are denoted as C_g^+ and C_g^-, respectively. The punishment cost of the imbalance in electricity C_e^p and natural gas C_g^p can be written by,

$$C_e^p(S_e^a, S_e^f) = \begin{cases} C_e^+(S_e^a - S_e^f), & S_e^a \geq S_e^f, \\ (C_e^- - C_e^0)(S_e^f - S_e^a), & S_e^a < S_e^f. \end{cases}$$
$$C_g^p(S_g^a, S_g^f) = \begin{cases} C_g^+(S_g^a - S_g^f), & S_g^a \geq S_g^f, \\ (C_g^- - C_g^0)(S_g^f - S_g^a), & S_g^a < S_g^f. \end{cases} \tag{5.62}$$

Therefore, the total operating cost for the multi-vector energy system is given by considering both (5.58a) and (5.62) as follows,

$$C^{all}(S_e^a, S_g^a, S_e^f, S_g^f) = C_e^0 S_e^f + C_g^0 S_g^f + C_e^p(S_e^a, S_e^f) + C_g^p(S_g^a, S_g^f). \tag{5.63}$$

It can be seen from (5.63), the total cost is depending on both the day-ahead market and the balancing service market. Forecasting errors in renewable energy and load demands will cause the system to make a deviated decision on the day-ahead market and result in the punishment cost in electricity and natural gas (5.62).

However, in real-world energy systems, it is not possible to know forecasting errors when making the day-ahead forecasts. As illustrated in Figures 5.19 and 5.20 with British electricity load demands and on-shore wind energy generation in 2017 [100], it can be seen that the electricity load demands' forecasting errors show a much smaller variance range than that of the wind energy generations' forecasting errors.

In this section, we present a deep learning embedded multi-vector energy scheduling scheme consisting of two stages: day-ahead scheduling stage and real-time dispatching stage as shown in Fig 5.22.

At the day-ahead scheduling stage, the day-ahead scheduling model makes the day-ahead scheduling based on forecasts via DNN, and then commits purchases with the day-ahead market. At the real-time dispatching stage, the multi-vector energy system loads the DNN for the optimal dispatching model, then makes an optimal energy dispatching according to the actual load demands and renewable energy consumption. The energy adjustment is achieved via the balancing service on the balancing market.

5.10.2.3 A Brief Revisit on DNN

A typical feedforward DNN can be defined as $\mathbf{x}_{R+1} = \Theta(\mathbf{x}_0; \boldsymbol{\theta})$, where inputs and outputs are $\mathbf{x}_0$ and $\mathbf{x}_{R+1}$, respectively [115]. The parameter set of the R hidden layers is $\boldsymbol{\theta} = \{\mathbf{W}_r, \mathbf{b}_r, r = 1, \ldots, R\}$, where the computation process at each layer r is given by $\mathbf{x}_r = \psi_r(\mathbf{W}_r \mathbf{x}_{r-1} + \mathbf{b}_r)$. Here $\psi_r(\cdot)$ denotes the activation function at layer r, e.g., ReLU activation function $\psi_{\text{ReLU}}(x) = \max\{x, 0\}$ [116]. It has been proved that a DNN can be used to approximate a wide range of functions with a proper design of the DNN structure with hidden layer number R and dimensions of the parameters $\mathbf{W}_r$ and $\mathbf{b}_r$ [117].

5.10.2.4 Optimal Dispatching Model via DNN

The optimal dispatching model is responsible to make energy dispatching decisions on the balancing market, which is based on actual load demands, renewable energy generations and economic considerations. In this part, we will present how to train a DNN as the optimal dispatching model. Here we define an optimal energy dispatching DNN model with inputs $\{L_e, L_h, S_r\}$ and outputs $\{S_e, S_g, \mathbf{v}\}$ as follows:

$$\{S_e, S_g, \mathbf{v}\} = \Theta_f(L_e, L_h, S_r; \boldsymbol{\theta}_f), \tag{5.64}$$

where $\boldsymbol{\theta}_f$ denotes the parameter set of $\Theta_f(\cdot)$.

Without uncertain energy sources, the optimal outputs can be known for arbitrarily given inputs via exhaustive search. Hence we can generate an arbitrary large training set $\mathbb{T}_f$ with known forecasts and corresponding optimal energy dispatching decisions. This feature is exploited to provide supervised training for the DNN, where the optimal energy dispatching is used to supervise the training procedure and optimize the network parameter set $\boldsymbol{\theta}_f$.

If define the loss function $\mathcal{L}_f(\boldsymbol{\theta}_f)$ as follows:

$$\mathcal{L}_f(\boldsymbol{\theta}_f) = f_{\boldsymbol{\eta}}(L_e, L_h, S_r) - \Theta_f(L_e, L_h, S_r; \boldsymbol{\theta}_f), \tag{5.65}$$

Then the supervised training procedure can be formulated as finding the DNN parameter set $\boldsymbol{\theta}_f$ as follows:

$$\max_{\{L_e, L_h, S_r\} \in \mathbb{T}_f} |\mathcal{L}_f(\boldsymbol{\theta}_f)| \leq \epsilon_f, \tag{5.66}$$

where ϵ_f is a sufficiently small approximation error.

Gradient descent (GD) method is used to update $\boldsymbol{\theta}_f$ as follows [118]:

$$\boldsymbol{\theta}_f^{(t)} = \boldsymbol{\theta}_f^{(t-1)} - \beta_f \nabla_{\boldsymbol{\theta}_f} \{\mathcal{L}_f(\boldsymbol{\theta}_f^{(t-1)})\}, \tag{5.67}$$

where the positive value β_f is the learning rate and the gradient operator is defined by $\nabla_{\boldsymbol{\theta}_f} = \left(\frac{\partial}{\partial \theta_1}, \ldots, \frac{\partial}{\partial \theta_n}\right)$ and $\boldsymbol{\theta}_f = \{\theta_1, \ldots, \theta_n\}$ [119, eq.1.6.20].

With the trained DNN for the optimal dispatching model (5.64), the optimal dispatching can be obtained directly as the DNN output. As the inference of a DNN is with deterministic computational complexity, it has an advantage over the extensive searching-based solution.

5.10.2.5 Day-Ahead Scheduling Model via DNN

The day-ahead scheduling model is to make day-ahead energy scheduling decisions on the day-ahead market, which is based on day-ahead forecasts containing forecasting errors.

Due to the underlying nonlinear relations between the cost and forecasting errors, the decisions of optimal dispatching model using forecasts are usually not optimal in the day-ahead market. The operating cost can be further reduced by the decisions considering the statistical characteristics of forecasting errors, which is referred to as the day-ahead scheduling model. In this section, a DNN will be trained for the energy scheduling under the uncertainties. This day-ahead scheduling model's inputs are forecasts $\{L_e^f, L_h^f, S_r^f\}$, while its outputs are the day-ahead scheduling decisions $\{S_e^0, S_g^0, \mathbf{v}^0\}$.

Suppose this day-ahead scheduling function is defined as follows:

$$\{S_e^0, S_g^0, \mathbf{v}^0\} = g_{\boldsymbol{\eta}}(L_e^f, L_h^f, S_r^f). \tag{5.68}$$

By using (5.68), the total multi-vector energy system operating cost can be rewritten as follows:

$$\begin{aligned} C^{\text{all}}(S_e^{\text{a}}, S_g^{\text{a}}, S_e^0, S_g^0) &= C_e^0 S_e^{\text{f}} + C_g^0 S_g^{\text{f}} + C_e^{\text{p}}(S_e^{\text{a}}, S_e^0) + C_g^{\text{p}}(S_g^{\text{a}}, S_g^0) \\ &= C^{\text{all}}(f_{\boldsymbol{\eta}}((1+\delta_e)L_e^{\text{f}}, (1+\delta_h)L_h^{\text{f}}, (1+\delta_r)S_r^{\text{f}}), g_{\boldsymbol{\eta}}(L_e^f, L_h^f, S_r^f)). \end{aligned} \tag{5.69}$$

For given forecasts, the total cost depends on the day-ahead scheduling described by $g_{\boldsymbol{\eta}}(L_e^f, L_h^f, S_r^f)$ and the corresponding forecasting errors $\{\delta_e, \delta_r, \delta_h\}$. Note that the forecasts $\{L_e^f, L_h^f, S_r^f\}$ are made by the forecasting services, whose values are unknown in advance and could be different between settlements. Here one settlement means a period of 30 minutes ending on the hour or half hour in each hour during a day.

In this way, the optimal day-ahead scheduling decision problem for given forecasts $\{L_e^f, L_h^f, S_r^f\}$ with the least total cost C^{all} can be defined as follows:

$$\begin{aligned} \min_{S_e^0, S_g^0, \mathbf{v}^0} \quad & \mathbb{E}_{\delta_e, \delta_h, \delta_r} \left\{ C^{\text{all}}(f_{\boldsymbol{\eta}}((1+\delta_e)L_e^{\text{f}}, (1+\delta_h)L_h^{\text{f}}, (1+\delta_r)S_r^{\text{f}}), g_{\boldsymbol{\eta}}(\{L_e^f, L_h^f, S_r^f\})) \right\} \\ \text{s.t.:} \quad & \mathbf{L}^{\text{f}} - H_{\boldsymbol{\eta}}(S_e^0, S_g^0, S_r^f, \mathbf{v}^0) = 0, \\ & \sum_{S_n^{\text{f}} \in \mathbf{S}^{\text{f}}} v_{kn}^0 S_n^{\text{f}} \le P_k^{\max}, \forall k, \end{aligned}$$

$$\tag{5.70}$$

where $\mathbb{E}_x\{z(x)\} = \int_{-\infty}^{\infty} z(x)p(x)dx$ is the expectation operation and $p(x)$ is the probability distribution function (PDF) of x.

Traditionally, the problem (5.70) is solved by exhaustive search methods, which should give the optimal energy dispatching corresponding to a specific set of forecasts

$\{L_e^f, L_h^f, S_r^f\}$. However, exhaustive search methods are not applicable in this case because:

- the exact distribution of forecasting errors δ_e, δ_h and δ_r are unknown, where (5.70) cannot be computed directly.

- even with accurate approximations of the objective function, solving (5.70) gives only one day-ahead scheduling result, which is only valid to a specific forecast. For a different forecast, it needs to repeat the searching from the beginning.

In this section, we present a deep learning-based method to address these two challenges. Define a DNN $\Theta_g(\cdot)$ that can achieve the same day-ahead scheduling (5.68) as follows,

$$\{S_e^0, S_g^0, \mathbf{v}^0\} = \Theta_g(L_e^f, L_h^f, S_r^f; \boldsymbol{\theta}_g), \qquad (5.71)$$

where $\boldsymbol{\theta}_g$ is the parameter set of the DNN for the day-ahead scheduling model. If we can train an optimal DNN parameter set $\boldsymbol{\theta}_g$, then the day-ahead scheduling can be directly obtained as the DNN outputs in (5.71).

Note that the supervised training method of the DNN for the optimal dispatching model in Section 5.10.2.4 cannot be used here. This is because the real distribution of the uncertainty sources is unknown for generating a supervised training set.

In the following, we propose a new unsupervised method to train the DNN for the day-ahead scheduling model $\boldsymbol{\theta}_g$. By substituting day-ahead scheduling function (5.68) with the DNN in (5.71), (5.70) can be rewritten as follows:

$$\begin{aligned} \min_{\boldsymbol{\theta}_g} \quad & \mathbb{E}_{L_e^f, L_h^f, S_r^f, \delta_e, \delta_h, \delta_r} \Big\{ C^{\text{all}} \Big(\Theta_f((1+\delta_e)L_e^f, (1+\delta_h)L_h^f, \\ & (1+\delta_r)S_r^f; \boldsymbol{\theta}_f), \Theta_g(L_e^f, L_h^f, S_r^f; \boldsymbol{\theta}_g) \Big) \Big\} \\ \text{s.t.:} \quad & \mathbb{E}_{L_e^f, L_h^f, S_r^f} \{ \mathbf{L}^f - H_{\boldsymbol{\eta}}(\Theta_g(L_e^f, L_h^f, S_r^f; \boldsymbol{\theta}_g), S_r^f \mathbf{v}^0) \} = 0, \\ & \sum_{S_n^f \in \mathbf{S}^f} v_{kn}^0 S_n^f \le P_k^{\max}, \forall k. \end{aligned} \qquad (5.72)$$

Remarks: It should be noted that in (5.72), the objective function and optimization values are different from those of (5.70), with the following reasons.

- After training, the DNN for the day-ahead scheduling model should take all potential forecasts as inputs. Therefore the expectation in (5.70) also considers all potential forecasts L_e^f, L_h^f, S_r^f together with forecasting errors $\delta_e, \delta_h, \delta_r$.

- The optimization goals are changed from finding optimal day-ahead scheduling decision $S_e^0, S_g^0, \mathbf{v}^0$ in (5.70), to the search of optimal DNN for the day-ahead scheduling model parameters $\boldsymbol{\theta}_g$ in (5.72). This transforms the problem of solving an optimization problem to the problem of training a DNN network to minimize the loss function (i.e., the objective function in the optimization problem) in an unsupervised method.

- The training procedure involves frequent evaluation of optimal energy dispatch under different forecasting values, which has now been fulfilled by the trained DNN for the optimal dispatching model in the first training stage instead of exhaustive search based solving method.

As constraints are not supported during unsupervised training, we further transform the original constrained problem (5.72) to an unconstrained problem. This is achieved by formulating a loss function $\mathcal{L}_g(\boldsymbol{\theta}_g)$ as follows:

$$
\begin{aligned}
\mathcal{L}_g(\boldsymbol{\theta}_g) = \\
\mathbb{E}_{L_e^f, L_h^f, S_r^f, \delta_e, \delta_h, \delta_r} & \left\{ C^{\text{all}} \left(\Theta_f((1+\delta_e)L_e^f, (1+\delta_h)L_h^f, (1+\delta_r)S_r^f; \boldsymbol{\theta}_f), \Theta_g(L_e^f, L_h^f, S_r^f; \boldsymbol{\theta}_g)) \right) \right\} \\
& + \lambda \left\{ \left| \mathbb{E}_{L_e^f, L_h^f, S_r^f} \{ \mathbf{L}^f - H_{\boldsymbol{\eta}}(\Theta_g(L_e^f, L_h^f, S_r^f; \boldsymbol{\theta}_g), S_r^f, \mathbf{v}^0) \} \right| \right. \\
& \left. + \max\{ \sum_{\forall k, S_N \in \mathbf{S}^f} v_{kn}^0 S_N^f - P_k^{\max}, 0 \} \right\}
\end{aligned}
$$

(5.73)

where constraints are multiplied with a penalty factor λ and appended to the objective given in (5.73).

Then the training procedure of the DNN for the day-ahead scheduling model is given by the following unconstraint optimization problem,

$$
\min_{\boldsymbol{\theta}_g} \mathcal{L}_g(\boldsymbol{\theta}_g).
$$

(5.74)

To obtain the optimal DNN parameter set $\boldsymbol{\theta}_g$, the training procedure will aim to reduce the value of $\mathcal{L}_g(\boldsymbol{\theta}_g)$. Specifically, the update of $\boldsymbol{\theta}_g$ at training step t based on the step $t-1$ can be calculated by the GD method [118] as follows:

$$
\boldsymbol{\theta}_g^{(t)} = \boldsymbol{\theta}_g^{(t-1)} - \beta_g \nabla_{\boldsymbol{\theta}_g} \mathcal{L}_g(\boldsymbol{\theta}_g^{(t-1)}),
$$

(5.75)

where β_g is the learning rate of the training procedure.

5.10.2.6 *Addressing Multiple Uncertainties via Deep Learning*

In (5.73), the expectation should consider not only the uncertainties $\{\delta_e, \delta_h, \delta_r\}$, but also the potential forecasts $\{L_e^f, L_h^f, S_r^f\}$. However, the PDFs of these random variables are unknown in real-world systems.

Traditional solutions are to first approximate the PDF of each random variable (RV) and then substitute to (5.73) [104]. But approximations are prone to errors and the errors can accumulate if there are multiple RVs. In this section, we take advantage of the deep learning training features to address the uncertainties, which are based on historical data and detailed as follows.

Forecasting errors $\{\delta_e, \delta_h, \delta_r\}$ are the uncertainty sources of the multi-vector energy system, which should cover all potential uncertainty situations. In fact, the historical data for each forecast error are samples from their real distributions, while the forecast errors can be regarded as mutually independent. Therefore given historical datasets Δ_e, Δ_h and Δ_r, the augmented forecasting error training set can be generated as all possible combinations from each set as $\{\delta_e, \delta_h, \delta_r | \delta_e \in \Delta_e, \delta_h \in \Delta_h, \delta_r \in \Delta_r\}$. With a large historical dataset from each forecasting error, the augmented forecasting error training set can cover all potential forecasting error scenarios that the DNN for the day-ahead scheduling model can encounter.

Table 5.4 Parameters in the case studies

Parameter	Description	Value
η_e^{TF}	Transformer efficiency	0.980
η_e^{CHP}	CHP power generation efficiency	0.404
η_h^{CHP}	CHP thermal generation efficiency	0.566
η_h^{Boiler}	Boiler thermal efficiency	0.900
$P_{\mathrm{TF}}^{\max}$	Max. transformer power	1000MW
$P_{\mathrm{WT}}^{\max}$	Max. wind turbine power	200MW
$P_{\mathrm{CHP}}^{\max}$	Max. combined heat & power system power	300MW
$P_{\mathrm{B}}^{\max}$	Max. boiler power	800MW

Given the fact that the unsupervised DNN performance is not universal across all inputs, it is essential that the DNN output is more accurate for the common situation in practice than in rare cases. Therefore the training datasets for the forecasts L_e^f, L_h^f, S_r^f also follow the augmented method based on historical data. Given historical datasets $\mathbb{L}_e$, $\mathbb{L}_h$ and $\mathbb{S}_r$, the training datasets can be generated as the all possible combinations from each set as $\{L_e^f, L_h^f, S_r^f | L_e^f \in \mathbb{L}_e, L_h^f \in \mathbb{L}_h, S_r^f \in \mathbb{S}_r\}$.

In this way, the augmented training set can be generated as the sample set from all potential combinations of forecasting errors $\{\delta_e, \delta_h, \delta_r\}$ and forecasts $\{L_e^f, L_h^f, S_r^f\}$ regarding their real distributions. By training the DNN with the augmented training set and taking the average of all outputs, the expectation operation in (5.73) can be achieved.

In practice, the augmented training set is very large in size, which cannot be all put into the training procedure at one time. Therefore the mini-batch training method is used, where a subset of the whole training set is randomly selected as the input of each training step. Moreover, to prevent overfitting to the training set, parts of data are reserved for verification. In summary, the two-stage deep learning training algorithm can be given in Algorithm 3.

Note that instead of manually approximating an error-prone discrete probability distribution, the proposed training procedure exploits the deep learning training procedure to learn the actual distribution from the historical data automatically.

5.10.3 Multi-Vector Energy System Implementations

The multi-vector energy system in Figure 5.21 is used to evaluate the proposed deep learning scheme, whose parameters are given in Table 5.4.

The electricity load and wind energy are considered as the uncertainty sources, whose day-ahead forecasts and actual data are scaled from the historical data in the UK [100]. The average day-ahead prices and balancing prices in the 2018 UK markets [100][114] are used, where $C_e^0 = £35/\mathrm{MWh}$, $C_g^0 = £14/\mathrm{MWh}$, $C_e^+ = £58/\mathrm{MWh}$, $C_e^- = £21/\mathrm{MWh}$, $C_g^+ = £18/\mathrm{MWh}$ and $C_g^- = £10/\mathrm{MWh}$.

The forecasting error of the electricity load shows different performances regarding the user's behaviours as illustrated in Figure 5.23, which can be reflected by the day of a week and the time of a day. With an observation of the electricity load and

---

Algorithm 3 Two-Stage Deep Learning Training

Stage 1: DNN Training for the Optimal Dispatching Model

1: Generate the training set $\mathbb{T}$ with randomly sampled inputs and their optimal outputs using (5.58a)– (5.58c).

2: Initialize the optimal dispatching parameters $\boldsymbol{\theta}_f^{(0)}$ and learning rate β_f.

while $\epsilon > \epsilon_f$ **do**

   1: Calculate the loss function $\mathcal{L}_f(\boldsymbol{\theta}_f^{(t)}) = f_{\boldsymbol{\eta}}(L_e, L_h, S_r) - \Theta_f(L_e, L_h, S_r; \boldsymbol{\theta}_f^{(t)})$.

   2: Update the DNN parameter set $\boldsymbol{\theta}_f^{(t)} = \boldsymbol{\theta}_f^{(t-1)} - \beta_f \nabla_{\boldsymbol{\theta}_f} \mathcal{L}_f(\boldsymbol{\theta}_f^{(t-1)})$.

   3: Calculate the maximum error $\epsilon = \max_{\{L_e, L_h, S_r\} \in \mathbb{T}} |\mathcal{L}_f(\boldsymbol{\theta}_f^{(t)})|$.

end while

Stage 2: DNN Training for the Day-Ahead Scheduling Model

1: Generate the augmented datasets from historical data $\{L_e^f, L_h^f, S_r^f, \delta_e, \delta_h, \delta_r | L_e^f \in \mathbb{L}_e, L_h^f \in \mathbb{L}_h, S_r^f \in \mathbb{S}_r, \delta_e \in \Delta_e, \delta_h \in \Delta_h, \delta_r \in \Delta_r\}$.

2: Split the augmented datasets into training set $\mathbb{T}_g$ and validation set $\mathbb{V}_g$.

2: Initialize DNN parameter set for the day-ahead scheduling model $\boldsymbol{\theta}_g^{(0)}$ and learning rate β_g.

while $\epsilon > \epsilon_g$ **do**

   1: Randomly select a subset $T_g \in \mathbb{T}_g$ as the mini-batch set.

   2: Load the trained optimal dispatching parameter $\Theta_f(L_e, L_h, S_r; \boldsymbol{\theta}_f^{(t)})$.

   3: Calculate training loss $L_T^{(t)}$ using (5.73) with mini-batch T_g.

   4: Update the DNN parameter set for the day-ahead scheduling model $\boldsymbol{\theta}_g^{(t)} = \boldsymbol{\theta}_g^{(t-1)} - \beta_g \nabla_{\boldsymbol{\theta}_g} \mathcal{L}_g(\boldsymbol{\theta}_g^{(t-1)})$

   5: Calculate validation loss $L_V^{(t)}$ using (5.73) with validation set $\mathbb{V}_g$.

   6: Calculate the validation loss error $\epsilon = L_V^{(t)} - L_V^{(t-1)}$

end while

return DNN parameter set for the optimal dispatching model $\boldsymbol{\theta}_f$ and DNN parameter set for the day-ahead scheduling model $\boldsymbol{\theta}_g$

Table 5.5 Scenario list

	Scenario 1	Scenario 2	Scenario 3	Scenario 4	Scenario 5
Day	Monday	Tuesday	Wednesday,Thursday,Friday	Saturday	Sunday
Settlement	1–10	15–18	13–14,19–22	11–12,23–32	33–48

wind forecasting in 2017, we found that there are similarities in load profiles between the day of the week as well as settlements. Therefore according to the load profile similarity, we categorize the day of the week into 5 scenarios and settlements into 5 scenarios respectively as listed in Table 5.5, whose combination makes a total of 25 scenarios.

During the training of the DNN for the optimal dispatching model, the inputs and outputs are normalized to improve the model accuracy and the approximation error ϵ is set as 10^{-3}. One DNN for the day-ahead scheduling model is trained for each of the

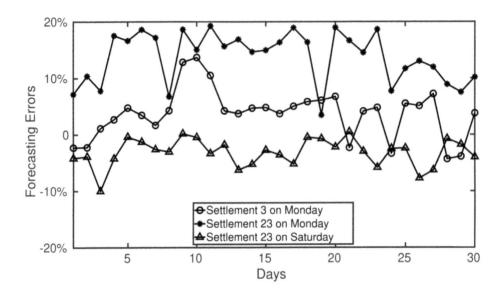

Figure 5.23 Electricity load forecasting errors in 2017, Monday Settlement 3, Monday Settlement 23, Saturday Settlement 23 [100].

25 scenarios, which makes a total of 25 models. During the evaluation, the proposed scheme in Figure 5.22 is applied, where a specific DNN for the day-ahead scheduling model is used according to the day-ahead forecasts. The DNNs are configured as 5 fully connected layers, where the width for each hidden layer is set as 30. Moreover, the ReLU activation is used for all hidden layers. The batch normalisation technique is used at every layer to speed up the training progress [120]. The models were trained with PyTorch [74] on GPU GeForce RTX 2080Ti.

For the validation purpose, the proposed deep learning-based method is compared against two benchmark methods. The first benchmark method is referred to as the discrete probability method, which exploits small sets of discrete probability distribution to approximate each forecasting error [104]. For each variable's forecasting error, five discrete sets and corresponding probabilities are generated by applying histogram analysis on the same training data used in the deep learning-based method.

Moreover, the second benchmark method uses forecasts directly for the day-ahead energy scheduling, namely without considering the uncertainties. The best performances are also calculated during the comparisons, which exploit the actual data instead of forecasted data for the day-ahead multi-vector energy scheduling. Although this best performance can never be achieved in real-world systems, it clearly indicates the minimum cost that the system can achieve in a given scenario.

5.10.3.1 A Case Study of Settlement Performance

In this case study, the proposed deep learning-based multi-vector scheduling method is evaluated with the data of Settlement 17 on 30 Wednesdays across 2017, whose results are shown in Figure 5.24. The extra cost is depicted for the proposed deep

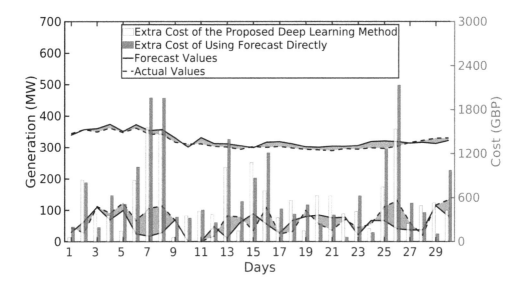

Figure 5.24 The evaluation of the proposed deep learning-based multi-vector scheduling method with the data of Settlement 17 on 30 Wednesdays across 2017. The differences between forecasts and corresponding actual values are also illustrated with colors, where red represents forecasts larger than actual values and blue represents the reverse relation.

learning method and that of using forecast directly method, which is defined by the difference from the best performance cost. Along with the cost performance, the uncertainty sources including electricity load demand and wind energy generation are also illustrated.

It can be seen that the extra cost spikes when a large mismatch between forecasts and actual value appears, which corresponds to large forecasting errors. Typical examples are the days 7, 8, 13 and 26. Under such conditions, using forecasts directly for the day-ahead scheduling results in very large extra costs. Although on these days the extra costs are also very high with the proposed deep learning-based method, it can be seen that the costs have been considerably reduced compared to the method using forecast directly. This is because during the training procedure, the day-ahead scheduling model has learned from the historical uncertainties and made the scheduling decision based on that.

Meanwhile, for the days with very small forecasting errors, e.g., the days 3, 22 and 29, the method of using the forecast directly shows an advantage over the proposed deep learning method. This is expected because the proposed deep learning-based method is to optimize the long-term performance of the multi-vector energy system, where the reduction in the long-run cost is at the expense of some short-term disadvantages.

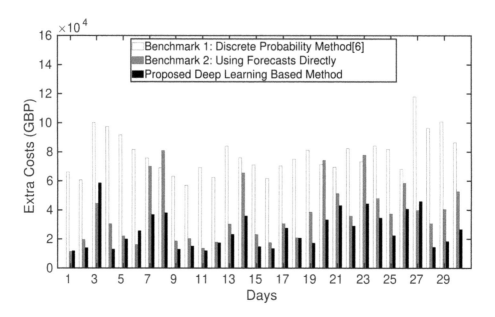

Figure 5.25 A comparison between the proposed Deep Learning based method against best performance, discrete probability method and using forecasts directly method, an example of the results for the full April month (30 days) in 2017.

5.10.3.2 A Case Study of 2017 UK Dataset

A case study is made using the 2017 dataset, whose data are used for the unsupervised training procedure of the multi-vector energy scheduling model. In Fig. 5.25, 30 days in April 2017 are used to evaluate the performances of each method, where the day costs are the summation of each settlement cost on that day. Due to the uncertainties caused by load forecasting errors and wind generation forecasting errors, it is seen from Figure 5.25 that both the proposed deep learning-based method and these two benchmark methods are fluctuating from day to day.

No single method can outperform every other method every day. This is because although the multiple uncertainty sources are simultaneously considered in the training procedure, the deep learning-based model is trained using (5.73), where the energy scheduling decisions are supposed to show good performances in a mathematical expectation manner. In other words, the performance is expected to be good for most cases, but it is normal to be sub-optimal for the rare but extreme events as illustrated in Figures 5.19 and 5.20. Therefore correspondingly, an observation from Figure 5.25 shows that the proposed deep learning-based method does make the most frequent least cost energy scheduling on a day-to-day basis (27 days out of 30 days in this studied case).

Since the performance of the methods is fluctuating according to the specific samples, the statistical analysis will provide more perceptions. Therefore the data of the whole calendar year 2017 is applied, whose costs are averaged according to the settlement periods as presented in Figure 5.26. It can be seen that the proposed deep

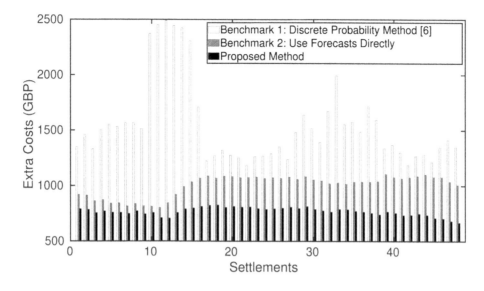

Figure 5.26 A comparison between the proposed Deep Learning based method against best performance, discrete probability method and using forecasts directly method, the averaged results in whole year 2017 are presented according to the full 48 settlements.

learning-based method outperforms all benchmark methods, from the view of yearly statistical average and in every settlement.

5.10.3.3 A Case Study of 2018 UK Dataset

As a part of the 2017 data is used for the training procedure, a further validation against the data in 2018, which are totally blind to the DNN training, will give a better evaluation of the proposed deep learning-based method. To this end, the data for the whole calendar year 2018 are used. The day-to-day performance in the 30 days in April 2018 and average settlement cost performance in whole 2018 is presented in Figures 5.27 and 5.28, respectively.

Similar to the case in 2017, the averaged daily costs in 2018 are varying with days in Figure 5.27, where no single method outperforms every other method as expected. Meanwhile, the average settlement costs across the whole calendar year 2018 in Figure 5.28 agree with that of Fig. 5.26. This verifies that the trained model has learned the distribution of forecasting errors of each uncertainty variable, and is able to make the expected optimal decision based on the forecasts.

For a better quantified comparison, here we define the improvement metric as follows. If one method with extra cost A and the other method with extra cost B, then the improvement metric is defined as $(A - B)/B$. A study on the settlements in 2017 reveals that the proposed deep learning-based method shows a maximum improvement of 76.68% against the discrete method and a 28.78% against the method of using forecasts directly.

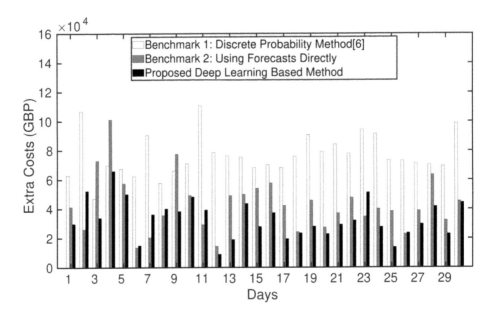

Figure 5.27 A comparison between the proposed Deep Learning based method against best performance, discrete probability method and using forecasts directly method, an example of the results for the full April month (30 days) in 2018.

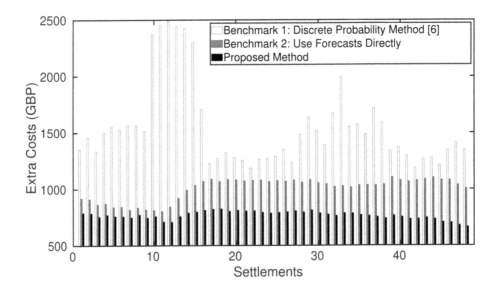

Figure 5.28 A comparison between the proposed Deep Learning based method against best performance, discrete probability method and using forecasts directly method, the averaged results in whole year 2018 are presented according to the full 48 settlements.

A further observation from Figures 5.25 – 5.28 is that, the proposed deep learning algorithm outperforms the discrete probability method for every scenario, in both day-to-day views and the average settlement views. This can be explained as follows. At the training stage, forecasting errors are input to the model, which minimizes the loss of the probability distribution information. The proposed method exploits the ability of the deep learning method to learn the distributions and their impacts on energy scheduling. On the contrary, the discrete sets based methods not only lose information about the true distributions, but also are prone to the accumulated inaccuracy due to multiple uncertainty scenarios.

Bibliography

[1] S. Javaid, M. Kaneko, and Y. Tan, "A linear programming model for power flow control problem considering controllable and fluctuating power devices," in *2019 IEEE 8th Global Conference on Consumer Electronics (GCCE)*, 2019, pp. 96–99.

[2] R. Fernandez-Blanco, J. M. Arroyo, N. Alguacil, and X. Guan, "Incorporating price-responsive demand in energy scheduling based on consumer payment minimization," *IEEE Transactions on Smart Grid*, vol. 7, no. 2, pp. 817–826, 2016.

[3] Y. Du, F. Li, J. Li, and T. Zheng, "Achieving 100x acceleration for n-1 contingency screening with uncertain scenarios using deep convolutional neural network," *IEEE Transactions on Power Systems*, vol. 34, no. 4, pp. 3303–3305, 2019.

[4] F. Meng and X. Zeng, "A profit maximization approach to demand response management with customers behavior learning in smart grid," *IEEE Transactions on Smart Grid*, vol. 7, no. 3, pp. 1516–1529, 2016.

[5] D. J. Olsen, Y. Dvorkin, R. Fernández-Blanco, and M. A. Ortega-Vazquez, "Optimal carbon taxes for emissions targets in the electricity sector," *IEEE Transactions on Power Systems*, vol. 33, no. 6, pp. 5892–5901, 2018.

[6] D. Li, W. Chiu, H. Sun, and H. V. Poor, "Multiobjective optimization for demand side management program in smart grid," *IEEE Transactions on Industrial Informatics*, vol. 14, no. 4, pp. 1482–1490, April 2018.

[7] H. Chang, W. Chiu, H. Sun, and C. Chen, "User-centric multiobjective approach to privacy preservation and energy cost minimization in smart home," *IEEE Systems Journal*, vol. 13, no. 1, pp. 1030–1041, 2019.

[8] C. Zhang, J. Li, Y. Angela Zhang, and Z. Xu, "Data-driven sizing planning of renewable distributed generation in distribution networks with optimality guarantee," *IEEE Transactions on Sustainable Energy*, vol. 11, no. 3, pp. 2003–2014, 2020.

[9] A. Belgana, B. P. Rimal, and M. Maier, "Open energy market strategies in microgrids: A stackelberg game approach based on a hybrid multiobjective evolutionary algorithm," *IEEE Transactions on Smart Grid*, vol. 6, no. 3, pp. 1243–1252, 2014.

[10] F.-L. Meng and X.-J. Zeng, "A stackelberg game-theoretic approach to optimal real-time pricing for the smart grid," *Soft Computing*, vol. 17, no. 12, pp. 2365–2380, 2013.

[11] A. Ghosh, V. Aggarwal, and H. Wan, "Exchange of renewable energy among prosumers using blockchain with dynamic pricing," 2018.

[12] H. Wang, T. Huang, X. Liao, H. Abu-Rub, and G. Chen, "Reinforcement learning in energy trading game among smart microgrids," *IEEE Transactions on Industrial Electronics*, vol. 63, no. 8, pp. 5109–5119, 2016.

[13] R. Mohammadi, H. R. Mashhadi, and M. Shahidehpour, "Enhancement of distribution system reliability: A framework based on cournot game model," *IEEE Transactions on Smart Grid*, vol. 11, no. 3, pp. 2172–2181, 2020.

[14] J. Vuelvas and F. Ruiz, "A novel incentive-based demand response model for cournot competition in electricity markets," *Energy Systems*, vol. 10, no. 1, pp. 95–112, 2019.

[15] M. I. Jordan and D. E. Rumelhart, "Forward models: Supervised learning with a distal teacher," *Cognitive Science*, vol. 16, no. 3, pp. 307–354, 1992.

[16] H. B. Barlow, "Unsupervised learning," *Neural Computation*, vol. 1, no. 3, pp. 295–311, 1989.

[17] R. S. Sutton and A. G. Barto, *Reinforcement Learning: An Introduction*. MIT press, 2018.

[18] M. Gasse, D. Chetelat, N. Ferroni, L. Charlin, and A. Lodi, "Exact combinatorial optimization with graph convolutional neural networks," in *Advances in Neural Information Processing Systems*, 2019, pp. 15 580–15 592.

[19] Z. Wen, D. O'Neill, and H. Maei, "Optimal demand response using device-based reinforcement learning," *IEEE Transactions on Smart Grid*, vol. 6, no. 5, pp. 2312–2324, 2015.

[20] F. Ruelens, B. J. Claessens, S. Vandael, S. Iacovella, P. Vingerhoets, and R. Belmans, "Demand response of a heterogeneous cluster of electric water heaters using batch reinforcement learning," in *2014 Power Systems Computation Conference*. IEEE, 2014, pp. 1–7.

[21] D. Zhang, S. Li, M. Sun, and Z. O'Neill, "An optimal and learning-based demand response and home energy management system," *IEEE Transactions on Smart Grid*, vol. 7, no. 4, pp. 1790–1801, 2016.

[22] N. Kalchbrenner, E. Grefenstette, and P. Blunsom, "A convolutional neural network for modelling sentences," *arXiv preprint arXiv:1404.2188*, 2014.

[23] M. Valueva, N. Nagornov, P. Lyakhov, G. Valuev, and N. Chervyakov, "Application of the residue number system to reduce hardware costs of the convolutional neural network implementation," *Mathematics and Computers in Simulation*, 2020.

[24] D. Owerko, F. Gama, and A. Ribeiro, "Optimal power flow using graph neural networks," in *ICASSP 2020-2020 IEEE International Conference on Acoustics, Speech and Signal Processing (ICASSP)*. IEEE, 2020, pp. 5930–5934.

[25] B. J. Claessens, P. Vrancx, and F. Ruelens, "Convolutional neural networks for automatic state-time feature extraction in reinforcement learning applied to residential load control," *IEEE Transactions on Smart Grid*, vol. 9, no. 4, pp. 3259–3269, 2018.

[26] A. Gupta, G. Gurrala, and P. S. Sastry, "An online power system stability monitoring system using convolutional neural networks," *IEEE Transactions on Power Systems*, vol. 34, no. 2, pp. 864–872, 2019.

[27] H. Choi, S. Ryu, and H. Kim, "Short-term load forecasting based on ResNet and LSTM," in *2018 IEEE International Conference on Communications, Control, and Computing Technologies for Smart Grids (SmartGridComm)*. IEEE, 2018, pp. 1–6.

[28] H. Liao, J. V. Milanović, M. Rodrigues, and A. Shenfield, "Voltage sag estimation in sparsely monitored power systems based on deep learning and system area mapping," *IEEE Transactions on Power Delivery*, vol. 33, no. 6, pp. 3162–3172, 2018.

[29] J. M. Morales, R. Minguez, and A. J. Conejo, "A methodology to generate statistically dependent wind speed scenarios," *Applied Energy*, vol. 87, no. 3, pp. 843–855, 2010.

[30] K. N. Hasan, R. Preece, and J. V. Milanović, "Existing approaches and trends in uncertainty modelling and probabilistic stability analysis of power systems with renewable generation," *Renewable and Sustainable Energy Reviews*, vol. 101, pp. 168–180, 2019.

[31] M. J. Santos, P. Ferreira, and M. Araújo, "A methodology to incorporate risk and uncertainty in electricity power planning," *Energy*, vol. 115, pp. 1400–1411, 2016.

[32] M. Hemmati, B. Mohammadi-Ivatloo, and A. Soroudi, "Uncertainty management in power system operation decision making," *arXiv preprint arXiv:1911.10358*, 2019.

[33] R. Preece and J. V. Milanović, "Efficient estimation of the probability of small-disturbance instability of large uncertain power systems," *IEEE Transactions on Power Systems*, vol. 31, no. 2, pp. 1063–1072, 2015.

[34] M. Liang, W. Li, J. Yu, and L. Shi, "Kernel-based electric vehicle charging load modeling with improved latin hypercube sampling," in *2015 IEEE Power Energy Society General Meeting*, 2015, pp. 1–5.

[35] Y. Chen, Y. Wang, D. Kirschen, and B. Zhang, "Model-free renewable scenario generation using generative adversarial networks," *IEEE Transactions on Power Systems*, vol. 33, no. 3, pp. 3265–3275, 2018.

[36] H. Xiao, W. Pei, Z. Dong, L. Kong, and D. Wang, "Application and comparison of metaheuristic and new metamodel based global optimization methods to the optimal operation of active distribution networks," *Energies*, vol. 11, no. 1, p. 85, 2018.

[37] G. Mavromatidis, K. Orehounig, and J. Carmeliet, "Design of distributed energy systems under uncertainty: A two-stage stochastic programming approach," *Applied Energy*, vol. 222, pp. 932–950, 2018.

[38] C. Huang, D. Yue, J. Xie, Y. Li, and K. Wang, "Economic dispatch of power systems with virtual power plant based interval optimization method," *CSEE Journal of Power and Energy Systems*, vol. 2, no. 1, pp. 74–80, 2016.

[39] H. V. D. Parunak, "Practical and industrial applications of agent-based systems," *Environmental Research Institute of Michigan (ERIM)*, 1998.

[40] J. M. G. de Durana, O. Barambones, E. Kremers, and L. Varga, "Agent based modeling of energy networks," *Energy Conversion and Management*, vol. 82, pp. 308–319, 2014.

[41] D. Divenyi and A. M. Dan, "Agent-based modeling of distributed generation in power system control," *IEEE Transactions on Sustainable Energy*, vol. 4, no. 4, pp. 886–893, 2013.

[42] R. Sangi and D. Muller, "A novel hybrid agent-based model predictive control for advanced building energy systems," *Energy Conversion and Management*, vol. 178, pp. 415–427, 2018.

[43] A. Mittal, C. C. Krejci, and M. C. Dorneich, "An agent-based approach to designing residential renewable energy systems," *Renewable and Sustainable Energy Reviews*, vol. 112, pp. 1008–1020, 2019.

[44] X. Fang, S. Misra, G. Xue, and D. Yang, "Smart grid—the new and improved power grid: A survey," *IEEE Communications Surveys & Tutorials*, vol. 14, no. 4, pp. 944–980, 2011.

[45] D. J. Gaushell and H. T. Darlington, "Supervisory control and data acquisition," *Proceedings of the IEEE*, vol. 75, no. 12, pp. 1645–1658, 1987.

[46] W. Hua, D. Li, H. Sun, P. Matthews, and F. Meng, "Stochastic environmental and economic dispatch of power systems with virtual power plant in energy and reserve markets." *International Journal of Smart Grid and Clean Energy.*, vol. 7, no. 4, pp. 231–239, 2018.

[47] D. Li, W. Hua, H. Sun, and W.-Y. Chiu, "Multiobjective optimization for carbon market scheduling based on behavior learning," *Energy Procedia*, vol. 142, pp. 2089–2094, 2017.

[48] W.-Y. Chiu, H. Sun, and H. V. Poor, "A multiobjective approach to multimicrogrid system design," *IEEE Transactions on Smart Grid*, vol. 6, no. 5, pp. 2263–2272, 2015.

[49] P.-H. Kuo and C.-J. Huang, "A high precision artificial neural networks model for short-term energy load forecasting," *Energies*, vol. 11, no. 1, p. 213, 2018.

[50] Apr. 2021. [Online]. Available: https://www.ofgem.gov.uk/energy-data-and-research/data-portal

[51] W. Hua, "Smart grid enabling low carbon future power systems towards prosumers era," Ph.D. dissertation, Durham University, 2020.

[52] D. Gielen, F. Boshell, D. Saygin, M. D. Bazilian, N. Wagner, and R. Gorini, "The role of renewable energy in the global energy transformation," *Energy Strategy Reviews*, vol. 24, pp. 38–50, 2019.

[53] W. Hua, Y. Chen, M. Qadrdan, J. Jiang, H. Sun, and J. Wu, "Applications of blockchain and artificial intelligence technologies for enabling prosumers in smart grids: A review," *Renewable and Sustainable Energy Reviews*, vol. 161, p. 112308, 2022.

[54] I. Dukovska, N. G. Paterakis, and H. J. Slootweg, "Local energy exchange considering heterogeneous prosumer preferences," in *2018 International Conference on Smart Energy Systems and Technologies (SEST)*. IEEE, 2018, pp. 1–6.

[55] H. Wei, Y. Zhang, Y. Wang, W. Hua, R. Jing, and Y. Zhou, "Planning integrated energy systems coupling v2g as a flexible storage," *Energy*, vol. 239, p. 122215, 2022.

[56] W. Seward, W. Hua, and M. Qadrdan, "Electricity storage in local energy systems," *Microgrids and Local Energy Systems*, vol. 1, p. 127, 2021.

[57] L. Gan, A. Wierman, U. Topcu, N. Chen, and S. H. Low, "Real-time deferrable load control: Handling the uncertainties of renewable generation," in *Proceedings of the Fourth International Conference on Future Energy Systems*, 2013, pp. 113–124.

[58] J. Ruan, G. Liu, J. Qiu, G. Liang, J. Zhao, B. He, and F. Wen, "Time-varying price elasticity of demand estimation for demand-side smart dynamic pricing," *Applied Energy*, vol. 322, p. 119520, 2022.

[59] C. Z. Mooney, *Monte Carlo Simulation.* Sage, 1997, no. 116.

[60] J. C. Helton and F. J. Davis, "Latin hypercube sampling and the propagation of uncertainty in analyses of complex systems," *Reliability Engineering & System Safety*, vol. 81, no. 1, pp. 23–69, 2003.

[61] Z. Liang and W. Su, "Game theory based bidding strategy for prosumers in a distribution system with a retail electricity market," *IET Smart Grid*, vol. 1, no. 3, pp. 104–111, 2018.

[62] S. Cui, Y.-W. Wang, and N. Liu, "Distributed game-based pricing strategy for energy sharing in microgrid with pv prosumers," *IET Renewable Power Generation*, vol. 12, no. 3, pp. 380–388, 2018.

[63] S. Choi and S.-W. Min, "Optimal scheduling and operation of the ess for prosumer market environment in grid-connected industrial complex," *IEEE Transactions on Industry Applications*, vol. 54, no. 3, pp. 1949–1957, 2018.

[64] K. O'Shea and R. Nash, "An introduction to convolutional neural networks," *arXiv preprint arXiv:1511.08458*, 2015.

[65] Y. Ju, G. Sun, Q. Chen, M. Zhang, H. Zhu, and M. U. Rehman, "A model combining convolutional neural network and lightgbm algorithm for ultra-short-term wind power forecasting," *IEEE Access*, vol. 7, pp. 28 309–28 318, 2019.

[66] L. Du, L. Zhang, and X. Tian, "Deep power forecasting model for building attached photovoltaic system," *IEEE Access*, vol. 6, pp. 52 639–52 651, 2018.

[67] H. Rinne, *The Weibull Distribution: A Handbook.* Chapman and Hall/CRC, 2008.

[68] M. Ahsanullah, B. Kibria, and M. Shakil, "Normal distribution," in *Normal and Student's t Distributions and Their Applications.* Springer, 2014, pp. 7–50.

[69] G. R. Terrell and D. W. Scott, "Variable kernel density estimation," *The Annals of Statistics*, pp. 1236–1265, 1992.

[70] W.-L. Loh, "On latin hypercube sampling," *The annals of Statistics*, vol. 24, no. 5, pp. 2058–2080, 1996.

[71] J. Manurung, "Application of fifo algorithm (first in first out) to simulation queue," *INFOKUM*, vol. 7, no. 2, Juni, pp. 44–47, 2019.

[72] J. Ngiam, A. Khosla, M. Kim, J. Nam, H. Lee, and A. Ng, "Bmultimodal deep learning,[in proc. 28th int. conf," *Mach. Learn*, vol. 2, 2011.

[73] Apr. 2021. [Online]. Available: https://www.gridwatch.templar.co.uk/

[74] A. Paszke, S. Gross, F. Massa, A. Lerer, J. Bradbury, G. Chanan, T. Killeen, Z. Lin, N. Gimelshein, L. Antiga *et al.*, "Pytorch: An imperative style, high-performance deep learning library," *Advances in Neural Information Processing Systems*, vol. 32, 2019.

[75] S.-H. Wang, K. Muhammad, J. Hong, A. K. Sangaiah, and Y.-D. Zhang, "Alcoholism identification via convolutional neural network based on parametric relu, dropout, and batch normalization," *Neural Computing and Applications*, vol. 32, no. 3, pp. 665–680, 2020.

[76] L. Bottou, "Stochastic gradient descent tricks," in *Neural Networks: Tricks of the Trade*. Springer, 2012, pp. 421–436.

[77] M. Majidi, S. Nojavan, and K. Zare, "A cost-emission framework for hub energy system under demand response program," *Energy*, vol. 134, pp. 157–166, 2017.

[78] Z. Zhou, C. Sun, R. Shi, Z. Chang, S. Zhou, and Y. Li, "Robust energy scheduling in vehicle-to-grid networks," *IEEE Network*, vol. 31, no. 2, pp. 30–37, 2017.

[79] M. You, W. Hua, M. Shahbazi, and H. Sun, "Energy hub scheduling method with voltage stability considerations," in *2018 IEEE/CIC International Conference on Communications in China (ICCC Workshops)*. IEEE, 2018, pp. 196–200.

[80] P. S. Georgilakis and N. D. Hatziargyriou, "Unified power flow controllers in smart power systems: Models, methods, and future research," *IET Smart Grid*, vol. 2, no. 1, pp. 2–10, 2019.

[81] S. Vandael, B. Claessens, D. Ernst, T. Holvoet, and G. Deconinck, "Reinforcement learning of heuristic ev fleet charging in a day-ahead electricity market," *IEEE Transactions on Smart Grid*, vol. 6, no. 4, pp. 1795–1805, 2015.

[82] E. C. Kara, M. Berges, B. Krogh, and S. Kar, "Using smart devices for system-level management and control in the smart grid: A reinforcement learning framework," in *2012 IEEE Third International Conference on Smart Grid Communications (SmartGridComm)*. IEEE, 2012, pp. 85–90.

[83] V. Davatgaran, M. Saniei, and S. S. Mortazavi, "Optimal bidding strategy for an energy hub in energy market," *Energy*, vol. 148, pp. 482–493, 2018.

[84] H. Shahinzadeh, J. Moradi, G. B. Gharehpetian, S. H. Fathi, and M. Abedi, "Optimal energy scheduling for a microgrid encompassing drrs and energy hub paradigm subject to alleviate emission and operational costs," in *2018 Smart Grid Conference (SGC)*. IEEE, 2018, pp. 1–10.

[85] M. C. Bozchalui, S. A. Hashmi, H. Hassen, C. A. Canizares, and K. Bhattacharya, "Optimal operation of residential energy hubs in smart grids," *IEEE Transactions on Smart Grid*, vol. 3, no. 4, pp. 1755–1766, 2012.

[86] J. Lafferty, A. McCallum, and F. C. Pereira, "Conditional random fields: Probabilistic models for segmenting and labeling sequence data," 2001.

[87] Y. Huang, F. Z. Peng, J. Wang, and D.-w. Yoos, "Survey of the power conditioning system for pv power generation," in *2006 37th IEEE Power Electronics Specialists Conference*. IEEE, 2006, pp. 1–6.

[88] J. H. Horlock, "*Combined Heat and Power*," Pergamon Press, 1987.

[89] S. G. Dukelow, "*The Control of Boilers*," 1986.

[90] J. W. Bialek and P. A. Kattuman, "Proportional sharing assumption in tracing methodology," *IEE Proceedings-Generation, Transmission and Distribution*, vol. 151, no. 4, pp. 526–532, 2004.

[91] C. Sutton, A. McCallum *et al.*, "An introduction to conditional random fields," *Foundations and Trends ® in Machine Learning*, vol. 4, no. 4, pp. 267–373, 2012.

[92] P. Carmona and L. Coutin, "Fractional Brownian motion and the markov property," *Electronic Communications in Probability*, vol. 3, pp. 95–105, 1998.

[93] S. Skarvelis-Kazakos, P. Papadopoulos, I. G. Unda, T. Gorman, A. Belaidi, and S. Zigan, "Multiple energy carrier optimisation with intelligent agents," *Applied Energy*, vol. 167, pp. 323–335, 2016.

[94] R. Fowler, O. Elmhirst, and J. Richards, "Electrification in the United Kingdom: A case study based on future energy scenarios," *IEEE Power and Energy Magazine*, vol. 16, no. 4, pp. 48–57, 2018.

[95] M. Moeini-Aghtaie, P. Dehghanian, M. Fotuhi-Firuzabad, and A. Abbaspour, "Multiagent genetic algorithm: An online probabilistic view on economic dispatch of energy hubs constrained by wind availability," *IEEE Transactions on Sustainable Energy*, vol. 5, no. 2, pp. 699–708, 2013.

[96] Y. Xiang, H. Cai, C. Gu, and X. Shen, "Cost-benefit analysis of integrated energy system planning considering demand response," *Energy*, vol. 192, p. 116632, 2020.

[97] H. Wang, C. Gu, X. Zhang, and F. Li, "Optimal chp planning in integrated energy systems considering network charges," *IEEE Systems Journal*, vol. 14, no. 2, pp. 2684–2693, 2019.

[98] M. You, J. Jiang, A. M. Tonello, T. Doukoglou, and H. Sun, "On statistical power grid observability under communication constraints," *IET Smart Grid*, vol. 1, no. 2, pp. 40–47, 2018.

[99] "National grid eso," http://fes.nationalgrid.com/fes-document/.

[100] "Balancing mechanism reporting service," https://www.bmreports.com/.

[101] S. Ge, Z. Xu, H. Liu, C. Gu, and F. Li, "Flexibility evaluation of active distribution networks considering probabilistic characteristics of uncertain variables," *IET Generation, Transmission & Distribution*, vol. 13, no. 14, pp. 3148–3157, 2019.

[102] H. Liu, Y. Zhang, S. Ge, C. Gu, and F. Li, "Day-ahead scheduling for an electric vehicle pv-based battery swapping station considering the dual uncertainties," *IEEE Access*, vol. 7, pp. 115 625–115 636, 2019.

[103] F. Liu, Z. Bie, and X. Wang, "Day-ahead dispatch of integrated electricity and natural gas system considering reserve scheduling and renewable uncertainties," *IEEE Transactions on Sustainable Energy*, vol. 10, no. 2, pp. 646–658, 2018.

[104] A. Ghasemi, M. Banejad, and M. Rahimiyan, "Integrated energy scheduling under uncertainty in a micro energy grid," *IET Generation, Transmission & Distribution*, vol. 12, no. 12, pp. 2887–2896, 2018.

[105] S. R. Salkuti, "Day-ahead thermal and renewable power generation scheduling considering uncertainty," *Renewable Energy*, vol. 131, pp. 956–965, 2019.

[106] C. Fan, J. Wang, W. Gang, and S. Li, "Assessment of deep recurrent neural network-based strategies for short-term building energy predictions," *Applied Energy*, vol. 236, pp. 700–710, 2019.

[107] S. Ryu, J. Noh, and H. Kim, "Deep neural network based demand side short term load forecasting," *Energies*, vol. 10, no. 1, p. 3, 2016.

[108] L. Benali, G. Notton, A. Fouilloy, C. Voyant, and R. Dizene, "Solar radiation forecasting using artificial neural network and random forest methods: Application to normal beam, horizontal diffuse and global components," *Renewable Energy*, vol. 132, pp. 871–884, 2019.

[109] A. Sharifian, M. J. Ghadi, S. Ghavidel, L. Li, and J. Zhang, "A new method based on type-2 fuzzy neural network for accurate wind power forecasting under uncertain data," *Renewable Energy*, vol. 120, pp. 220–230, 2018.

[110] T. Teo, T. Logenthiran, W. L. Woo, and K. Abidi, "Near-optimal day-ahead scheduling of energy storage system in grid-connected microgrid," in *2018 IEEE Innovative Smart Grid Technologies-Asia (ISGT Asia)*. IEEE, 2018, pp. 1257–1261.

[111] M. Zhou, B. Wang, and J. Watada, "Deep learning-based rolling horizon unit commitment under hybrid uncertainties," *Energy*, vol. 186, p. 115843, 2019.

[112] F. Watanabe, T. Kawaguchi, T. Ishizaki, H. Takenaka, T. Y. Nakajima, and J.-i. Imura, "Machine learning approach to day-ahead scheduling for multiperiod energy markets under renewable energy generation uncertainty," in *2018 IEEE Conference on Decision and Control (CDC)*. IEEE, 2018, pp. 4020–4025.

[113] ——, "Day-ahead strategic marketing of energy prosumption: A machine learning approach based on neural networks," in *2019 18th European Control Conference (ECC)*. IEEE, 2019, pp. 3910–3915.

[114] "National grid gas," https://www.nationalgridgas.com/balancing.

[115] Y. LeCun, Y. Bengio, and G. Hinton, "Deep learning," *Nature*, vol. 521, no. 7553, pp. 436–444, 2015.

[116] V. Nair and G. E. Hinton, "Rectified linear units improve restricted Boltzmann machines," in *International Conference on Machine Learning*, 2010.

[117] Z. Lu, H. Pu, F. Wang, Z. Hu, and L. Wang, "The expressive power of neural networks: A view from the width," *Advances in Neural Information Processing Systems*, vol. 30, 2017.

[118] I. Sutskever, J. Martens, G. Dahl, and G. Hinton, "On the importance of initialization and momentum in deep learning," in *International Conference on Machine Learning*. PMLR, 2013, pp. 1139–1147.

[119] F. W. Olver, D. W. Lozier, R. F. Boisvert, and C. W. Clark, *NIST Handbook of Mathematical Functions Hardback and CD-ROM*. Cambridge university press, 2010.

[120] S. Ioffe and C. Szegedy, "Batch normalization: Accelerating deep network training by reducing internal covariate shift," in *International Conference on Machine Learning*. PMLR, 2015, pp. 448–456.

CHAPTER 6

Implementation of Blockchain in Local Energy Markets

I n this chapter, the research and practices for implementing the blockchain technology in energy systems are introduced. Section 6.1 provides an introduction of the categories of potential applications of blockchain and smart contracts in energy systems. Section 6.2 reviews the research and innovations on the application of blockchain and smart contracts in energy systems and energy markets. Section 6.3 provides an example research of blockchain based peer-to-peer trading coupling energy and carbon markets while Section 6.4 extends this example research from the perspective of fundamental mechanism of blockchain technologies.

6.1   INTRODUCTION

Blockchain and smart contract technologies have a range of potential applications in energy systems. The key applications are summarised as follows:

- *Peer-to-peer energy trading:* With blockchain and smart contracts, consumers can sell the excess renewable energy they generate to other consumers in the same local grid, without relying on a utility grid as a middleman. By using a distributed ledger to record transactions and smart contracts to automate the settlement process, peer-to-peer energy trading can be made more secure and efficient. This can also enable the creation of local energy communities that share and trade energy among themselves.

- *Decentralised grid management:* Microgrids are decentralized energy systems that can operate independently or in parallel with the larger grid. Blockchain technology can be used to manage these microgrids by enabling the peer-to-peer energy trading, ensuring the energy supply and demand balance, and facilitating the exchange of information between different microgrids. Smart contracts

DOI: 10.1201/9781003170440-6

can be used to automate the process of energy trading, enabling seamless transactions without the need for intermediaries.

- *Carbon credit trading:* Carbon credits are certificates that represent the right to emit a certain amount of carbon dioxide or other greenhouse gases. Blockchain technology can be used to create a transparent and auditable carbon credit trading system. By using a distributed ledger to record transactions and smart contracts to automate the issuance, tracking, and trading of carbon credits, this system can help to reduce emissions and incentivise the adoption of clean energy technologies.

- *Electric vehicle charging:* Blockchain and smart contracts can be used to facilitate secure and transparent payments for electric vehicle charging stations. By using a distributed ledger to record transactions and smart contracts to automate the payment process, electric vehicle charging can be made more efficient and cost-effective. This can also enable the creation of charging networks that span multiple locations and providers, reducing the need for consumers to sign up for multiple accounts and payment systems.

- *Renewable energy certificates:* Renewable energy certificates are credits that represent the environmental attributes of renewable energy generation. Blockchain can be used to create a transparent and tamper-proof renewable energy certificates trading system. Smart contracts can automate the verification and tracking of these certificates, making the process more efficient and cost-effective. This can help to incentivise the adoption of renewable energy technologies and enable more transparent and traceable reporting of renewable energy generation.

- *Energy supply chain management:* The energy supply chain involves the production, distribution, and consumption of energy. Blockchain technology can be used to track the flow of energy throughout the supply chain, enabling more efficient and transparent management. By using a distributed ledger to record information and smart contracts to automate the exchange of information and settlement of transactions, the energy supply chain can be made more secure and efficient, reducing waste and increasing efficiency. This can help to reduce the environmental impact of energy production and consumption, while also reducing costs and increasing reliability.

Overall, the use of blockchain and smart contract technologies in energy systems has the potential to increase efficiency, reduce costs, and promote the transition to a more sustainable energy future.

6.2 BLOCKCHAIN ENABLING DECENTRALISED ENERGY MARKETS

In this section, research and innovations on the blockchain technologies including smart contracts, as enabling technologies of decentralised energy systems and local energy markets, are reviewed. Subsection 6.2.1 provides the concept and benefits of the peer-to-peer energy trading. Subsection 6.2.2 reviews the work on applying blockchain and smart contracts in energy systems. Subsection 6.2.3 compares the difference between the conventional centralised energy trading and blockchain based peer-to-peer energy trading.

6.2.1 Peer-to-Peer Energy Trading

The term 'peer-to-peer energy trading [1]' (similar to the terms of 'transactive energy [2]' and 'community self-consumption [3]') refers to a novel approach in which individual consumers who also generate their own electricity can trade excess energy with each other, without the need for a centralised agent. This concept has been gaining attention in recent years, and researchers have explored its potential benefits and challenges.

The goal of peer-to-peer energy trading is to create a more efficient and sustainable energy system, in which energy is generated, consumed and shared locally. This has several benefits, including the reduction of transmission losses and carbon emissions, increased energy resilience and supply–demand balance, and cost savings for both the community and individual prosumers.

- *Promoting local economy:* Peer-to-peer energy trading can lead to the development of an economically stronger community by utilising the sharing economy model. When profits from supplying energy are maintained locally, the community becomes as a whole benefits. This approach creates opportunities for training, education, and work, which can further strengthen the local economy. By encouraging community members to participate in the energy system, there is a greater sense of ownership and responsibility towards its success, leading to a more collaborative and supportive community.

- *Local energy resilience:* Peer-to-peer energy trading can enhance local energy resilience and supply–demand balance by facilitating the integration of small and independent prosumers with their distributed energy sources into power grids. This approach allows for a more diverse and distributed energy system that is less reliant on centralised power plants. By integrating small-scale energy sources into the grid, the system becomes more flexible and responsive to changes in energy demand. This can help avoid power outages during peak demand periods and ensure a reliable supply of electricity for the community.

- *Net zero energy transitions:* The increasing penetration of distributed renewable energy sources, facilitated by peer-to-peer energy trading, can reduce carbon emissions caused by long-distance power transmission and fossil fuel-based power generation. By generating and consuming energy locally, the need for long-distance transmission of electricity is reduced. In addition, distributed renewable energy sources such as solar panels and wind turbines produce clean energy that does not contribute to greenhouse gas emissions. This approach can help communities reduce their carbon footprint and contribute to global efforts to combat climate change.

- *Increasing prosumers' interests:* Peer-to-peer energy trading can also lead to bill savings, profit improvements, and cost savings for prosumers. By strategically deciding on their bidding or selling prices, local generation, and consumption behaviours, prosumers can increase their profits while reducing their energy costs. This approach allows for more control over energy bills and can reduce the reliance on centralised energy providers. By participating in the energy market, prosumers can benefit financially from their investment in renewable energy sources, which can lead to a faster return on investment and increased financial stability.

To achieve these benefits, a decentralised framework and mechanism for peer-to-peer energy trading must be designed. Several approaches have been proposed in the literature. For example, a two-stage aggregated control framework was designed in [4] for the peer-to-peer energy sharing in microgrids. This allowed prosumers to manage their distributed energy sources through the energy sharing coordinator, resulting in significant cost savings for the community and bill savings for individual prosumers.

Another approach was proposed by Morstyn et al. [5], who introduced a federated power plant that combines virtual power plants with the peer-to-peer energy trading. This incentivises coordination among individual prosumers and addresses the social, institutional, and economic issues that can arise with top-down strategies of the conventional trading framework.

In addition, bilateral contracting networks were developed in [6] for the peer-to-peer energy trading on real-time and forward markets. These networks coordinated upstream larger-scale power plants with downstream small-scale distributed energy sources, considering uncertainties in forward markets, to ensure an agreed market pricing for market participants.

In summary, peer-to-peer energy trading is a promising concept that can revolutionise the energy industry by empowering consumers to take control of their energy production, consumption and sharing. With the development of decentralised frameworks and mechanisms, the potential benefits of this approach can be fully realised.

6.2.2 Potential Applications of Blockchain Technologies

Blockchain technologies are emerging as a potential solution for establishing a decentralised trading platform with automated negotiation procedures, reduced transactional costs, secured information infrastructure, and protected residential privacy [7]. In the energy markets, blockchain can support a platform for energy trading, where residential privacy, such as address, load patterns, and pricing patterns, can be protected through the hash encryption of blockchain networks [8]. The blockchain also allows for the collective verification of transactions, which overcomes issues of double spending and the same energy being supplied twice. By removing intermediaries, prosumers can trade with each other, and the role of the market operator becomes that of a neutral facilitator towards open and accessible local energy markets.

The most promising technology to be explored in the energy market design and energy trading is smart contracts. These executable programs allow for self-enforcing settlement and negotiation procedures, thus securely automating trading procedures with standardised contracts, and reducing the costs of processing information flows from transactions of a large number of prosumers [9]. The features of replicability, security, and verifiability of smart contracts ensure that trading, negotiation, and agreement become more trustworthy without the interference of centralised authorities.

A fundamental principle of smart contracts is that if an event happens, the smart contracts transfer payments to the appropriate receivers. In the context of energy trading, the event could be the supply of energy or ancillary services, which is monitored by smart meters of prosumers. The pay function is executed in a self-enforcing manner. Therefore, the trustworthiness of energy trading is dependent on the trustworthiness of smart meters and programs executed by smart contracts.

Smart contracts have the potential to transform the energy market by making trading more secure, transparent, and automated. By reducing transactional costs and allowing for the efficient trading of energy, smart contracts could encourage the growth of decentralised energy systems and accelerate the transition to renewable energy. However, challenges remain, such as the need for interoperability and standardisation of smart contracts across different blockchain platforms. As the technology continues to evolve, the full potential of smart contracts in energy trading may yet to be realised.

Overall, the blockchain technologies including smart contracts provide a transaction and control foundation for the prosumers to participate in the peer-to-peer energy trading, with the following advantages:

- *Asset accounting:* The blockchain can prevent double spending attacks in energy trading by accounting for the ownership of digital and physical assets [10]. This ensures that the same energy or digital currency cannot be sold or spent twice, which increases the security of energy trading. The ownership of assets is transparently recorded in the blockchain, which makes it easy to trace the history of transactions and detect any attempt at fraud.

- *Collective verification:* The distributed feature of blockchain enables all participants in the energy market, including generators, consumers, prosumers, energy retailers, power system operators, and market operators, to hold a copy of the ledger [11]. Changes to the ledger require the consensus of all participants, which makes the blockchain network open and accessible to all. This ensures that the ledger is tamper-proof, transparent, and trustworthy, as every participant has a say in the decision-making process.

- *Disintermediation:* The disintermediating property of blockchain avoids domination by centralised authorities, such as energy retailers or suppliers [12]. Instead of supplying energy, the role of these intermediaries becomes that of a neutral facilitator to encourage passive customers to become both producers and consumers. This ensures that the market is not dominated by one or more participants and prevents market manipulation.

- *Security and privacy:* From a cryptographic perspective, the public/private key encryption guarantees the privacy and information security of a prosumer [13]. The computational difficulty of block mining and the collective validation of transactions through reaching a consensus ensure the security of local energy markets and energy trading [14]. The security of the blockchain is further enhanced by the use of hashing algorithms that make it practically impossible to reverse-engineer the original data.

- *Interoperability:* The blockchain supports smart control architectures that enable the interoperability of the smart grid [15]. Interoperability is defined as multiple agents collectively performing a function through exchanging information [16]. Automatically executed control functions written in smart contracts interact with smart meters, distributed computing, and fog computing to minimise latency and enhance computational efficiency and security.

- *Self-enforcement:* Smart contracts support a trading platform that minimises or eliminates the costs of handling information flows from transactions through automatically self-enforcing settlement and negotiation procedures [17]. This enables energy trading to be more efficient and cost-effective.

- *Standardisation:* Smart contracts with standardised auction procedures have the potential to prevent unforeseen trading behaviours in local energy markets [18]. This ensures that all participants in the market understand the rules and procedures for trading, which increases the transparency and fairness of the market. Standardised auction procedures also reduce transactional costs and promote efficiency in the market.

While blockchain and smart contracts hold promise for peer-to-peer energy trading, there are several challenges that need to be addressed to fully realise their potentials. Some of the main challenges include:

- *Technical complexity:* Blockchain technology and smart contracts are complex and require specialised knowledge to develop and maintain. Implementation and integration of these technologies into existing energy systems may require significant technical expertise.

- *Scalability:* The current scalability limitations of blockchain technology pose a significant challenge for large-scale energy systems. As the number of participants and transactions increases, the blockchain can become slower and less efficient, leading to increased transaction times and costs.

- *Energy demand and supply uncertainty:* The energy market is dynamic, and demand and supply can be unpredictable. Smart contracts require specific conditions to be met before execution, which can be challenging to implement in such an uncertain environment.

- *Consistency:* Different blockchain platforms have different technical specifications and standards, which can make it difficult to achieve interoperability between them. This can limit the ability to integrate with other systems and technologies, creating challenges for scaling and collaboration.

- *Regulatory challenges:* The regulatory environment for energy systems is complex, and blockchain and smart contract technologies must comply with regulations. Ensuring compliance while maintaining the decentralised and transparent nature of blockchain is a significant challenge.

  Energy storage and distribution: Peer-to-peer trading of energy requires energy storage and distribution systems to support the efficient and effective use of the available energy. The integration of these systems with blockchain technology and smart contracts is a significant challenge.

  Cybersecurity: Blockchain technology and smart contracts are not immune to cybersecurity threats. With energy systems being a critical infrastructure, the security and protection of these systems is of paramount importance.

Addressing these challenges will be critical to unlocking the full potential of blockchain and smart contracts in energy systems. However, continued innovation and development in the field may overcome these challenges in the future.

The use of blockchain technologies in power systems and energy markets is a subject of active research and industrial practice. Several studies have explored the potential benefits of applying blockchain and smart contracts to energy trading and management, and some of these studies are summarized below.

Thomas et al. [19] proposed a general form of smart contracts for controlling energy transfer processes between separated distribution networks. The designed negotiation framework and use case on a DC-link provided the means of applying smart contracts into power systems.

In another study, real-time power losses caused by transactions in microgrids were accounted for by the blockchain, and prosumers were considered as negotiators of energy transactions. Distribution system operators were responsible for computing losses [20].

Li et al. [21] applied smart contracts to distributed hybrid energy systems to facilitate energy exchange among end-users. The framework considered demand-side management and uncertainties caused by renewable generation.

Mihaylov et al. [22] designed a paradigm for energy trading with a virtual currency generated by the energy supply of prosumers. The designed currency incentivized prosumers to achieve demand response and supply–demand balance, as demonstrated by case studies.

Saxena et al. [23] proposed a blockchain-based transactive energy system to address the incentivizing, contract auditability, and enforcement of the voltage regulation service. Smart contracts were used to enforce the validity of each transaction and automate the negotiation and bidding process.

Finally, a transparent and safe power trading algorithm was executed on the Ethereum blockchain platform for prosumers to trade energy [24]. The platform leveraged the security and transparency features of blockchain technology to enable efficient and secure energy trading.

While these studies have demonstrated the potential of blockchain and smart contracts in energy trading and management, there are still several challenges that need to be addressed to achieve widespread adoption of these technologies in the energy sector. Some of these challenges include scalability, interoperability, regulatory compliance, and cybersecurity. Overcoming these challenges will require continued innovation and development in the field.

Furthermore, the blockchain has been developed for various purposes related to the decarbonisation of the energy sector, including trading carbon allowances and allocating monetary incentives for emission reduction. Several studies have explored the potential benefits of applying blockchain and smart contracts to carbon trading and management, as summarised below.

Khaqqi et al. [25] customized the trading of carbon allowances for industries using a reputation-based blockchain, in which reputation signified performances and commitments for carbon reduction from network participants. The reputation system was maintained by the consensus of blockchain networks to guarantee fairness and security.

Pan et al. [26] implemented blockchain technology into an emission trading scheme to reduce the entry threshold for the carbon market and improve the reliability of information exchange. The blockchain was used to ensure the accuracy and transparency of data, thereby increasing the reliability of the trading system.

Similarly, Richardson and Xu [27] proposed a blockchain-based emission trading scheme to ensure transparency, tamper-resistance, and high liquidity. The blockchain technology ensured that the trading process was secure and reliable, thereby promoting the growth of the carbon market.

In terms of the application of smart contracts, a distributed carbon ledger system was designed in [28] to strengthen the corporate accounting system for carbon asset management. The ledger system was integrated with existing market-based emission trading schemes and enabled more efficient and secure trading of carbon assets.

6.2.3 Comparison Remark

The difference between the conventional centralised trading and blockchain based peer-to-peer trading in the energy sector is summarised in Table 6.1. This table describes advantages of using the blockchain based the peer-to-peer energy trading, with details explained as follows.

- *Primary energy supplier:* In conventional energy markets, the primary energy supplier is the retailers in retail energy markets. In blockchain-based decentralised energy markets, the primary energy supplier is the prosumers with distributed energy sources. This shift in the primary supplier reflects the increased emphasis on decentralisation and the ability of individuals to participate in the energy market.

- *Pricing schemes:* In conventional energy markets, the pricing schemes are centralised, with a centralised wholesale energy price determined by the wholesale

Table 6.1 Comparison between the conventional centralised energy trading and blockchain-based peer-to-peer energy trading

	Conventional Centralised Trading	Blockchain Based Peer-to-Peer Trading
Primary Energy Supplier	Retailers	Prosumers with Distributed Energy Sources
Pricing Scheme	Centralised Pricing	Bidding/Selling Pricing from Prosumers
Contract Type	Idiosyncratic Contract	Standardised Smart Contracts
Settlement Enforcement	Legal Restraint	Self-Enforcement
Trustee	Third Party	Smart Meters and Smart Contracts
Incentive Supplier	Policy Maker	Consensus of Network

market and a retail energy price determined by energy retailers in retail markets [29]. In contrast, the pricing scheme in decentralised local energy markets is decentralised, with individual prosumers able to determine their own bidding or selling prices for sharing energy based on their real-time situation of the supply–demand balance. This pricing flexibility allows for a more tailored approach to energy pricing.

- *Negotiation and contracting:* The process of negotiation and contracting in conventional energy markets is idiosyncratic, with each large-scale generator signing a contract with the transmission system operator individually, and the content of each contract varying according to the specific agreement. In blockchain-based peer-to-peer energy trading, a standardised contract and negotiation can be formulated using smart contracts, reducing complexity when large amounts of prosumers formulate their own contracts.

- *Settlement:* The settlement of conventional centralised energy trading is enforced by legal restraints. If energy is not delivered at the agreed time, retailers will be accused or receive penalties from the power system operator afterwards. In contrast, in blockchain-based peer-to-peer energy trading, the self-enforcing nature of smart contracts enables the prevention of contract violations beforehand by querying smart meters to ensure that prosumers have enough capacity to supply.

- *Third-party reliance:* In conventional centralised energy trading, energy trading relies on a third party, such as an auditing institution or market operator. In blockchain-based peer-to-peer networks, prosumers' trust relies on the automatic interactions between smart contracts and smart meters, under the consensus of blockchain networks. This reduces the reliance on third parties and promotes a more open and accessible energy market.

Overall, the use of blockchain and smart contracts in energy trading represents a shift towards decentralisation, increased flexibility, and reduced reliance on third parties. While challenges remain in terms of scalability, interoperability, regulatory compliance, and cybersecurity, continued innovation and development in the field may overcome these challenges and unlock the full potential of these technologies in the energy market.

6.3 EXAMPLE RESEARCH 1: PEER-TO-PEER TRADING INTEGRATING ENERGY AND CARBON MARKETS

Prosumers, individuals who both produce and consume energy, are poised to play a crucial role in the future of energy systems. However, the rise of prosumers has brought about challenges in monitoring carbon emissions and pricing individual energy behaviours. To address these challenges, this example research proposes a revolutionary blockchain-based peer-to-peer trading framework for the trading of energy and carbon allowances.

This innovative framework utilises the bidding and selling prices of prosumers to incentivise the reshaping of energy behaviours, with the ultimate goal of achieving regional energy balance and carbon emissions reduction. The mechanism is decentralised, creating a low-carbon incentive structure that targets specific energy behaviours.

To test the efficacy of this proposed trading framework, the modified IEEE 37-bus test feeder was utilized in a series of case studies. The results showed that the new trading framework was capable of exporting 0.99 kWh of daily energy and saving 1465.90 g of daily carbon emissions. These findings indicate that the blockchain-based peer-to-peer trading framework outperforms existing centralised and aggregator-based trading systems.

Overall, the proposed trading framework has the potential to revolutionise the way energy is traded and consumed, creating a more efficient and environmentally friendly system. By incentivising prosumers to adjust their behaviours and reduce carbon emissions, this new system offers a promising path toward a sustainable future.

The rest of this example research is organised as follows: Sub-Section 6.3.2 introduces the proposed three-layer trading framework coupling energy and carbon markets. Corresponding to each layer, the details of problem formulation and the smart contract based auction mechanism are described in Sub-Section 6.3.3. Section 6.3.4 provides case studies to verify the proposed framework and demonstrate the trading platform. Section 6.3.5 draws the conclusion of this chapter.

6.3.1 Introduction

In achieving the net zero energy transition, once the peer-to-peer trading decentralises the local energy markets, what is the next we should decentralise? The answer to this question is the carbon market. In the carbon markets, the pollutant emitters are enforced to compensate the environmental damage in a monetary manner. Two classic mechanisms in conventional carbon markets are emission trading scheme (also called cap and trade) and carbon tax, which are centralised market-based carbon pricing schemes [30]. The carbon tax levies fixed price on carbon emissions whereas the emissions trading scheme assigns certain amount of carbon emissions. Compared to the carbon tax which is implemented based on the existing tax systems, the emissions trading scheme is more flexible and extendable to financial tools, e.g., the peer-to-peer energy trading. For this reason, our research focuses on the emissions trading scheme, and this background motivates us to couple both the energy market and carbon market in a decentralised manner.

The blockchain technology [31], as one of the distributed ledger technologies, has the potential of establishing a decentralised trading platform with automated trading procedures and protected residential privacy. The smart contracts [32], as one of the key blockchain technologies, enable prosumers to proceed the trading in a manner of self-enforcing settlement and setting out negotiation.

The research questions identified by this example research are summarised as follows:

- How to trace carbon emissions at the micro scope when prosumers trade energy and carbon allowance simultaneously?

- How to design the decentralised pricing of energy and carbon allowance targeting on the behaviours of individual prosumers?

- How to design a decentralised framework enabling prosumers to trade energy and carbon simultaneously?

This example research proposes a novel blockchain-based peer-to-peer trading framework. This framework enables prosumers to jointly exchange the energy and carbon allowance, since purchasing carbon allowance is a part of generating costs. The biding/offering prices of individual prosumers in energy and carbon markets are able to directly incentivise the reshaping of energy behaviours of prosumers for local energy balance and carbon saving. Additionally, when prosumers exchange energy as both generators and consumers, they need to know how much carbon allowance would be required. The carbon emissions tracing approach is developed to identify the carbon emissions caused by a prosumer's generation for self-consumption, consumption from self-generation, and generation (or consumption) for (or from) energy exchange with other prosumers. A low-carbon incentive mechanism is subsequently designed for individual prosumers. Case studies based on the modified IEEE 37-bus distribution network testify the proposed trading framework, in comparison with the centralised trading scheme and aggregator-based trading scheme. The execution of smart contracts on the Ethereum blockchain networks, and the interface between scheduling algorithms and smart contracts are demonstrated.

A conceptual graph of the proposed peer-to-peer trading framework is presented in Figure 6.1. Overall, this chapter offers the following key contributions:

- A novel trading framework is designed enabling the exchange of energy and carbon allowance at both prosumer level and microgrid level, using a blockchain smart contract based trading platform. The proposed energy scheduling algorithms interact with the self-enforcing nature of smart contract to automate the standardised auction procedure.

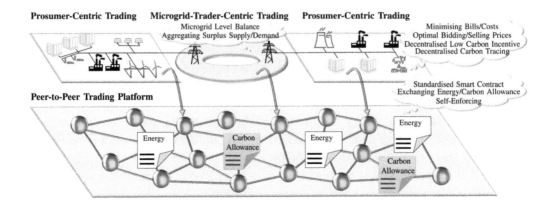

Prosumer-Centric Trading   Microgrid-Trader-Centric Trading   Prosumer-Centric Trading

Figure 6.1 Conceptual graph of peer-to-peer trading framework coupling energy and carbon markets.

- A carbon emissions tracing approach targeting on individual prosumers' energy behaviours is developed to ensure a fair allocation of low carbon incentives.

- Case studies show that the proposed trading framework achieves better energy balance and carbon saving than those approaches of centralised trading and aggregator-based trading. The interface between scheduling algorithms and smart contract, and the execution of smart contract are demonstrated.

6.3.2   Trading Framework

In this subsection, we introduce the trading framework for exchanging both energy and carbon allowances within distribution networks. This framework is designed based on the commercial relations between market participants, namely prosumers and microgrid-traders. The trading procedure is divided into three hierarchical layers: prosumer-centric trading, microgrid-trader-centric trading, and peer-to-peer trading platform, as illustrated in Figure 6.2.

The proposed framework is implemented in the day-ahead market, where energy behaviours are scheduled and trading is performed for the following day. Prosumers, who are considered masters of energy exchange seeking both personal benefits, such as bill or cost savings, and environmental goals, such as carbon emissions reduction, participate in both energy and carbon markets using their distributed renewable energy sources.

To facilitate this trading framework, the Ethereum blockchain [33] is used, consisting of full nodes and light nodes. The market operator acts as full nodes, providing and managing the trading platform by offering computing power for block mining, storing all blocks, and earning rewards for mined blocks. Prosumers and microgrid-traders act as light nodes, storing header chain and verifying transactions. As light nodes, they do not require powerful computers, and the trading process can be supported by smart meters or mobile phones.

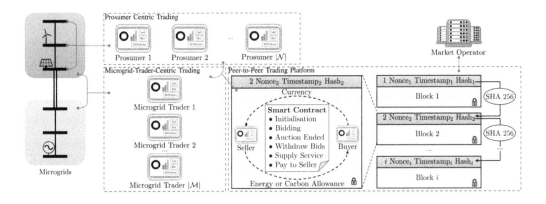

Figure 6.2 Architecture and information flows of the proposed framework for peer-to-peer trading of energy and carbon allowance. Individual prosumers trade energy or carbon allowance on the layer of prosumer-centric trading. The residual supply and demand for an ensemble of prosumers in the same microgrid are aggregated and traded by microgrid-traders on the layer of microgrid-trader-centric trading. The trading of energy or carbon allowance is proceeded on the layer of the peer-to-peer trading platform.

Each layer of the trading framework is designed to address specific challenges in peer-to-peer energy trading, with the problem formulation discussed in Sub-Section 6.3.2. The prosumer-centric layer is responsible for energy trading between individual prosumers and the microgrid-traders. The microgrid-trader-centric layer handles energy trading between microgrid-traders and the market operator. The peer-to-peer trading platform layer provides a mechanism for direct energy and carbon allowance trading between prosumers.

6.3.2.1 Prosumer-Centric Trading

The prosumer-centric trading layer is designed to help individual prosumers make optimal decisions regarding their energy behaviours and trading prices. The collected metering data is used to solve optimisation problems, with the objective of minimising electricity bills for buyers or maximising profits for sellers. The optimal decisions are then implemented by controllers, and the optimal decisions of bidding prices are sent to smart contracts for auctions. The blockchain automatically updates monetary incentives for individual prosumers by evaluating their carbon emissions behaviours.

To achieve regional energy balance and reduce transmission losses, prosumer-centric trading is only applicable to an ensemble of prosumers geographically located in the same microgrid, with a designated microgrid index. The prosumer-centric trading layer offers several advantages, including:

- Direct incentivisation of reshaping energy behaviours through prosumers' bidding or selling prices, rather than relying on central authorities such as aggregators or energy retailers.

- Direct linking of monetary incentives for carbon reduction with individual pro-sumers, considering their carbon emissions behaviours.

- Through the prosumer-centric trading layer, prosumers are empowered to take control of their energy usage and become active participants in the energy mar-ket. By offering financial incentives and promoting carbon emissions reduction, this layer can encourage prosumers to make environmentally responsible choices that benefit both themselves and the broader community.

Overall, the prosumer-centric trading layer provides a novel mechanism for incentivise energy behaviour reshaping and carbon emissions reduction. It achieves this by empowering individual prosumers to make optimal decisions regarding their bidding prices and energy behaviours. This layer offers significant advantages over traditional centralised approaches, such as direct incentivisation and linked monetary incentives. Ultimately, the prosumer-centric trading layer has the potential to create a more sustainable and efficient energy market by promoting responsible energy behaviours and reducing carbon emissions.

6.3.2.2 *Microgrid-Trader-Centric Trading*

The microgrid-trader-centric trading layer manages a group of physically connected prosumers under the management of a virtual entity, microgrid-trader. Within this layer, microgrid-trader aggregates the residual supply and demand of energy and carbon allowance for its ensemble of prosumers to trade with other microgrid-traders. Similar to the prosumer-centric trading layer, the optimal decisions of bidding prices are also yielded by solving optimisation problems with the objective of minimising electricity bills for buyers or maximising profits for sellers.

The primary aim of the microgrid-trader-centric trading layer is to assist an ensemble of prosumers within the same microgrid in balancing supply and demand by exchanging energy and carbon allowances with other microgrids. By leveraging the collective bargaining power of multiple prosumers, microgrid-traders can negotiate more favourable terms for energy and carbon allowance trading than would be possible for individual prosumers. This layer also enables microgrid-traders to optimise their bidding prices, taking into account the energy and carbon allowance trading activity of their ensemble of prosumers.

The microgrid-trader-centric trading layer provides several advantages, including:

- Efficient balancing of supply and demand within a microgrid, by leveraging the collective bargaining power of microgrid-traders to negotiate more favourable terms for energy and carbon allowance trading.

- Optimisation of bidding prices, taking into account the trading activity of an microgrid-trader's ensemble of prosumers.

- The ability to trade with other microgrid-traders, enabling the exchange of energy and carbon allowances between microgrids, promoting efficient and sustainable energy usage across regions.

Overall, the microgrid-trader-centric trading layer is a critical component of the proposed trading framework, enabling the efficient balancing of energy supply and demand within a microgrid, while promoting sustainable energy usage across regions. By leveraging the collective bargaining power of microgrid-traders and optimising bidding prices, this layer can enable the more efficient and cost-effective trading of energy and carbon allowances, benefiting both prosumers and the broader community.

6.3.2.3 Peer-to-Peer Trading Platform

The peer-to-peer trading platform layer is designed to provide a secure and standardised way for buyers and sellers to trade energy and carbon allowances. This is achieved through the use of smart contracts, which allow for the automatic execution of trades based on predefined conditions.

The smart contracts are based on the 'if-then' principle, where a specific event triggers a specific action. In the case of energy and carbon trading, the event is the delivery of energy or carbon allowance, which is verified by querying the smart meter. Once the delivery is confirmed, the smart contract automatically executes the trade by transferring the agreed amount of currency from the buyer to the seller.

The execution of the smart contract involves several steps, including initialisation, bid matching, bidding, winner selection, and ownership exchange. The seller initiates the smart contract by specifying the conditions of the offer. Buyers who meet these conditions can deposit their bids on the smart contract for auction. The buyer with the highest bidding price is selected as the winner and their deposited bid is transferred to the seller. The rest of the buyers can withdraw their deposits from the smart contract.

All transactions on the peer-to-peer trading platform are stored on a blockchain, which is a distributed ledger that is shared and audited by all nodes in the network. The blockchain is a series of blocks, with each block containing a record of all transactions that have taken place since the previous block. Each block is secured through a proof-of-work consensus mechanism that uses a secure hash algorithm to protect the integrity of the blockchain.

The use of blockchain technology ensures that all transactions are traceable, verifiable, and resistant to tampering. This provides greater transparency and security in the trading process, as all parties can see the history of transactions and trust that they are valid.

Overall, the peer-to-peer trading platform layer provides a secure and efficient way for buyers and sellers to trade energy and carbon allowances, with smart contracts and blockchain technology ensuring the integrity and transparency of the trading process.

6.3.3 Problem Formulation

This sub-section provides details on the problem formulations corresponding to each layer of the proposed framework of the peer-to-peer trading of energy and carbon allowances.

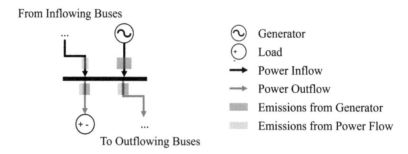

From Inflowing Buses

To Outflowing Buses

Generator
Load
Power Inflow
Power Outflow
Emissions from Generator
Emissions from Power Flow

Figure 6.3 Schematic illustration for the distribution of carbon emissions flow. The carbon emissions caused by all power outflows from a bus equals to the carbon emissions caused by all power inflows to this bus (indicated by the size of carbon emissions flow box). The proportion of the carbon emissions from each power inflow keeps unchanged in each power outflow (indicated by the colour of carbon emissions flow box), resulting in the same carbon intensity in power outflows.

6.3.3.1 Carbon Emissions Flow

For the conventional power systems, the large scale generators report their annual fuel usage and electricity supply to evaluate the efficiency of energy supply. With the information of the efficiency of energy supply by various technologies and carbon intensities of fuels, the carbon intensities of each generation technology can be traced.

To trace the carbon emissions from the energy behaviours of prosumers or consumers, first, we need to find an approach to evaluate carbon emissions from each component of power systems. In our research, we implemented an approach called the carbon emissions flow [34]. The carbon emissions flow is a virtual network flow concurrent with the power flow. It is ejected from the outflowing buses, including generators, and delivered to the inflowing buses, including loads. Since it is a concurrent flow with the power flow, the proportional sharing principle [35] which applies to the power flow analysis also works on the carbon emissions flow. As shown in Figure 6.3, we have

- *Conservation*: The carbon emissions caused by all power outflows from a bus equals to the carbon emissions caused by all power inflows to this bus.

- *Proportional sharing*: The proportion of the carbon emissions from each power inflow keeps unchanged in each power outflow, resulting in the same carbon intensity in power out flows.

This carbon emissions tracing approach is subsequently implemented to evaluate the carbon emissions from generation, transmission, transmission loss, and consumption. By contrast, with the integration of distributed energy sources in distribution networks, prosumers play a role as both generators and consumers. This carbon emissions tracing approach needs to be extended to distinguish the following portions of carbon emissions:

- Carbon emissions resulting from a prosumer's self-consumption of electricity generated through their own renewable energy sources, such as solar panels or wind turbines. The use of this on-site generation reduces the prosumer's reliance on electricity from the grid and can lead to a decrease in overall carbon emissions. However, the carbon emissions associated with the manufacturing and installation of the renewable energy system, as well as the emissions from any backup generators or grid-supplied electricity used during periods of low generation, must also be considered.

- Carbon emissions caused by a prosumer's distribution of excess electricity generated through their own renewable energy sources to other prosumers, either through a peer-to-peer energy trading system or by feeding it back into the grid. This practice can help to reduce overall carbon emissions by decreasing the need for energy from non-renewable sources. However, there may be carbon emissions associated with the transportation of electricity to the other prosumers, as well as any grid losses or inefficiencies.

- Carbon emissions resulting from a prosumer's consumption of electricity that was generated by other prosumers and supplied through a peer-to-peer energy trading system or through the grid. This approach can be beneficial for reducing carbon emissions by encouraging the use of renewable energy sources and improving energy efficiency. However, there may still be carbon emissions associated with the generation of the electricity, which could come from non-renewable sources if the prosumers supplying the electricity do not have renewable energy sources. Additionally, there may be carbon emissions associated with the transmission and distribution of the electricity, as well as any grid losses or inefficiencies.

The carbon emissions tracing approach in micro scope has been designed to evaluate the carbon emissions in microgrids, taking into account the bidirectional power flows that arise from energy trading between prosumers. The index sets for the generators and loads of a prosumer are denoted as $\mathcal{I}$ and $\mathcal{K}$, respectively. The carbon emissions rates caused by the power generation of a generator $i \in \mathcal{I}$ and the power consumption of load $k \in \mathcal{K}$ at scheduling time t are represented by $r_{i,t}$ and $r_{k,t}$, respectively.

To illustrate the different portions of carbon emissions, a schematic diagram is presented in Figure 6.4. Consider the case where prosumer A generates surplus energy after meeting its own demand and supplies this excess energy to prosumer B, who is unable to generate enough energy to meet its own demand. In this scenario, the portion of carbon emissions caused by using prosumer A and prosumer B's own generation for meeting their own demand can be quantified by $\sum_{k \in \mathcal{K}} r_{k,t}^{\mathrm{A}}$ and $\sum_{k \in \mathcal{K}} r_{k,t}^{\mathrm{B}}$, respectively.

In addition, the portion of carbon emissions caused by using a prosumer's own generation for supplying other prosumers' demand can be expressed as:

$$r_{\mathrm{net},t} = \sum_{i \in \mathcal{I}} r_{i,t} - \sum_{k \in \mathcal{K}} r_{k,t}, \tag{6.1}$$

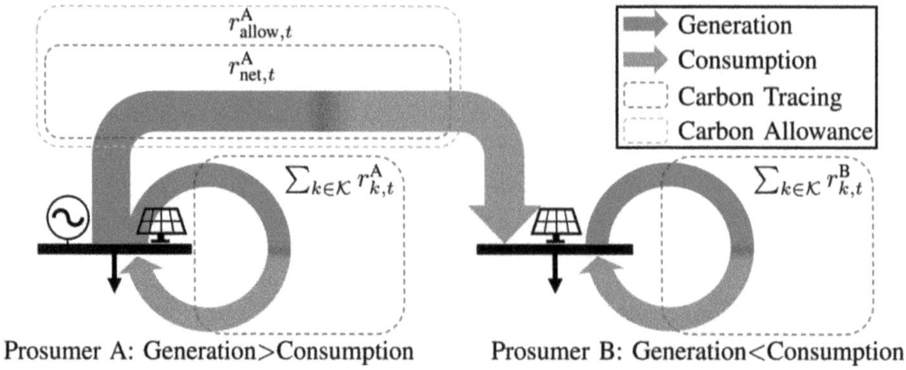

Figure 6.4 Schematic illustration of carbon emissions tracing for prosumers. Prosumer A supplies surplus energy to prosumer B. The prosumer A needs to have the carbon allowance when supplying energy to the prosumer B.

Here, $r_{\text{net},t}$ represents the carbon emissions rate caused by using a prosumer's own generation to meet the demand of other prosumers at scheduling time t.

The above carbon emissions tracing approach can provide valuable insights into the carbon footprint of microgrids, especially in the context of energy trading. By quantifying the different portions of carbon emissions, it can help identify opportunities for reducing carbon emissions and improving the overall sustainability of microgrids.

As shown in Figure 6.4, the portion of carbon emissions resulting from prosumer A' s own generation being used to meet the energy demand of prosumer B can be evaluated using $r_{\text{net},t}^{\text{A}}$. It is important to note that the same amount of carbon emissions will also be generated when prosumer B' s energy demand is met by prosumer A' s generation.

By using the carbon emissions tracing approach, it becomes possible to quantify the carbon emissions associated with energy trading between prosumers. In this context, it is important to consider not only the carbon emissions resulting from the generation and consumption of energy but also the carbon emissions associated with the transmission and distribution of the energy. Through the quantification of carbon emissions, it is possible to identify opportunities to reduce the carbon footprint of energy trading and improve the sustainability of microgrids.

Once the different portions of carbon emissions have been identified and traced, it becomes possible to design a decentralised low carbon incentive mechanism for individual prosumers. This mechanism targets prosumers as individuals rather than as a group. The principle of carbon accounting is critical in this incentive mechanism. When a prosumer supplies energy to other prosumers, they need to have a carbon allowance as a permission to emit pollutants. The carbon allowance for a prosumer at scheduling time t is denoted as $r_{\text{allow},t}$. The function $\gamma(\cdot)$ represents the monetary compensation for carbon reduction by a prosumer.

To formulate the prosumer-centric low carbon incentive, several assumptions need to be considered as follows:

- *Assumption 1*: If the prosumer's net carbon emissions rate $r_{\text{net},t}$ exceeds their carbon allowance $r_{\text{allow},t}$, the prosumer will have to purchase the additional carbon allowance from other prosumers. On the other hand, if the prosumer's net carbon emissions rate is lower than their carbon allowance, they can sell the surplus carbon allowance to other prosumers and receive monetary compensation through the consensus of peer-to-peer trading networks. This assumption ensures that prosumers are incentivised to reduce their carbon emissions while also facilitating the exchange of carbon allowances between prosumers in a transparent and efficient manner.

  Assumption 2: The monetary compensation for carbon reduction should be non-negative. When the net carbon emissions rate of a prosumer reduces to zero, the monetary compensation should be maximised, i.e., $\gamma = \gamma^{\max}$, where $\gamma^{\max}$ represents the maximum monetary compensation for carbon reduction that a prosumer can receive. This assumption incentivises prosumers to reduce their carbon emissions as much as possible and rewards those who achieve zero net carbon emissions.

  Assumption 3: When the carbon allowance is assigned to a prosumer, the monetary compensation should decrease monotonically with the net carbon emissions rate. This can be expressed as:

$$\frac{\partial \gamma \left(r_{\text{net},t}, r_{\text{allow},t} \right)}{\partial r_{\text{net},t}} < 0. \tag{6.2}$$

This assumption ensures that prosumers are incentivised to reduce their net carbon emissions rate, as a higher net carbon emissions rate will result in a lower monetary compensation for carbon reduction. In other words, prosumers who emit more carbon will receive lower monetary compensation than those who emit less carbon, creating a fair and effective incentive mechanism for reducing carbon emissions.

- *Assumption 4*: Let $\mathcal{N}$ be the index set of prosumers in the same microgrid. The initial carbon allowance for each prosumer is assigned by the blockchain system based on the carbon emissions intensities of the prosumers and a carbon reduction target for the entire group of prosumers. This is given by:

$$r_{\text{allow},t} = \frac{\rho_n}{\sum_{n \in \mathcal{N}} \rho_n} \cdot \bar{e}, \tag{6.3}$$

where ρ_n represents the carbon emissions intensity of prosumer n, and $\bar{e}$ is the targeted total carbon emissions for the group of prosumers. This allocation ensures that each prosumer is allocated a carbon allowance that is proportional to their carbon emissions intensity and contributes to the overall carbon reduction target of the group.

To further incentivise prosumers to reduce their carbon emissions, the marginal monetary compensation should be monotonically increasing with the assigned carbon allowance. This means that prosumers with higher carbon emissions intensity will receive more monetary compensation than those with lower carbon emissions intensity, as they require more urgent carbon mitigation. This condition can be expressed as:

$$\frac{\partial^2 \gamma \left(r_{\text{net},t}, r_{\text{allow},t} \right)}{\partial r_{\text{allow},t}^2} > 0. \tag{6.4}$$

Based on these conditions, the following function is proposed as the prosumer-centric monetary compensation for carbon reduction:

$$\gamma \left(r_{\text{net},t}, r_{\text{allow},t} \right) := \begin{cases} \alpha_t \cdot \sqrt{\left(r_{\text{allow},t} \cdot \Delta t \right)^2 - \left(r_{\text{net},t} \cdot \Delta t \right)^2}, & r_{\text{allow},t} > r_{\text{net},t}, \\ 0, & r_{\text{allow},t} \leq r_{\text{net},t}, \end{cases} \tag{6.5}$$

where α_t is the monetary compensation rate at scheduling time t, and Δt is the scheduling interval. This function ensures that prosumers receive monetary compensation for carbon reduction that is proportional to the difference between their carbon allowance and their net carbon emissions rate.

By implementing this prosumer-centric monetary compensation mechanism, it is possible to incentivise prosumers to reduce their carbon emissions and contribute to the overall carbon reduction target of the microgrid. This can help to foster the development of a more sustainable and resilient energy system, while also promoting the efficient and transparent exchange of carbon allowances and monetary compensation between prosumers.

6.3.3.2 Prosumer-Centric Algorithm

The prosumer-centric trading approach enables prosumers in the same microgrid to exchange energy or carbon allowance with neighbouring prosumers for the purpose of achieving local energy balance and carbon reduction. However, this trading process is not without its losses. Power losses are a significant consideration in this regard, and they consist of transmission losses and distribution losses.

Transmission losses refer to the losses that occur during the transmission of power from generators to distribution networks. These losses account for approximately 2-6% of total power generation from a whole power systems perspective [36]. On the other hand, distribution losses refer to the losses that occur within the distribution networks, such as power losses within a community. However, the distribution losses are relatively small compared to transmission losses because the generators and loads are nearby, and the amount of distributed generation from prosumers is smaller relative to the amount of large-scale generation in power systems. Therefore, distribution losses are neglected in this work.

Recall that $p_{i,t}$ and $p_{k,t}$ denote the power generation of generator $i \in \mathcal{I}$ and the power consumption of load $k \in \mathcal{K}$ at scheduling time t, respectively. The prosumer-centric algorithm is discussed below when a prosumer is either an energy buyer or an energy seller.

- *Prosumer as an energy buyer:*

When a prosumer is unable to generate enough energy to meet their own demand ($\sum_{i \in \mathcal{I}} p_{i,t} < \sum_{k \in \mathcal{K}} p_{k,t}$), they must purchase energy from other prosumers as an energy buyer. The objective function for a prosumer as an energy buyer is given by

$$f_b \left(p_{i,t}, p_{k,t}, b_{\text{energy},t} \right) := \sum_{t \in \mathcal{T}_{\text{buyer}}} \left(\sum_{k \in \mathcal{K}} p_{k,t} - \sum_{i \in \mathcal{I}} p_{i,t} \right) \cdot \Delta t \cdot b_{\text{energy},t}, \qquad (6.6)$$

where $f_b \left(\cdot \right)$ represents the prosumer's electricity bill. The bidding price of a prosumer at scheduling time t for buying energy is denoted by $b_{\text{energy},t}$ and the index set of scheduling time when a prosumer is an energy buyer is given by $\mathcal{T}_{\text{buyer}}$.

To participate in the peer-to-peer trading network as an energy buyer, the smart contract requires the prosumer to have enough account balance for purchasing the bided energy. This balance constraint is expressed by Equation (6.7), where b_{balance} is the account balance of the buyer.

$$\sum_{t \in \mathcal{T}_{\text{buyer}}} \left(\sum_{k \in \mathcal{K}} p_{k,t} - \sum_{i \in \mathcal{I}} p_{i,t} \right) \cdot \Delta t \cdot b_{\text{energy},t} \leq b_{\text{balance}}, \qquad (6.7)$$

In addition to this, the smart contract requires a buyer's bidding price to be higher than the currently highest bidding prices submitted by other energy buyers for the same offer. This constraint is represented by Equation (6.8), where $b_{\text{energy},t}^{\text{highest}}$ is the currently highest bidding price for the energy being sold at scheduling time t over all energy buyers updated by the blockchain network. The set of bidding prices submitted by all energy buyers for the offer of selling energy at scheduling time t is denoted by $\mathcal{B}_{\text{energy},t}$. Equation (6.9) represents the computation of the highest bidding price.

$$b_{\text{energy},t}^{\text{highest}} < b_{\text{energy},t}, \qquad (6.8)$$

$$b_{\text{energy},t}^{\text{highest}} = \max : \mathcal{B}_{\text{energy},t}. \qquad (6.9)$$

Thus, the objective of a prosumer as an energy buyer is to minimize their electricity bills by strategically deciding the bidding prices of energy and reshaping their energy behaviours. This is represented by the optimisation problem given in Equation (6.10), subject to the balance constraint given in Equation (6.7), the bidding price constraint given in Equation (6.8), and the computation of the highest bidding price as given in Equation (6.9).

$$\min_{p_{i,t}, p_{k,t}, b_{\text{energy},t}} : f_b \left(p_{i,t}, p_{k,t}, b_{\text{energy},t} \right), \qquad (6.10)$$

s.t.:(6.7), (6.8), and (6.9).

- *Prosumer as an energy seller:*

When a prosumer generates surplus energy after meeting its own demand ($\sum_{k \in \mathcal{K}} p_{k,t} < \sum_{i \in \mathcal{I}} p_{i,t}$), this prosumer can sell the surplus energy to other prosumers as an energy seller. However, the prosumer must have the necessary carbon allowance assigned by the blockchain system in order to sell energy. If the prosumer's net carbon emissions $r_{net,t}$ exceeds the assigned carbon allowance $r_{allow,t}$, then the prosumer must buy the extra carbon allowance as part of its generating costs. Conversely, if $r_{net,t}$ is less than $r_{allow,t}$, then the prosumer can sell the extra carbon allowance and be compensated as part of its revenue.

To calculate the cost or revenue of buying or selling carbon allowance, we use the function $c_{carbon}(\cdot)$, which describes the carbon cost/revenue of an energy seller. This function takes into account the highest bidding price $b_{carbon,t}^{highest}$ submitted by carbon allowance buyers for the energy seller's carbon allowance at scheduling time t, and is defined as follows:

$$c_{carbon}(r_{net,t}) := \begin{cases} (r_{net,t} - r_{allow,t}) \cdot \Delta t \cdot b_{carbon,t}^{highest} - \gamma(r_{net,t}, r_{allow,t}), & r_{net,t} < r_{allow,t}, \\ (r_{net,t} - r_{allow,t}) \cdot \Delta t \cdot b_{carbon,t}, & r_{net,t} > r_{allow,t}, \end{cases}$$
$$(6.11)$$

In the case where $r_{net,t} < r_{allow,t}$, the energy seller is compensated with the bidding price $b_{carbon,t}$ for each unit of surplus carbon allowance sold. On the other hand, if $r_{net,t} > r_{allow,t}$, the energy seller must pay for the extra carbon allowance it needs to purchase at the highest bidding price $b_{carbon,t}^{highest}$, and this cost is subtracted from the revenue gained from selling energy.

To participate in the auction as an energy seller, the prosumer must submit a bidding price $b_{carbon,t}$ for the carbon allowance it wishes to sell at scheduling time t. Furthermore, the smart contract requires that the energy seller's bidding price is higher than the highest bidding prices submitted by other carbon allowance sellers for the same offer, which is updated by the blockchain network. This can be expressed as:

$$b_{carbon,t}^{highest} = \max : \mathcal{B}_{carbon,t}, \qquad (6.12)$$

where $\mathcal{B}_{carbon,t}$ denotes the set of bidding prices submitted by all the carbon allowance buyers for the offer of selling carbon allowance at scheduling time t.

In addition to the carbon cost, there are other operating costs that need to be considered in our dynamic scheduling problem. These include costs of operation, maintenance, fuel, and carbon capture and storage. However, costs of pre-development, construction, decommissioning, and waste are not considered in our analysis.

The coefficients of operating costs for each energy source can be evaluated by the levelised cost of energy (LCoE) [37]. The LCoE represents the total cost of generating energy over the lifetime of a power plant, including initial capital costs, operating costs, and fuel costs, as well as costs associated with decommissioning and any required environmental controls.

The function of operating costs of a prosumer can be modelled as follows:

$$c\left(p_{i,t}\right) := \sum_{i \in \mathcal{I}} p_{i,t} \cdot \Delta t \cdot \delta_i, \tag{6.13}$$

where $c\left(\cdot\right)$ is the function of operating costs of a prosumer excluding the carbon cost, and δ_i is the coefficient of the total operating costs of generator i. This equation represents the total operating cost incurred by a prosumer at scheduling time t, given the power generation of generator i during that time.

The objective function of a prosumer as an energy seller can be modelled as

$$f_p\left(p_{i,t}, p_{k,t}, b_{\text{carbon},t}\right)$$
$$:= \sum_{t \in \mathcal{T}_{\text{seller}}} \left[\left(\sum_{i \in \mathcal{I}} p_{i,t} - \sum_{k \in \mathcal{K}} p_{k,t}\right) \cdot \Delta t \cdot b_{\text{energy},t}^{\text{highest}} - c_{\text{carbon}}\left(r_{\text{net},t}\right) - c(p_{i,t})\right], \tag{6.14}$$

where $f_p\left(\cdot\right)$ is the objective function of profits of a prosumer, and $\mathcal{T}_{\text{seller}}$ is the index set of scheduling time when a prosumer is an energy seller.

When a prosumer participates in the peer-to-peer trading network as a carbon allowance buyer, they need to have enough account balance to purchase the bided carbon allowance. This is enforced by the smart contract, which checks that the sum of carbon costs over all scheduling times when the prosumer is a buyer is less than or equal to the prosumer's account balance:

$$\sum_{t \in \mathcal{T}_{\text{seller}}} c_{\text{carbon}}\left(r_{\text{net},t}\right) \le b_{\text{balance}}, \tag{6.15}$$

In addition, the smart contract also requires that a buyer's bidding price for carbon allowance be higher than the currently highest bidding prices submitted by other carbon allowance buyers for the same offer. This is necessary to ensure that the buyer obtains the carbon allowance and that the market operates efficiently:

$$b_{\text{carbon},t}^{\text{highest}} < b_{\text{carbon},t}, \tag{6.16}$$

The objective of a prosumer as a carbon allowance buyer is to minimise their carbon costs by strategically deciding the bidding prices of carbon allowance and reshaping their energy consumption behaviours. This is formulated as the maximisation of profits:

$$\max_{p_{i,t}, p_{k,t}, b_{\text{carbon},t}} : f_p\left(p_{i,t}, p_{k,t}, b_{\text{carbon},t}\right), \tag{6.17}$$

$$\text{s.t.:}(6.12), (6.15), \text{ and } (6.16)$$

It is important to note that the decision variable $b_{\text{carbon},t}$, as well as constraints (6.15) and (6.16), only apply when a prosumer is buying carbon allowance. When a prosumer generates surplus energy and is selling carbon allowance, they do not need to bid for the allowance, and the carbon revenue is automatically calculated based on the market price.

6.3.3.3 *Microgrid-Trader-Centric Algorithm*

After the completion of the prosumer-centric trading, there might be residual supply or demand which cannot be met inside the microgrid due to the surplus or scarcity generation of all prosumers in the same microgrid. The microgrid-trader-centric trading aims to help an ensemble of prosumers in the same microgrid aggregate the residual supply and demand. Through solving the prosumer-centric algorithm, the optimal power generation of generator i and optimal power consumption of load k for a prosumer at each scheduling time t are yielded, denoted as $p_{i,t}^*$ and $p_{k,t}^*$, respectively. The residual power of prosumer $n \in \mathcal{N}$ can be described as:

$$p_{n,t} = \sum_{i \in \mathcal{I}} p_{i,t}^* - \sum_{k \in \mathcal{K}} p_{k,t}^* \tag{6.18}$$

where $p_{n,t}$ is the residual power of prosumer $n \in \mathcal{N}$ at scheduling time t.

After the prosumer-centric trading is completed, there may still be some residual supply or demand that cannot be met within the microgrid due to the surplus or scarcity generation of all prosumers within that microgrid. In order to address this issue, the microgrid-trader-centric trading comes into play. This approach aims to help a group of prosumers within the same microgrid aggregate their residual supply and demand, by trading with external entities such as the utility grid or other microgrids.

To implement this approach, the prosumer-centric algorithm is first executed to determine the optimal power generation of generator i and optimal power consumption of load k for each prosumer at each scheduling time t, which are denoted as $p_{i,t}^*$ and $p_{k,t}^*$, respectively. With these results, the residual power of prosumer $n \in \mathcal{N}$ can be computed as:

$$p_{n,t} = \sum_{i \in \mathcal{I}} p_{i,t}^* - \sum_{k \in \mathcal{K}} p_{k,t}^* \tag{6.19}$$

where $p_{n,t}$ is the residual power of prosumer n at scheduling time t.

The residual power can either be positive or negative, representing surplus power or demand, respectively. If the residual power is positive, it means that the prosumers have excess power that can be sold to other entities. On the other hand, if the residual power is negative, it means that the prosumers need additional power to meet its demand and can buy it from other entities. In either case, the residual power can be traded in the microgrid-trader-centric trading to balance the supply and demand between the microgrids.

The cases when the microgrid-trader acts as the energy buyer or energy seller are discussed as follows:

- *Microgrid-trader as energy buyer:*

  When an ensemble of prosumers in the same microgrid is unable to generate enough energy to meet their own demand, i.e. $\sum_{n \in \mathcal{N}} p_{n,t} < 0$, the microgrid-trader needs to help its prosumers buy energy from other microgrids or import from the main grid. In this case, the microgrid-trader acts as an energy buyer in the microgrid-trader-centric trading. The objective of the microgrid-trader as an energy buyer is to minimise the overall electricity bills for its prosumers by strategically deciding the optimal bidding price of energy. The objective function of the microgrid-trader as an energy buyer can be modelled as follows:

$$f_B\left(b_{\text{energy},t}\right) = \sum_{t \in \mathcal{T}\text{buyer}} \sum n \in \mathcal{N}\left(-p_{n,t}\right) \cdot \Delta t \cdot b_{\text{energy},t}, \tag{6.20}$$

where $f_B\left(\cdot\right)$ is the objective function of electricity bills of a microgrid-trader, $b_{\text{energy},t}$ is the bidding price of a microgrid-trader at scheduling time t for buying energy, and $\mathcal{T}_{\text{buyer}}$ is the index set of scheduling time when the microgrid-trader is an energy buyer.

Similar to the prosumer-centric trading, the microgrid-trader has to satisfy account balance and highest bidding constraints when it is an energy buyer. The account balance constraint is expressed as follows:

$$\sum_{t \in \mathcal{T}\text{buyer}} \sum n \in \mathcal{N}\left(-p_{n,t}\right) \cdot \Delta t \cdot b_{\text{energy},t} \leq b_{\text{balance}}, \tag{6.21}$$

where b_{balance} is the account balance of the microgrid-trader.

The highest bidding constraint for the microgrid-trader is given as follows:

$$b_{\text{energy},t}^{\text{highest}} < b_{\text{energy},t}, \tag{6.22}$$

where $b_{\text{energy},t}^{\text{highest}}$ is the currently highest bidding price for the energy selling at scheduling time t over all energy buyers updated by the blockchain network. Let $\mathcal{B}_{\text{energy},t}$ denote the set of bidding prices submitted by all energy buyers for the offer of selling energy at scheduling time t. We have

$$b_{\text{energy},t}^{\text{highest}} = \max : \mathcal{B}_{\text{energy},t}. \tag{6.23}$$

Therefore, the decision variable for the microgrid-trader as an energy buyer is $b_{\text{energy},t}$, and the constraints to be satisfied are (6.21) and (6.22). The objective of the microgrid-trader is to minimise overall electricity bills for its prosumers as expressed in (6.24).

$$\min_{b_{\text{energy},t}} : f_B\left(b_{\text{energy},t}\right), \tag{6.24}$$

$$\text{s.t.: (6.21), and (6.22).}$$

- *Microgrid-trader as energy seller:* When an ensemble of prosumers in the same microgrid generates surplus energy after meeting their own demand ($\sum_{n \in \mathcal{N}} p_{n,t} > 0$), microgrid-trader can help its prosumers sell energy to other microgrids. Meanwhile, microgrid-trader can help its energy sellers trade residual carbon allowance with other microgrids. If the net carbon emissions of an ensemble of prosumers in the same microgrid exceed the carbon allowance of this microgrid, microgrid-trader has to help its prosumers buy carbon allowance from other microgrids. If the net carbon emissions of an ensemble of prosumers in the same microgrid are less than the carbon allowance of this microgrid, microgrid-trader can help its prosumers sell the extra carbon allowance and earn monetary compensation for its prosumers.

To achieve these goals, microgrid-trader as an energy seller aims to maximise the overall profits for its prosumers by strategically deciding optimal bidding prices of carbon allowance as well as electricity prices for energy trading. The objective function of a microgrid-trader as an energy seller can be modelled as

$$f_P\left(b_{\text{carbon},t}\right) := \sum_{t \in \mathcal{T}_{\text{seller}}} \sum_{n \in \mathcal{N}} \left[p_{n,t} \cdot \Delta t \cdot b_{\text{energy},t}^{\text{highest}} - \left(c_{\text{carbon},n} + c_n\right)\right], \qquad (6.25)$$

where $f_P\left(\cdot\right)$ is the objective function of profits of a microgrid-trader, $c_{\text{carbon},n}$ is the carbon cost/revenue of prosumer n, c_n is the operating costs excluding the carbon cost of prosumer n, and $b_{\text{energy},t}^{\text{highest}}$ is the highest bidding price for energy selling at scheduling time t over all energy buyers updated by the blockchain network. The decision variable $b_{\text{carbon},t}$ and $b_{\text{energy},t}^{\text{highest}}$ only hold when a microgrid-trader buys the carbon allowance.

Similar to the prosumer-centric trading, there are account balance constraints and the highest bidding constraints for microgrid-trader as an energy seller. The account balance constraint is defined as

$$\sum_{t \in T_{\text{seller}}} \sum_{n \in \mathcal{N}} c_{\text{carbon},n} \leq b_{\text{balance}}, \qquad (6.26)$$

The highest bidding constraint for energy selling is defined as

$$b_{\text{energy},t}^{\text{highest}} < b_{\text{energy},t}. \qquad (6.27)$$

The highest bidding constraint for carbon allowance selling is defined as

$$b_{\text{carbon},t}^{\text{highest}} < b_{\text{carbon},t}. \qquad (6.28)$$

Therefore, the objective of a microgrid-trader as an energy seller is to maximise the overall profits for its prosumers by strategically deciding optimal bidding prices of carbon allowance as

$$\max_{b_{\text{carbon},t}} : f_P\left(b_{\text{carbon},t}\right), \qquad (6.29)$$

$$\text{s.t.: (6.26), (6.27) and (6.28).}$$

The decision variable $b_{\text{carbon},t}$ and this optimisation problem only hold when a microgrid-trader buys the carbon allowance.

Remark: The optimisation problems for both the prosumer-centric and microgrid-trader-centric algorithms can be quite complex and computationally intensive, which can lead to long computation times and scalability issues. To address these challenges, a learning approach has been proposed, which utilises artificial neural networks to predict optimal scheduling decisions as introduced in the previous chapter which can significantly reduce computation times while maintaining high accuracy in the predicted results.

The learning approach involves training artificial neural networks to predict optimal scheduling decisions based on historical data. In particular, the artificial neural networks are trained on data from past scheduling periods to learn the relationship between the inputs and the optimal scheduling decisions. The inputs include various parameters such as energy demand, available energy sources, carbon allowance, and bidding prices. The artificial neural networks can then be used to predict the optimal scheduling decisions for future periods, which can significantly reduce computation times and allow for near-real-time decision making.

The learning approach can be especially useful for large-scale microgrid systems with many prosumers and complex trading networks. With a large number of prosumers, the number of possible scheduling decisions can become very large, which can be difficult to solve using traditional optimisation methods. By using artificial neural networks to predict the optimal scheduling decisions, the computational burden can be significantly reduced, and the system can operate more efficiently.

In addition, the learning approach can also be used to improve the accuracy of the predicted results. By training the artificial neural networks on historical data, the model can learn the patterns and relationships between the inputs and outputs, which can lead to more accurate predictions. This is especially useful for systems with high variability and uncertainty, such as renewable energy sources, which can be difficult to predict accurately using traditional methods.

Therefore, the learning approach provides a powerful tool for optimising microgrid systems, which can help to improve the scalability, computational efficiency, and accuracy of the system. By combining traditional optimisation methods with machine learning techniques, it is possible to create highly efficient and effective microgrid systems that can help to reduce energy costs, improve sustainability, and promote a more efficient use of energy resources.

6.3.3.4 Smart Contract-Based Auction Mechanism

The proposed smart contract-based auction mechanism in the peer-to-peer trading platform is a versatile solution for both prosumers and microgrid-traders to trade energy or carbon allowance. This mechanism ensures standardised negotiation and self-enforcing of the smart contract, providing a transparent and efficient trading platform. The auction process involves several steps, namely initialisation, matching, bidding, withdrawal, and pay-to-seller, each of which is performed by a specific function in the smart contract. These functions are denoted as $f_{\text{init}}(\cdot)$, $f_{\text{match}}(\cdot)$, $f_{\text{bid}}(\cdot)$, $f_{\text{withdraw}}(\cdot)$, and $f_{\text{pay}}(\cdot)$, respectively. The smart contract-based auction mechanism is designed to ensure the integrity and transparency of the auction process. By using this mechanism, participants can trade energy or carbon allowance without intermediaries, reducing transaction costs and increasing the efficiency of the trading platform. Let $\mathcal{U}$ denote the index set of sellers, and $\mathcal{V}$ denote the index set of buyers.

Algorithm 4 Smart Contract Based Auction Procedure

1: **function: initialisation** $f_{\text{init}}(\cdot)$
2: input: $\text{id}_u, \varepsilon, \beta, m_u, s_u, b_{u,t}^{\min}, b_{u,t}^{\text{highest}}, \tau_u$
3: output: $\mathcal{O}_u$

4: **function: matching** $f_{\text{match}}(\cdot)$
5: **for** $v \in \mathcal{V}$ **do**
6:    find optimal offers combination $\mathcal{U}_v^*$ by (6.31) and (6.32)
7: **end for**

8: **function: bidding** $f_{\text{bid}}(\cdot)$
9: input: $\tau_{\text{now}}, b_v^*, m_v, b_{\text{balance},v}$
10: **while** $\tau_{\text{now}} \leq \tau_u, m_v = m_u, b_{u,t}^{\text{highest}} \cdot s_u < b_v^* \cdot s_u \leq b_{\text{balance},v}$ **do**
11:    submit bids and update the highest bidding price by (6.34)
12: **end while**
13: output: $b_{u,t}^{\text{highest}'}$

14: **function: withdrawal** $f_{\text{withdraw}}(\cdot)$
15: input: $\tau_{\text{now}}, b_v^*, b_{\text{balance},v}$
16: **while** $\tau_{\text{now}} > \tau, v \in \mathcal{V}, v \neq v^*$ **do**
17:    unsuccessful buyers withdraw their bids by (6.36)
18: **end while**
19: output: $b_{\text{balance},v}'$

20: **function: pay-to-seller** $f_{\text{pay}}(\cdot)$
21: input: $\tau_{\text{now}}, b_v^*, b_{\text{balance},u}$
22: **while** $\tau_{\text{now}} > \tau, v = v^*$ **do**
23:    pay the deposited highest bid to seller by (6.38)
24: **end while**
25: output: $b_{\text{balance},u}'$

The trading algorithm, as shown in Algorithm 4, is written in the Solidity language [38] and stored in the Ethereum blockchain [39]. Detailed steps of executing the auction are explained as:

Step 1: In the proposed smart contract-based auction mechanism, each seller initiates the auction process by calling the initialisation function $f_{\text{init}}(\cdot)$ from the smart contract. This function allows the seller to specify the necessary details about the auction, such as the seller address, trading type (energy or carbon allowance), seller type (prosumer or microgrid-trader), microgrid number, selling amount, minimal accepted bidding price, the currently highest bid, and the time of auction ended.

More specifically, for seller $u \in \mathcal{U}$, the offer $\mathcal{O}_u$ is initialised as follows:

$$\mathcal{O}_u = f_{\text{init}}\left(\text{id}_u, \varepsilon, \beta, m_u, s_u, b_{u,t}^{\min}, b_{u,t}^{\text{highest}}, \tau_u\right), \tag{6.30}$$

where $\text{id}u$ is the encrypted address of seller u, $\varepsilon \in 0, 1$ is a binary value indicating if the trading type is energy ($\varepsilon = 0$) or carbon allowance ($\varepsilon = 1$), $\beta \in 0, 1$ is a binary value indicating if the seller type is prosumer ($\beta = 0$) or microgrid-trader ($\beta = 1$),

m_u is the microgrid index of seller u, which enables buyers to find sellers in the same microgrid. s_u is the amount of energy or carbon allowance to be supplied by seller u, while $b_{u,t}^{\min}$ is the minimal accepted bidding price specified by seller u for the energy or carbon allowance to be provided at scheduling time t. At initialisation, $b_{u,t}^{\text{highest}}$ is equal to $b_{u,t}^{\min}$, as there are no bids yet. τ_u is the time of auction end specified by seller u. All the offers from sellers are stored and updated in the blockchain network, enabling buyers to access the information easily.

This initialisation step is crucial in setting up the auction and ensuring that all sellers provide the necessary information for buyers to make informed decisions. The use of a smart contract ensures that the information provided is standardised and that the auction process is transparent and secure. This step corresponds to the line 1-3 in Algorithm 4.

Step 2: In the proposed auction mechanism, each buyer is required to bid with a higher price than the currently highest bidding price over all the buyers. To help buyers submit their bids optimally, the matching function $f_{\text{match}}(\cdot)$ automatically matches the optimal offers combination according to certain criteria.

The matching function has two main objectives. First, it aims to ensure that the demand of energy or carbon allowance for a buyer can be met by the summation of selected offers. Second, it selects the optimal offers with the minimal summation of the currently highest bidding prices, allowing buyers to bid with minimal bidding prices.

The optimal offers combination for a buyer v can be obtained by solving the following optimisation problem:

$$\mathcal{U}v^* = \arg\min u : \sum_{u\in\mathcal{U}} b^{\text{highest}}u, t \cdot su, \tag{6.31}$$

subject to the constraint that the summation of the selling amounts of the selected optimal offers is greater than or equal to the demand of energy or carbon allowance of buyer v:

$$\sum_{u\in\mathcal{U}} s_u \geq d_v, \tag{6.32}$$

where $\mathcal{U}v^*$ is the set of optimal offers combination that can meet buyer v's demand with minimal required bidding prices, and dv is the demand of energy or carbon allowance of buyer v.

In other words, the matching function ensures that the selected optimal offers meet the buyer's demand while minimising the required bidding prices. By doing so, buyers can make informed decisions and participate in the auction process with minimal bidding prices, increasing the efficiency and fairness of the trading platform. This step corresponds to the line 4-7 in Algorithm 4.

Step 3: The bidding function $f_{\text{bid}}(\cdot)$ in the proposed auction mechanism enables buyers to submit their bids after fulfilling certain conditions. These conditions ensure that the auction is fair and transparent for all participants.

First, the bidding function checks that the auction is not ended, i.e., the current time τ_{now} is less than or equal to the time of auction end specified by the seller u, denoted as τ_u. Second, the function checks that the microgrid index of the buyer v, denoted as m_v, matches that of the seller u, i.e., $m_v = m_u$.

Finally, the function checks that the buyer has enough balance to provide a bid higher than the currently highest bidding price. This condition is expressed as:

$$b_{u,t}^{\text{highest}} \cdot s_u < b_v^* \cdot s_u \leq b_{\text{balance},v}, \tag{6.33}$$

where b_v^* is the optimal bidding price of buyer v obtained by solving the optimisation problems in the prosumer-centric algorithm or microgrid-trader-centric algorithm. $b_{\text{balance},v}$ is the account balance of the buyer v.

If all conditions are met, the buyer can successfully submit their bid, and the highest bidding price of the seller u's offer is updated using the following equation:

$$b_{u,t}^{\text{highest}'} = f_{\text{bid}}\left(\tau_{\text{now}}, b_v^*, m_v, b_{\text{balance},v}\right), \tag{6.34}$$

where $b_{u,t}^{\text{highest}'}$ is the updated currently highest bidding price for the energy or carbon allowance to be provided by the seller u at scheduling time t.

Before the auction ends, all the bids are frozen by the smart contract, which means that buyers are unable to withdraw their bids back to their account. This ensures that the auction process is fair and transparent, with all participants having an equal opportunity to bid on the available offers. This step corresponds to the line 8-13 in Algorithm 4.

Step 4: When the auction ends, i.e., the current time τ_{now} is greater than the time of auction end specified by the seller u, denoted as τ, the buyer with the highest bidding price wins the auction. This is expressed as:

$$v^* = \arg\max_v : \mathcal{B}_t, \tag{6.35}$$

where v^* is the buyer with the highest bidding price, and $\mathcal{B}_t$ is the set of bidding prices submitted by all buyers for the energy or carbon allowance provided at scheduling time t.

The rest of the unsuccessful buyers, denoted as $v \in \mathcal{V}, v \neq v^*$, can withdraw their previously submitted bids by calling the withdrawal function $f_{\text{withdraw}}(\cdot)$. The withdrawal function updates the account balance of the unsuccessful buyer by adding back the bidding price for the offer that was not won. This is expressed as:

$$b_{\text{balance},v}' = f_{\text{withdraw}}\left(\tau_{\text{now}}, b_v^*, b_{\text{balance},v}\right), \tag{6.36}$$

where $b_{\text{balance},v}'$ is the updated account balance of buyer v after withdrawing the bid for the seller u's offer. The updated account balance is calculated as:

$$b_{\text{balance},v}' = b_{\text{balance},v} + b_v^* \cdot s_u, \tag{6.37}$$

where b_v^* is the optimal bidding price of the buyer v and s_u is the selling amount of the offer from the seller u.

Overall, the withdrawal function allows unsuccessful buyers to withdraw their bids, increasing the transparency and fairness of the auction process. By doing so, buyers who did not win the auction can recover their account balance and participate in future auctions with minimal financial losses. This step corresponds to the line 14-19 in Algorithm 4.

Step 5: After the energy or carbon allowance is delivered and confirmed by querying the smart meter, the final highest bid for the offer u, denoted as $b_{u,t}^{\text{highest}*}$, is paid to the seller. The payment is made by calling the pay-to-seller function $f_{\text{pay}}(\cdot)$.

The pay-to-seller function updates the account balance of the seller u by adding the amount received for the winning bid. This is expressed as:

$$b'_{\text{balance},u} = f_{\text{pay}}\left(\tau_{\text{now}}, b_{u,t}^{\text{highest}*}, b_{\text{balance},u}\right), \qquad (6.38)$$

where $b'_{\text{balance},u}$ is the updated account balance of the seller u after receiving the payment. The updated account balance is calculated as:

$$b'_{\text{balance},u} = b_{\text{balance},u} + b_{u,t}^{\text{highest}*} \cdot s_u, \qquad (6.39)$$

where s_u is the selling amount of the offer from the seller u.

This step corresponds to the line 20-25 in Algorithm 4.

6.3.4 Case Studies

The proposed blockchain-based peer-to-peer trading framework has been evaluated through a series of case studies. The aim of these case studies is to demonstrate the effectiveness and efficiency of the proposed framework in supporting energy and carbon allowance trading between prosumers and microgrid-traders.

6.3.4.1 Simulation Setup and Data Availability

The proposed prosumer-centric algorithm and microgrid-trader-centric algorithm have been implemented using MATLAB. The proposed smart contract has been implemented in Solidity 0.6.0 and executed on the Remix-IDE. To ensure secure and transparent trading, individual deposit accounts have been created for each prosumer and microgrid-trader.

The testing environment for the proposed blockchain-based peer-to-peer trading framework is shown in Fig. 6.5. The simulations were conducted on a machine with an Intel^R Core^{TM} i9-9900K CPU running at 3.60 GHz.

The simulation results demonstrate the effectiveness of the proposed framework in supporting secure, transparent, and efficient energy and carbon allowance trading in microgrid systems. The simulation results show that the proposed framework can handle a large number of transactions and can effectively incentivise prosumers to generate and sell excess renewable energy.

In this study, we have used a modified version of the IEEE 37-bus distribution network, as depicted in Figure 6.6. The network has been partitioned into five interconnected microgrids, with each bus representing a prosumer. To create a more realistic simulation, we have replaced the static default data of generation and consumption from the IEEE 37-bus distribution network with dynamic data.

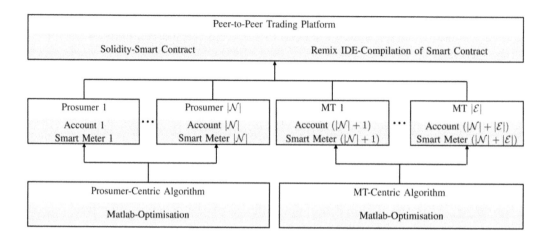

Figure 6.5 Overview of the testing environment for the blockchain-based peer-to-peer trading framework. The smart contract is written in the Solidity language and executed on the Remix-IDE. The prosumer-centric algorithm and microgrid-trader-centric algorithm are written in the MATLAB®. Individual deposit accounts are created for each prosumer and microgrid-trader.

Specifically, 7 solar photovoltaics, 4 diesel generators, 4 wind turbines, and 2 biomass generators have been arbitrarily assigned to each microgrid, and 33 loads are assigned to each bus. The demand data of residential loads has been collected using an EFERGY monitor hub and allocated to each prosumer, as shown in Figure 6.7. The solar generation data has been obtained from the UK rooftop solar generation of endpoint consumers.

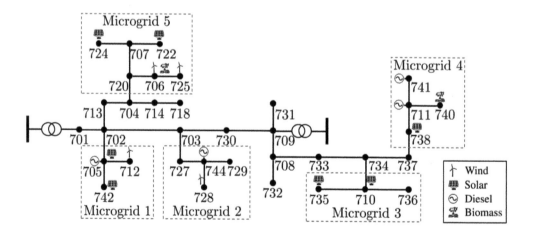

Figure 6.6 Modified IEEE 37-bus distribution network. The network is partitioned into 5 microgrids. Each bus represents a prosumer. 7 solar photovoltaics, 4 diesel generators, 4 wind turbines, and 2 biomass generators are arbitrarily assigned to each microgrid by connecting to prosumers' buses. 33 loads are assigned to each bus.

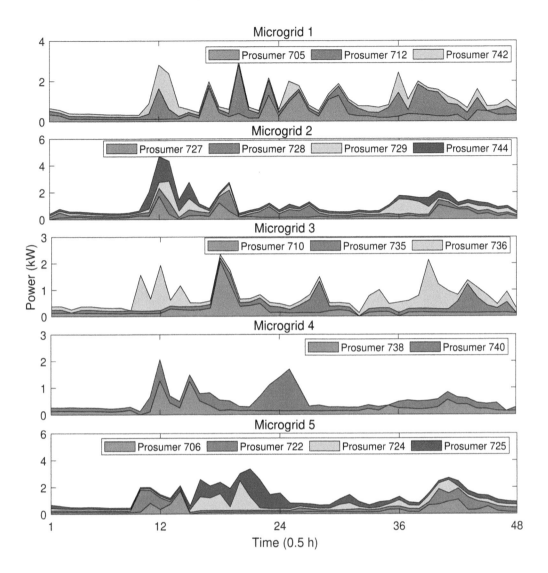

Figure 6.7 Demand allocation for prosumers and microgrids in the modified IEEE 37-bus distribution network.

To incorporate real-time states of GB generation, we have used data from the GridWatch. To scale down the generation of diesel, wind, and biomass, we have used the ratio of peak real-time demand to the peak static demand from the IEEE 37-bus distribution network. The total power outputs of each generation source have been equally allocated to the corresponding generators. The generation allocation for prosumers and microgrids in the modified IEEE 37-bus distribution network is shown in Figure 6.8.

The use of this modified IEEE 37-bus distribution network, along with the dynamic generation and consumption data, allows for a more accurate simulation of energy and carbon allowance trading in microgrids. By incorporating realistic data and scenarios, the simulation results provide a more comprehensive understanding

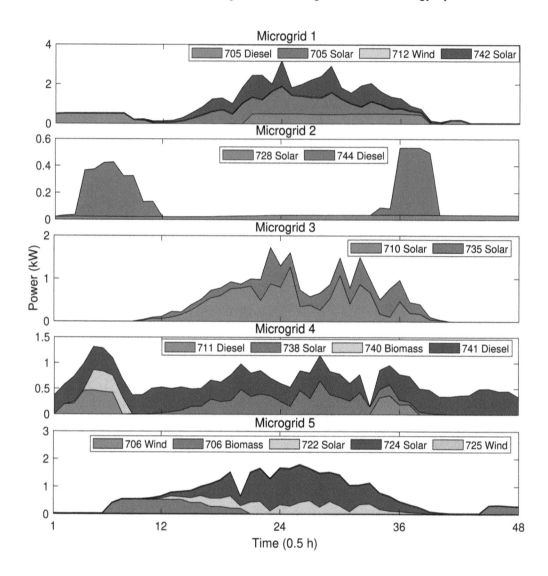

Figure 6.8 Generation allocation for prosumers and microgrids in the modified IEEE 37-bus distribution network.

of the performance and effectiveness of the proposed blockchain-based peer-to-peer trading framework.

To provide a more realistic simulation of the proposed blockchain-based peer-to-peer trading framework, we have obtained data on centralised prices of energy and carbon allowance from the UK energy retail market and the UK carbon market, respectively. Specifically, the centralised prices have been set as the minimal accepted bidding price of each seller during the auction process.

This design encourages more prosumers to sell their surplus energy or carbon allowance, and reduces the need to import from central markets. By allowing buyers to provide a higher price than the centralised prices through solving their own objective functions, the proposed mechanism promotes competition and encourages efficient trading. The use of realistic centralised prices adds to the accuracy and applicability of the simulation results, allowing for a more comprehensive evaluation of the proposed trading framework.

6.3.4.2 Balancing Performances of Energy and Carbon Allowance

To evaluate the performance of the proposed blockchain-based peer-to-peer trading framework, we compared it with two other trading schemes: centralised trading and aggregator-based trading as follows:

- The first scheme, known as centralised trading, involves trading of energy or carbon allowance exclusively on centralised markets. This scheme utilises the prices of energy and carbon allowance obtained from central markets.

- The second scheme, known as aggregator-based trading, follows the approach presented in [40]. In this scheme, the reshaping of energy behaviours is managed by relatively decentralised agents, known as aggregators, who aim to minimise the bills for buyers or maximise the profits for sellers. Aggregators then pay prosumers the monetary compensation for the reshaping, and the trading of energy or carbon allowance is only performed by these aggregators.

By comparing the proposed framework with these two trading schemes, we can evaluate the extent to which blockchain-based peer-to-peer trading can improve the efficiency and effectiveness of energy and carbon allowance trading. The results of these comparisons will provide insights into the potential benefits of blockchain technology for the energy and carbon markets.

Figure 6.9 shows the net power of the modified IEEE 37-bus distribution network, which is defined as the difference between the total power generation and the total power consumption. A positive net power indicates that the total generation is greater than the total demand, while a negative net power indicates the opposite, where the network has to import power from the main grid. The proposed peer-to-peer trading framework has resulted in a daily net energy summation of 0.99 kWh, which demonstrates a more balanced energy distribution, compared to the aggregator-based trading and centralised trading with net energy summations of -4.50 kWh and -46.44 kWh, respectively. These results indicate that the proposed framework is capable of achieving a better balance between energy generation and consumption, which is an essential aspect of modern power systems.

The surplus of carbon allowance in the modified IEEE 37-bus distribution network, which represents the total assigned carbon allowance minus the total carbon emissions, is presented in Figure 6.10. A positive surplus of carbon allowance indicates that the network's total carbon emissions are less than the total assigned carbon allowance, while a negative surplus indicates that the total carbon emissions exceed the total assigned carbon allowance.

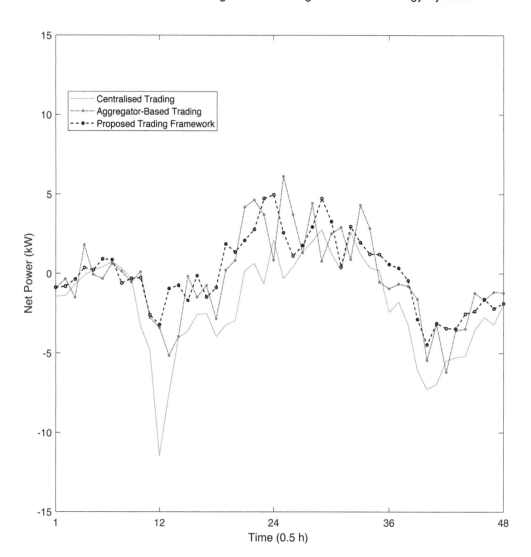

Figure 6.9 Net power of the modified IEEE 37-bus distribution network. The positive value of y-axis indicates the total generation is greater than the total demand. The negative value of y-axis indicates the total generation is less than the total demand. The x-axis indicates the scheduling time of day.

Through the proposed peer-to-peer trading framework, the total daily carbon emissions saved from the carbon allowance is 1465.90 g, which is approximately 6 times higher than the aggregator-based trading (385.91 g) and 9 times higher than the centralised trading (168.65 g). In particular, during the period from the thirty-sixth scheduling time to the forty-eighth scheduling time, the proposed framework achieves significantly more carbon savings compared to the other two trading schemes.

In contrast, while the aggregator-based trading also achieves carbon savings during this period, it results in carbon emissions exceeding the carbon allowance during the period from the twenty-second scheduling time to the thirty-fifth scheduling

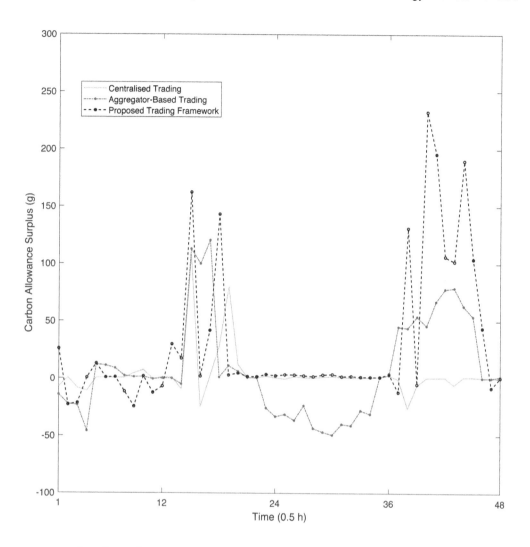

Figure 6.10 Surplus of carbon allowance of the modified IEEE 37-bus distribution network. The positive value of y-axis indicates the total carbon emissions are less than the total assigned carbon allowance. The negative value of y-axis indicates the total carbon emissions exceed the total assigned carbon allowance.

time. This highlights the advantage of the proposed peer-to-peer trading framework in achieving a more balanced and sustainable carbon emissions reduction across different periods.

6.3.4.3 Demonstration of Interface between Scheduling Algorithms and Smart Contract

The prosumer-centric algorithm yields the optimal energy scheduling and bidding prices for each individual prosumer in the microgrid. Figure 6.11 illustrates the comparison between the original net consumption and the scheduled net consumption, along with the optimal bidding prices for the scheduled energy trading.

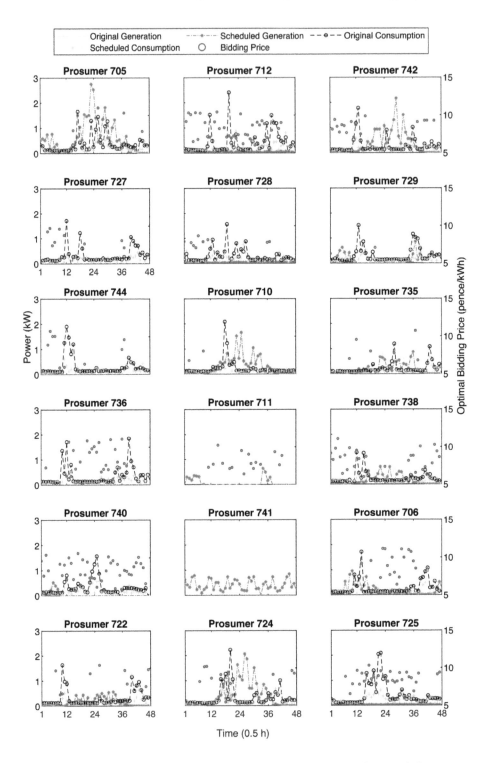

Figure 6.11 Optimal energy scheduling and bidding prices obtained by prosumer-centric algorithm. The left y axes indicate the power of original prosumption and scheduled prosumption of individual prosumers, and the right y axes indicate the optimal bidding prices. The x axes indicate the scheduling time of day.

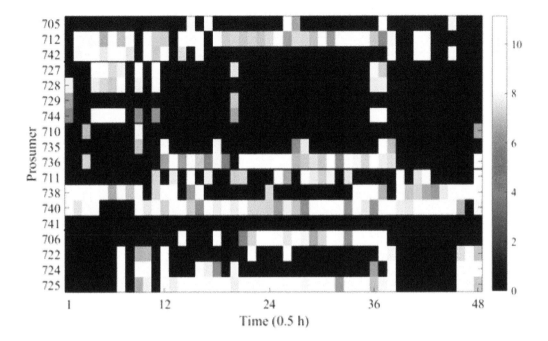

Figure 6.12 Optimal bidding prices of energy buyers as inputs of smart contract. The y-axis indicates the bus number of prosumers, assigned to corresponding microgrids. The x-axis indicates the scheduling time of day. The colourbar indicates the optimal bidding prices from each prosumer for a given 0.5 h scheduling interval. The scheduling interval without bidding price means there is no surplus energy on the microgrid to trade.

In scheduling intervals where all prosumers of a particular microgrid cannot generate surplus energy, there are no sellers or bidding prices.

The comparison between the scheduled net consumption and the original net consumption reveals that during peak demand periods, i.e., from the twelfth to the thirty-sixth scheduling time, generation is scheduled to increase while consumption is shifted to the off-peak periods, i.e., the remaining scheduling time. This ensures a balance between energy supply and demand and reduces the reliance on the main grid during peak demand periods.

Furthermore, during periods where prosumers experience high power consumption and low power generation and become energy buyers, the scheduling algorithm shifts the energy purchasing to off-peak demand periods, and the bidding prices are stabilised at around 10 pence/kWh without any significant increase. The slight fluctuations of the bidding prices dynamically reflect the actual supply–demand balance in energy markets, ensuring that energy prices remain stable and affordable for prosumers.

The interaction between the scheduling decisions and the smart contract is illustrated in Figure 6.12. After obtaining the optimal bidding prices for prosumers as buyers through solving the prosumer-centric algorithm, the bidding prices are automatically sent to the smart contract for auction. It can be observed that auctions are conducted over all scheduling intervals of the day in microgrid 4, whereas they are conducted only at a few scheduling intervals in microgrid 2. This is due to the limited generation capacity of microgrid 2, which is not sufficient to meet its demand, and thus the microgrid-trader 2 has to help its prosumers buy energy from other microgrid-traders. In addition, the proposed peer-to-peer trading framework helps to stabilise selling prices between 6 pence/kWh and 10 pence/kWh over all scheduling intervals, which is different from the aggregator-based trading scheme with dramatic peak and off-peak prices. The auction prices determined by individual prosumers accurately reflect the actual supply–demand relationship of prosumers.

6.3.4.4 Demonstration of Smart Contract Execution

The proposed auction mechanism, which is executed as a smart contract on the Ethereum blockchain, is illustrated in Figure 6.13 for microgrid 5. In this scenario, prosumers at bus 706 and bus 724 act as energy sellers, providing 319 Wh and 109 Wh of energy, respectively. Additionally, prosumers at bus 706, bus 724, and bus 725 are carbon allowance sellers, providing 7 g, 113 g, and 123 g of carbon allowances, respectively. To participate in the auction, the sellers first call the initialisation function from the full node to specify the offer conditions. On the other hand, prosumers at bus 722 and bus 725 are energy buyers with a demand of 419 Wh and 202 Wh, respectively. Prosumer at bus 722 is a carbon allowance buyer with a demand of 117 g. The proposed matching criteria are then used to match the bids and offers. The auction is carried out according to the proposed auction rules, and the buyer with the highest bidding price wins the auction. The unsuccessful buyers then withdraw their bids by calling the withdrawal function, and the winning buyer pays the seller by calling the pay-to-seller function.

Through the proposed blockchain-based peer-to-peer trading framework, energy and carbon allowances can be traded simultaneously, and individual prosumers can set their own bidding/selling prices, leading to a more accurate and efficient supply-demand balance. The use of smart contracts on the Ethereum blockchain ensures transparency, security, and self-executing of the auction process, reducing the need for intermediaries and ensuring trust between the participating parties.

To fulfil the carbon allowance demand of 117 g from the prosumer at bus 722, there are two available options. The first option is to purchase 123 g of carbon allowance from the prosumer at bus 725 at a bidding price of 4 pence/kg. The second option is to purchase 113 g of carbon allowance from the prosumer at bus 724 at a bidding price of 3 pence/kg and another 7 g of carbon allowance from the prosumer at bus 706 at a bidding price of 3 pence/kg. The matching criteria is applied to select the most suitable option, which in this case is the second option.

In the auction of energy, there were multiple buyers with multiple sellers. Prosumers at bus 725 and bus 722 attempted to bid as buyers for the offer of selling 109 Wh energy by the prosumer at bus 724. The prosumer at bus 725 won the auction with the highest bidding price of 7 pence/kWh. The unsuccessful buyer at bus 722 then called the withdrawal function from the full node to withdraw its bid. Once the

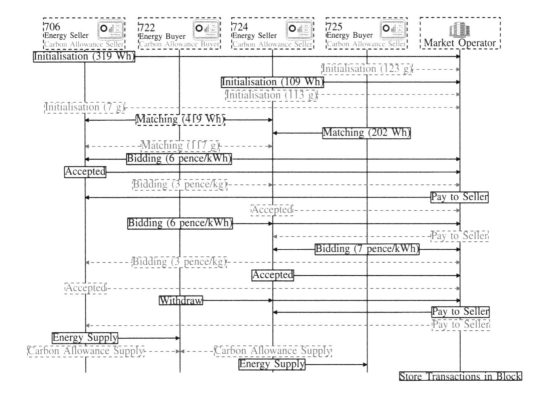

Figure 6.13 Execution of smart contract-based auction on the peer-to-peer trading platform. The black line is the execution of the energy trading, and the dashed blue line is the execution of the carbon allowance trading.

smart contract confirmed that the energy or carbon allowance was supplied, it paid the sellers with the highest bids.

For the auction of carbon allowance, there was a single buyer with multiple sellers. To meet the demand of 117 g carbon allowance from the prosumer at bus 722, the prosumer had two options: either buy 123 g carbon allowance from the prosumer at bus 725 with a bidding price of 4 pence/kg, or buy 113 g from the prosumer at bus 724 with a bidding price of 3 pence/kg and buy another 7 g from the prosumer at bus 706 with a bidding price of 3 pence/kg. According to the matching criteria, the second option was selected.

After the auction of energy and carbon allowance, there was residual 123 g carbon allowance from the prosumer at bus 725, 93 Wh energy demand from prosumer at bus 725, and 100 Wh energy demand from prosumer at bus 722. These were aggregated by the microgrid-trader 5 to trade with other microgrid-traders.

6.3.5 Research Summary

This example research proposes a blockchain-based peer-to-peer trading framework that aims to achieve local energy balance and reduce carbon emissions on distribution networks. By allowing for simultaneous energy and carbon allowance exchange, this framework enables prosumers to participate in low-carbon energy trading and benefit from the incentive allocation mechanism, which is based on specific energy behaviours. To generate optimal bidding/selling prices of prosumers and make energy reshaping decisions, the chapter proposes two algorithms: the prosumer-centric algorithm and microgrid-trader-centric algorithm.

The auction takes place under a standardised and self-enforcing smart contract, which ensures transparency, security, and fairness in the trading process. To demonstrate the effectiveness of the proposed trading framework, case studies were conducted on the modified IEEE 37-bus distribution network. The results show that the proposed trading framework can export 0.99 kWh of daily energy to the main grid and save 1465.90 g of daily carbon emissions.

Compared to other trading schemes such as centralised trading and aggregator-based trading, the proposed framework outperforms in balancing energy and carbon allowance. The proposed scheduling algorithms drive up prosumers' self-generation, shift away peak demand, and stabilise energy prices below 10 pence/kWh. The auction prices of individual prosumers are accurately targeted to the actual supply-demand relationship of prosumers.

Finally, this example research demonstrates the execution of the smart contract on the Ethereum blockchain and the interface between the scheduling algorithms and the smart contract. In summary, the proposed framework provides an effective and scalable solution for low-carbon energy trading on distribution networks.

6.4 EXAMPLE RESEARCH 2: BLOCKCHAIN-SECURED PEER-TO-PEER ENERGY TRADING

Climate change is driving the urgent need to transition to a sustainable energy system based on distributed renewable energy sources. As energy is increasingly generated and traded among distributed prosumers, it is crucial to develop a carbon pricing scheme to incentivise carbon reduction efforts. However, the transfer of carbon responsibilities and allowances from large-scale energy suppliers to prosumers presents significant challenges, including energy imbalance, uneven carbon reduction, and privacy concerns in centralised trading markets.

To address these challenges, we propose a fully decentralised blockchain-based peer-to-peer trading scheme that combines energy and carbon markets. Our approach utilises a pay-to-public-key-hash with multiple signatures as a transaction standard to enhance transaction security and reduce storage burdens on senders. Furthermore, we incorporate a script that is hashed during the wallet address generation for each new transaction to protect residential privacy.

To promote carbon reduction among distributed prosumers, the carbon accounting method and corresponding incentive mechanism used in the previous example research are implemented to evaluate emission behaviours. This approach incentivises prosumers to reduce their carbon footprint and contributes to overall carbon reduction efforts.

Case studies demonstrate that our proposed scheme leads to reduced costs and carbon emissions compared to centralised trading systems and existing blockchain-based trading schemes. Our approach offers a more secure, efficient, and privacy-preserving trading platform that enables distributed prosumers to trade energy and carbon credits seamlessly. Overall, this paper provides a comprehensive solution to address the challenges of the energy and carbon markets' integration and promotes the transition to a sustainable energy system.

The rest of this example research is organised as follows: Sub-Section 6.4.1 introduces the backgrounds, challenges and contributions of the proposed work. The peer-to-peer trading framework and specific details on transaction standard and address generation are discussed in Sub-Section 6.4.2. Sub-Section 6.4.3 introduces the procedures of the peer-to-peer trading coupling energy and carbon markets. Sub-Section 6.4.4 provides case studies to demonstrate the effectiveness of the proposed model. Sub-Section 6.4.6 concludes this example research.

6.4.1 Introduction

In today's energy sector, majority of the world's power demand is met through centralised power plants that rely on fossil fuels such as coal, gas, and oil. Centralised fossil fuel-based power plants account for roughly 80% of global energy supply. However, the combustion of fossil fuels leads to enormous carbon emissions, contributing to air pollution and irreversible climate change effects. Additionally, there are significant energy losses during long-distance transmission, further reducing transmission efficiency.

To address these environmental challenges, policymakers are taking a two-pronged approach. Firstly, they are facilitating the integration of distributed renewable energy sources into distribution systems to mitigate carbon emissions and improve transmission efficiency. This move towards distributed renewable energy sources includes various sources such as wind, solar, and hydroelectricity. With the advancement in technology, distributed renewable energy sources are becoming more efficient and cost-effective, making them a more practical solution for meeting energy demands. Secondly, policymakers are formulating a carbon pricing scheme to incentivise the reduction of carbon emissions [7]. Carbon pricing is a market-based climate policy that aims to charge carbon producers for the carbon allowances they require to operate. The goal is to phase out power plants with extreme high carbon intensities, while encouraging the use of low-emission technologies. This policy shift can drive innovation in clean energy technologies, create jobs in the renewable energy sector, and foster sustainable economic growth.

Hence, the integration of distributed renewable energy sources and the implementation of carbon pricing schemes are vital steps towards mitigating climate change and promoting sustainable development. By adopting a more sustainable approach to energy production and consumption, we can help reduce carbon emissions and mitigate the environmental impact of our energy use.

As distributed renewable energy sources become more prevalent, prosumers are playing an increasingly important role in reducing carbon emissions. This shift in responsibility requires an overhaul of traditional carbon accounting practices, particularly as new system structures emerge from peer-to-peer energy trading. To fully embrace this transition, there is an opportunity to reform carbon emission trading by incorporating it into decentralised energy trading.

The creation of a decentralised carbon market is desirable to facilitate local carbon pricing that can better incentivise carbon reduction among prosumers. Rather than designing separate energy and carbon markets, coupling both markets can be a more efficient approach. This is because the carbon cost is directly linked to the operational costs of energy generation, making it easier to incentivise the reduction of carbon emissions.

In this approach, the coupling of energy and carbon markets involves the use of blockchain technology to enable peer-to-peer energy and carbon trading between prosumers. This system provides a secure and efficient platform for prosumers to trade energy and carbon credits, while ensuring that the carbon cost is integrated into the operational costs of energy generation. Furthermore, the coupling of energy and carbon markets offers an opportunity to incentivise prosumers to reduce their carbon footprint by offering them incentives for adopting low-carbon technologies or practices. This can help to promote the widespread adoption of renewable energy technologies, and drive a sustainable transition to a low-carbon economy.

This example research approaches a novel peer-to-peer trading scheme for prosumers to exchange both energy and carbon allowance with contributions as follows:

- Our proposed model involves a decentralised, peer-to-peer energy and carbon trading scheme that utilises blockchain technologies to accommodate the engagement of prosumers. The integration of prosumers into the energy and carbon markets helps to improve local energy imbalance and reduce carbon emissions. This approach offers a more efficient and cost-effective solution compared to centralised energy and carbon markets, while also promoting the transition to a sustainable energy system.

- We have designed a monetary incentive mechanism for carbon reduction that helps to realise the revenue neutralisation of policymakers without intervention in the bidding process. This approach incentivises prosumers to reduce their carbon footprint, promotes the adoption of low-carbon technologies, and drives the transition towards a sustainable energy system. By incentivising carbon reduction efforts, we can help to mitigate climate change and promote sustainable economic growth.

- To ensure a fair allocation of carbon allowance and incentives, we evaluate the carbon responsibilities of prosumers in distribution systems. This approach helps to incentivise prosumers to reduce their carbon emissions, and contributes to overall carbon reduction efforts. By evaluating the carbon responsibilities of prosumers, we can ensure a fair allocation of carbon allowances and incentives, and help to promote the widespread adoption of low-carbon technologies.

- Our proposed trading scheme offers a more secure transaction and residential privacy compared to centralised trading systems. Case studies demonstrate that our approach can achieve a more secure transaction process, protecting the

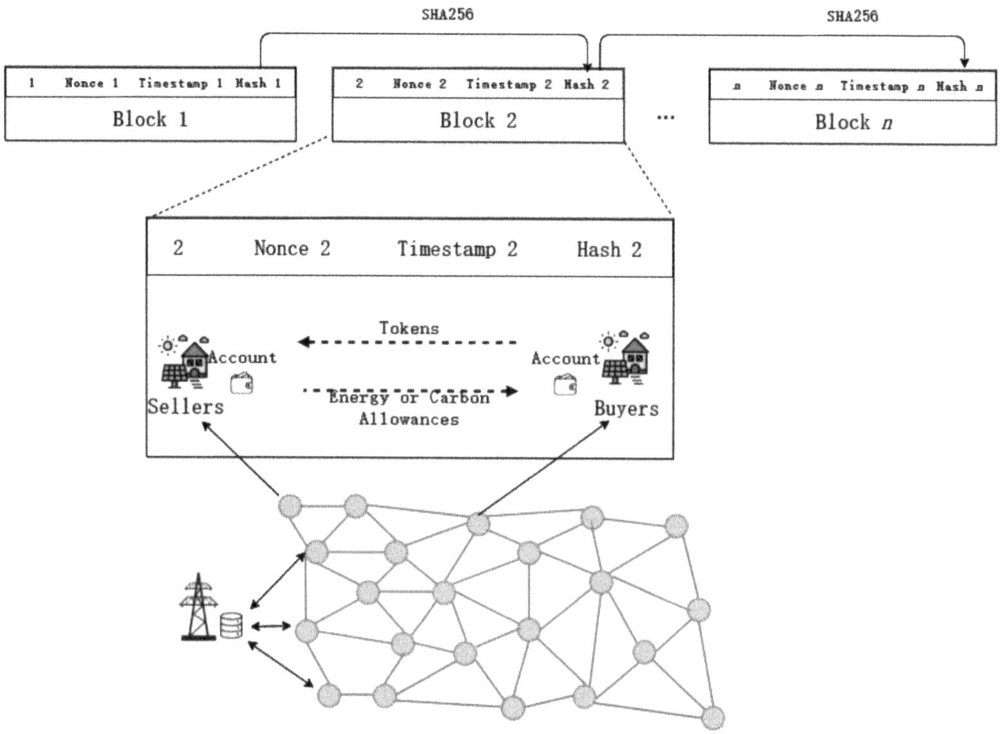

Figure 6.14 Schematic illustration of the blockchain-based peer-to-peer energy trading scheme. Prosumers directly communicate with each other to exchange energy and carbon allowance. During each transaction, the asset ownership in terms of both token and energy is exchanged after signing and broadcasting the encrypted trading outcomes to the network.

privacy of residential consumers. Additionally, the proposed trading scheme offers enhanced security features, such as pay-to-public-key-hash with multiple signatures, and a hashed script during wallet address generation to protect residential privacy.

6.4.2 System Model

In this section, we present the proposed scheme for a blockchain-based, peer-to-peer energy trading system, which leverages the unique features of blockchain technology to enable secure, decentralised transactions. In particular, we focus on the integration of distributed renewable energy sources and the evaluation of carbon responsibilities, which are key factors in the transition to a sustainable energy system.

6.4.2.1 Peer-to-Peer Trading Framework

Figure 6.14 provides a schematic illustration of our proposed blockchain-based, peer-to-peer energy trading scheme. In this framework, prosumers can directly communicate with each other to exchange energy and carbon allowances without the need for intermediaries.

To facilitate secure and transparent transactions, we use blockchain technology to store encrypted information, including energy and carbon allowance transactions, pricing signals, incentives, prosumer addresses, demand profiles, trading records, and timestamps. All participants in the blockchain network can access and validate this information, ensuring its authenticity and accuracy.

The use of blockchain technology also provides a more efficient and cost-effective way to manage transactions. With no intermediaries involved, transaction fees are minimised, and the process is streamlined, providing a more user-friendly experience for prosumers.

In addition to facilitating energy and carbon trading, the proposed system also promotes the adoption of low-carbon technologies and practices by incentivising carbon reduction efforts. The use of a novel carbon accounting method and corresponding incentive mechanism helps to evaluate the carbon responsibilities of prosumers and allocate carbon allowances and incentives fairly.

In the proposed blockchain-based, peer-to-peer energy trading scheme, transactions involve the exchange of asset ownership for energy and carbon credits. Each transaction is confirmed by the signature of the buyer and seller and then broadcast to the entire network. Multiple transactions are then structured into a block, with each block being chronologically chained to the previous block by incorporating the hash value of the previous block. This chain of blocks is commonly referred to as a blockchain.

The blockchain network is collectively secured through the consensus of the proof-of-work algorithm [41], which utilises the SHA 256 algorithm to solve a cryptographic puzzle. This algorithm takes the block header as an input and returns a fixed-length digest that serves as a unique identifier for the block. The digest is determined by the specially mined nonce, which adds an element of randomness to the algorithm.

The chaining of blocks and the computational difficulty of solving the mining puzzle provide robust security measures to protect against tampering and fraud in the blockchain network. If a malicious node attempts to modify the data in one block, it would result in an unverified block that cannot be accepted by subsequent blocks in the chain. If a malicious node attempts to tamper with data in one block and subsequently all following blocks, the computational difficulty of solving the mining puzzle makes it virtually impossible to carry out such an attack. This feature ensures that the data stored in blocks, including transactions, account addresses, and account states, is verifiable, traceable, and tamper-resistant. The use of blockchain technology provides a secure, efficient, and transparent platform for energy and carbon trading.

When prosumers communicate with each other for energy trading, the private key of the sender is used to sign the transaction or message. The receiving address is generated by the public key of the receiver. This approach ensures that the transactions are authenticated and authorised by the sender, and can be traced to their source.

To maintain the privacy of prosumers, our system uses hashed scripts during wallet address generation [42], which obfuscates the actual address of the prosumer.

This feature protects the residential privacy of prosumers and ensures that their sensitive information remains confidential.

In addition to protecting residential privacy, our system also prevents double spending attacks [43]. Each transaction is collectively voted on by every node in the blockchain network, which ensures that the transaction is valid and that the asset ownership of token and energy is transferred securely. By using a consensus mechanism that involves all nodes in the network, we can ensure that the transaction is authenticated and recorded accurately in the blockchain.

Therefore, the use of private and public keys, hashed scripts, and a consensus mechanism provides a secure, efficient, and privacy-preserving trading platform for energy and carbon credits. This approach enables prosumers to trade energy and carbon credits securely and transparently, while ensuring that their privacy is protected and that double spending attacks are prevented.

6.4.2.2 Transaction Standard

To enable prosumers to exchange energy or carbon credits in a secure and efficient manner, a standard transaction protocol has been designed based on existing blockchain technologies. The designed transaction protocol utilises several established transaction scripts, including pay-to-public-key-hash (P2PKH), pay-to-public-key (P2PK), multiple signatures (MS), and pay-to-script-hash (P2SH), which are commonly used in blockchain systems like Bitcoin [44]. Details of these scripts are explained as follows:

- The P2PKH script enables the sender of a transaction to specify the recipient's address by hashing their public key. This script is widely used in blockchain systems to provide a secure and efficient way of transferring funds without exposing the recipient's public key.

- The P2PK script allows the sender of a transaction to specify the recipient's public key directly, without hashing it. This script is used in certain blockchain systems to enable faster transaction processing times and lower transaction fees.

- The MS script enables multiple parties to sign a transaction, which adds an extra layer of security and reduces the risk of fraud. This script is often used in blockchain systems to enable more complex transactions that involve multiple parties.

- The P2SH script allows the sender of a transaction to specify a script that must be satisfied in order for the transaction to be executed. This script is used in certain blockchain systems to enable more advanced smart contract functionality, which can be used to implement more complex transactions and enable more sophisticated financial instruments.

When prosumers trade energy or carbon allowances, the buyer creates a script with the specified transferring amount and the account address of the seller. At the same time, the seller creates a script with their signature to authorise the transaction.

Once a transaction is initiated, it is broadcast over the blockchain network so that the scripts and transactions can be verified by every node in the network. The transaction is then included in a block and chronologically chained to the previous block, ensuring that the transaction is secure, transparent, and tamper-proof.

During the transaction process, the use of a public ledger ensures that every transaction is visible to all participants in the network, allowing for efficient and accurate auditing of energy and carbon credits. The scripts and transactions can be verified by every node in the network, ensuring that they are authentic and authorized. This feature ensures that every transaction is conducted with transparency, and that all parties have access to accurate and up-to-date information about their energy and carbon credit balances.

The proposed trading system utilises the transaction standard of P2PKH and P2SH to enable secure and efficient transactions between prosumers. In the case of P2PKH, the script of the buyer is generated using the public key of the seller, which is known as the scriptPubKey. This script is further hashed to create a unique address for the buyer. The script of the seller is generated using their private key, which serves as a unique identity of the seller, known as the scriptSig. On the other hand, the P2SH standard replaces the scriptPubKey of the buyer with a redeem script of the seller. This redeem script specifies the conditions under which a transaction can be redeemed [45]. If the hash of the seller matches the hash generated by the redeem script of the seller, and the signature is verified, the transaction is considered valid.

This design ensures a more secure transaction, as the seller, who is more concerned about receiving the token, monitors the success of the transaction instead of the buyer. This approach also reduces the storage requirement of the buyer, as there is no need to generate a script.

To ensure the security of transactions in the proposed trading system, all nodes in the network must collectively verify each transaction using the MS standard. This feature prevents malicious nodes from tampering with the transaction by requiring a minimum number of signatures to match the corresponding public keys.

The condition for a valid MS is that a minimum of $|\mathcal{P}|$ signatures must match $|\mathcal{Q}|$ public keys. This approach ensures that all transactions are authorised by the appropriate parties and that no unauthorized transactions can be executed.

To further simplify the encoding process, we use the Base 58 encoding standard [46] to encode the script under the P2SH transaction standard. This approach reduces the complexity of encoding the script, making it easier for prosumers to execute transactions in a timely and efficient manner.

6.4.2.3 Address Generation

The authenticity of transactions is guaranteed by generating a unique address for each prosumer's account using a public-private key pair. The public key is generated from the private key, and then used to generate the public key hash, which serves as the unique address for the account.

Figure 6.15 Flowchart of procedures for generating address of an account.

To ensure the security of the account address, the encryption process uses the SHA256 algorithm, which is a one-way cryptographic hash function that makes it irreversible. This means that given the public key or address, the private key cannot be decrypted, ensuring that each account is secure and tamper-proof.

By using public-private key pairs and the SHA256 algorithm, our blockchain-based, peer-to-peer energy trading system ensures that each account is authentic and secure. This approach provides a robust platform for energy and carbon trading that benefits both prosumers and the environment.

The procedures for generating address of an account are shown in Figure 6.15, with details described as follows:

- *Step 1*: A fixed-length private key is randomly generated using a cryptographic random number generator. This private key is then used to generate a corresponding public key using the elliptic curve digital signature algorithm, secp256k1 [47], from the asymmetric cryptography. This public key serves as a unique identifier for the prosumer's account.

- *Step 2*: According to the transaction standard P2SH, a redeem script is encrypted to generate a script hash through using the SHA256 and RIPEMD160 hash functions. This script hash is used to encode the transaction script, ensuring the security and authenticity of each transaction.

- *Step 3*: To ensure the validity and avoid typographical errors in the script, the result of double SHA256 is truncated to the first four bytes to generate a checksum. The version number and checksum are then concatenated with the hash of the script using the Base58 encoding standard to generate a unique address for the prosumer's account. This address serves as a secure and tamper-proof identifier for the prosumer's account, and is used in all transactions conducted within the system.

By using a public-private key pair and the P2SH transaction standard, our blockchain-based, peer-to-peer energy trading system ensures the security and authenticity of all transactions, while providing a robust platform for energy and carbon trading that benefits both prosumers and the environment.

6.4.3 Energy and Carbon Markets Coupling Theory

This subsection outlines the trading procedure for the proposed blockchain-based peer-to-peer energy and carbon markets. Unlike conventional markets, which require a central authority to match bids and offers and publish unique market clearing prices, the decentralised feature of blockchain technology allows prosumers to flexibly choose offers and conduct transactions directly with each other.

In addition, our proposed system incorporates the coupled emission trading mechanism, enabling incentive mechanisms to be applied to the trading process for carbon mitigation in the consumption side. This provides a more comprehensive approach to energy and carbon trading, encouraging prosumers to reduce their carbon footprint and promoting a more sustainable energy system.

To illustrate the trading procedure, we assume a peer-to-peer trading system for current transactions, and use G, K, U, V to denote the sets of energy sellers, energy buyers, carbon allowance sellers, and carbon allowance buyers, indexed by g, k, u, v, respectively, where $g, k, u, v \in N$. Prosumers can participate in the trading process by submitting their offers, which are then made available for other prosumers to review and accept.

Once a prosumer has found an offer they wish to accept, they can initiate a transaction using the standard transaction protocol, as described in earlier sub-sections. The transaction is then broadcast over the blockchain network, where it is verified and validated by other nodes in the network. Upon successful validation, the energy and/or carbon allowance is transferred from the seller's account to the buyer's account, completing the transaction.

Through this peer-to-peer trading procedure, prosumers have more control over their energy and carbon trading, enabling them to choose the most suitable offers and participate in a more sustainable energy system. By incorporating the incentive mechanism for carbon mitigation, our proposed system encourages prosumers to reduce their carbon footprint, promoting a more environmentally-friendly energy market.

The detailed trading procedures are explained as follows:

- *Step 1:* The trading process begins with the generation of unique addresses for each of the parties involved in the transaction, including the energy seller A_g, energy buyer A_k, carbon allowance seller A_u, and carbon allowance buyer A_v.

- *Step 2:* The energy buyer k announces their intended energy demand E_k and their bid π_k along with their address A_k to the blockchain network for verification of token ownership. The overall energy demand of the network is increased by the sum of all announced energy demands $\sum_{k=1}^{K} E_k$. Encrypted information key pairs for broadcasting include

$$I_{k,1} = \text{hash}\{\text{Rscript}_k|E_k|\pi_k|\text{timestamp}\}, \tag{6.40}$$

and

$$I_{k,2} = \text{hash}\{I_{k,1}|\delta_k\}, \tag{6.41}$$

where $I_{k,1}$ is a static key for verifying the ownership of tokens, Rscript_k is the redemption script of buyer k, and δ_k is a random number for generating $I_{k,2}$.

- *Step 3:* The system servers perform power flow tracing, carbon flow tracing, and reduction incentive calculation as introduced in Section 6.3. The required carbon allowances C_n and the amount of monetary compensation M_n are quantified and transmitted to the specific prosumer n.

- *Step 4:* The carbon allowance buyer v announces their required allowances C_v and their bid π_v along with its address to the blockchain network. Encrypted information key pairs include

$$I_{v,1} = \text{hash}\{\text{Rscript}_v|C_v|\pi_v|\text{timestamp}\}, \tag{6.42}$$

and

$$I_{v,2} = \text{hash}\{I_{v,1}|\delta_v\}. \tag{6.43}$$

- *Step 5:* The energy and carbon allowance sellers announce their intended supplies E_g and C_u, respectively, along with their offers π_g and π_u and their addresses to the blockchain network for verification. Encrypted information key pairs include

$$I_{g,1} = \text{hash}\{\text{Rscript}_g|E_g|\pi_g|\text{timestamp}\}, \tag{6.44}$$

or

$$I_{u,1} = \text{hash}\{\text{Rscript}_u|C_u|\pi_u|\text{timestamp}\}, \tag{6.45}$$

and

$$I_{g,2} = \text{hash}\{I_{g,1}|\delta_g\}, \tag{6.46}$$

or

$$I_{u,2} = \text{hash}\{I_{u,1}|\delta_u\}. \tag{6.47}$$

The keys are used as locks that can only be unlocked by the senders and receivers' identities to prevent double spending of tokens or energy and carbon allowances.

- *Step 6:* The system servers update the database of the offers π'_u and bids π'_v of carbon allowances by adding the monetary compensation to the sellers' original offers and the buyers' original bids, such that

$$\pi'_u = \pi_u + M_u, \tag{6.48}$$

and

$$\pi'_v = \pi_v - M_v, \tag{6.49}$$

before sorting them in sequence and publishing them to the auction board.

- *Step 7:* The buyers receive a list of filtered offers and corresponding addresses relevant to their queries by conditions: $E_g \geq E_k$ (or $C_u \geq C_v$), and select potential suppliers.

- *Step 8:* Each of the potential suppliers opens a transmission channel and feeds energy into the peer-to-peer network, before generating MS redemption scripts to note the amount and receivers. Upon receiving the redemption scripts, the buyers hash them and specify purchasing tokens. These MS transactions are broadcast to all nodes of networks for signing. If the signature script matches P2SH address, the transaction are validated by networks to transfer the ownership.

6.4.4 Case Studies

Case studies have been conducted to demonstrate the performance of the proposed trading scheme. Proposed scheme is applied in adjusted IEEE 14-bus test system which consists of 7 prosumers with DRESs including 4 solar, 2 wind, 1 biomass, and 4 vehicle-to-grid. The proposed peer-to-peer energy and carbon trading scheme has been evaluated through case studies, in which the scheme is implemented on the adjusted IEEE 14-bus test system. This test system is composed of 7 prosumers with distributed renewable energy sources, including 4 solar panels, 2 wind turbines, 1 biomass, and 4 vehicle-to-grid. These prosumers are assumed to be able to participate in the energy and carbon markets, either as buyers or sellers, and are expected to optimise their profits while meeting their energy demands and carbon emission reduction targets.

The case studies aim to evaluate the performance of the proposed trading scheme in terms of its ability to achieve efficient energy and carbon trading among prosumers, as well as to evaluate the accuracy and security of the proposed blockchain-based system. The simulation results will demonstrate the effectiveness of the proposed scheme in achieving carbon emission reduction and improving energy efficiency in a decentralised system, as well as in ensuring the privacy and security of the transactions.

6.4.5 Evaluation of Decentralised Trading Scheme

To evaluate the performance of the proposed fully decentralised peer-to-peer energy and carbon trading scheme, a comparison with conventional centralised trading was conducted in terms of environmental, economic, and security benefits. The results of the performance evaluation are presented in TABLE 6.2. The cost coefficients for each of the sources were multiplied with the power generation to evaluate the generating costs. The net carbon emissions were evaluated using the same method in Section 6.3. Power flow analysis was performed to obtain the transmission loss. The comparison showed that the decentralised trading system outperforms the centralised trading system in all dimensions, particularly in cost and carbon emissions reduction. Despite the fact that the power flow is not optimised in the peer-to-peer trading system, the transmission loss is still reduced as peers prefer to trade with neighbourhoods considering costs of using power grids.

Table 6.2 Multi-criteria evaluation of environmental, economic, and security benefits for proposed fully decentralised peer-to-peer energy and carbon trading scheme and conventional centralised trading

	Cost [£]	Net Carbon Emissions (kg)	Transmission Loss (kW)
Centralised Trading	331.63	104.84	302.78
Decentralised Peer-to-Peer Trading	142.98	42.23	298.13

The initial carbon allowances were distributed based on the carbon intensities of the prosumers. When their energy behaviours cause a positive net carbon emissions, they will purchase carbon allowances for the next half hour. The distribution of generation carbon emissions, transmission and distribution carbon emissions, consumption carbon emissions, net carbon emissions, carbon allowances, and compensation for 11 prosumers in half-hour intervals are presented in Figure 6.16. Each column represents the distribution of carbon emissions and monetary compensation in the peer-to-peer networks. The dark colour indicates a lower value, whereas the bright colour represents a higher value. The distribution of localized carbon emissions caused by energy behaviours in the blockchain network is reflected in this figure, and a fair allocation of monetary compensation can be formulated.

The proposed peer-to-peer energy and carbon trading system not only offers environmental and economic benefits but also ensures high levels of security and residential privacy. The use of encrypted residential addresses guarantees that only the trading type and amount can be traced and reviewed by the blockchain network, preventing any potential leakage of sensitive information. Moreover, the P2SH transaction standard with multi-signature effectively prevents double spending of tokens, carbon allowances, and energy, providing a more secure and reliable trading platform compared to the conventional centralised trading system. This robust security framework ensures the trust and confidence of prosumers in the blockchain network, encouraging their active participation and contribution towards a sustainable and efficient energy future.

6.4.5.1 Peer-to-Peer Trading

The data related to each transaction, including the encrypted trading addresses, trading type, and trading amount, are stored in the data cell of that particular transaction. An example of available bids and offers involving the proposed monetary incentive mechanism for the second settlement period is presented in TABLE 6.3. These bids and offers are published in the trading platform for participants to select from. Only participants who produce a positive net carbon emissions are allowed to participate in the carbon allowances trading, as their energy behaviours are the direct source of carbon emissions.

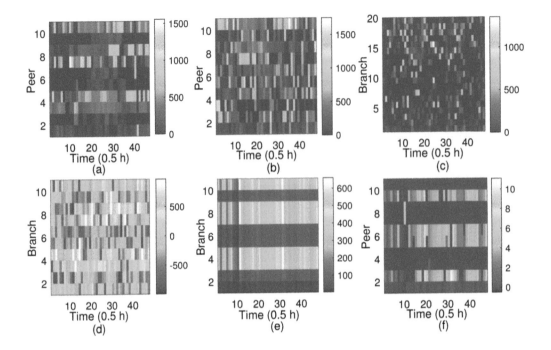

Figure 6.16 The distributions of generation carbon emissions (a), transmission and distribution carbon emissions (b), consumption carbon emissions (c), net carbon emissions (d), carbon allowances (e), and compensation (f) for 11 prosumers. The x axes denote each half-hour settlement period and y axes denote the number of peers or branches.

It is worth noting that the proposed monetary incentive mechanism enables prosumers to make more informed decisions when bidding or offering. As can be seen in Table 6.3, carbon allowances seller 10 provides a larger amount of allowance surplus than seller 7, but still has a lower offering price due to a higher monetary compensation. This shows that the proposed incentive mechanism can encourage sellers to provide more allowances at a lower cost, ultimately resulting in a more cost-effective and environmentally friendly trading scheme. Furthermore, the encrypted trading addresses ensure that only the trading type and amount can be traced and reviewed by the network, guaranteeing security and residential privacy. The use of the P2SH transaction standard with multi-signature also effectively prevents double spending of tokens, carbon allowances, and energy, further enhancing the security of the trading platform compared to a centralised trading system.

Table 6.3 Example of available bids and offers for one settlement period

Carbon Allowances Buyers			Carbon Allowances Sellers			Energy Buyers			Energy Sellers		
#	Amount [g]	Bid [GBP/kg]	#	Amount [g]	Offer [GBP/kg]	#	Amount [Wh]	Bid [GBP/kWh]	#	Amount [Wh]	Offer [GBP/kWh]
4	216	0.0308	7	173	0.0311	1	356	0.122	4	3617	0.108
11	16	0.0322	10	239	0.0302	2	16461	0.125	7	4432	0.133
						3	13272	0.123	8	48071	0.121
						5	14164	0.109	10	9449	0.126
						6	6687	0.123			
						9	4615	0.131			
						11	17508	0.110			

Table 6.4 Blockchain structure for peer-to-peer trading

Timestamp: t_1; Block index: 0; nonce: [];Genesis Block SelfHash: '075c27741a3506846368fa6e5b3477f85b31ceee71a5716e2'
Timestamp: t_2;Block index: 1;nonce: 224 Data:{Sender: '7b2891454769d57605dcfceaa9967121' Receiver: 'c9c940aec3ad22d7527863ecfc4cfc7c' Type: 'Carbon Allowances'; Amount: 216 } PreviousHash: '075c27741a3506846368fa6e5b3477f85b31ceee71a5716e2' SelfHash: '00c8091e1a5055e933f8498c6095ad44'

The process of trading and block generation in the proposed fully decentralised peer-to-peer energy and carbon trading scheme is presented in Table 6.4. As an example, the first two blocks are described in detail. The first block is a genesis block that does not contain any trading information.

In this trading system, buyers select sellers based on the lowest price principle. For instance, carbon allowances buyer 10 has two options:

- To purchase all 216 g of carbon allowances from seller 10 at a price of 0.0302 GBP/kg;

- To purchase 173 g of carbon allowances from seller 7 at a price of 0.0297 GBP/kg and 43 g from seller 10 at a price of 0.0309 GBP/kg.

Based on these options, buyer 10 will choose the first option and purchase all the carbon allowances from seller 10 at a price of 0.0302 GBP/kg.

Unlike a reputation-based trading system, all offers and bids are available for participants in the proposed scheme, and incentives are included without any market intervention. The trading and block generation procedures are fully automated and transparent, providing a fair and efficient trading platform.

In this system, once a trade is confirmed, a new block is generated and added to the blockchain. The encrypted information of trading addresses, trading type, and trading amount is stored in the data cell of each transaction, providing a secure and private trading platform. All nodes in the blockchain network collectively verify each transaction using MS, ensuring that double spending of tokens, carbon allowances, and energy is prevented.

Thus, the proposed fully decentralised peer-to-peer energy and carbon trading scheme offers significant advantages over traditional centralised trading systems, such as lower costs, reduced carbon emissions, improved security, and greater residential privacy. The availability of all offers and bids, along with the inclusion of incentives, creates a fair and transparent trading platform that benefits all participants.

6.4.6 Research Summary

This example research presents a blockchain-based peer-to-peer trading scheme that addresses the challenges posed by the integration of distributed renewable energy sources in traditional energy and carbon markets. The proposed scheme couples energy and carbon markets by using a carbon accounting method to evaluate the emission behaviours of prosumers, and a monetary compensation mechanism to incentivise carbon reduction. By doing so, the scheme encourages the localised energy and carbon emissions to be involved, resulting in a more efficient and sustainable trading process.

The proposed scheme leverages the decentralised features of blockchain to promote cost and carbon emissions reductions in a more secure and privacy-preserving manner. The use of P2SH with MS ensures that the double-spending attacks are prevented and residential privacy is maintained. Compared to centralised systems, the decentralised trading scheme promotes more reductions in costs and carbon emissions.

As a future direction, the effects of the proposed trading scheme on the long-term investment of low-carbon technologies should be investigated. By doing so, the proposed scheme can play a crucial role in incentivising the deployment of low-carbon technologies and promoting a sustainable energy and carbon market.

Bibliography

[1] C. Zhang, J. Wu, Y. Zhou, M. Cheng, and C. Long, "Peer-to-peer energy trading in a microgrid," *Applied Energy*, vol. 220, pp. 1–12, 2018.

[2] Z. Liu, Q. Wu, S. Huang, and H. Zhao, "Transactive energy: A review of state of the art and implementation," *2017 IEEE Manchester PowerTech*, pp. 1–6, 2017.

[3] D. Frieden, A. Tuerk, C. Neumann, S. d'Herbemont, and J. Roberts, "Collective self-consumption and energy communities: Trends and challenges in the transposition of the eu framework," *COMPILE Consortium: Novo mesto, Slovenia*, 2020.

[4] C. Long, J. Wu, Y. Zhou, and N. Jenkins, "Peer-to-peer energy sharing through a two-stage aggregated battery control in a community microgrid," *Applied Energy*, vol. 226, pp. 261–276, 2018.

[5] T. Morstyn, N. Farrell, S. J. Darby, and M. D. McCulloch, "Using peer-to-peer energy-trading platforms to incentivize prosumers to form federated power plants," *Nature Energy*, vol. 3, no. 2, pp. 94–101, 2018.

[6] T. Morstyn, A. Teytelboym, and M. D. McCulloch, "Bilateral contract networks for peer-to-peer energy trading," *IEEE Transactions on Smart Grid*, vol. 10, no. 2, pp. 2026–2035, 2018.

[7] W. Hua, Y. Chen, M. Qadrdan, J. Jiang, H. Sun, and J. Wu, "Applications of blockchain and artificial intelligence technologies for enabling prosumers in smart grids: A review," *Renewable and Sustainable Energy Reviews*, vol. 161, p. 112308, 2022.

[8] D. Li, W. Peng, W. Deng, and F. Gai, "A blockchain-based authentication and security mechanism for iot," in *2018 27th International Conference on Computer Communication and Networks (ICCCN)*. IEEE, 2018, pp. 1–6.

[9] D. Kirli, B. Couraud, V. Robu, M. Salgado-Bravo, S. Norbu, M. Andoni, I. Antonopoulos, M. Negrete-Pincetic, D. Flynn, and A. Kiprakis, "Smart contracts in energy systems: A systematic review of fundamental approaches and implementations," *Renewable and Sustainable Energy Reviews*, vol. 158, p. 112013, 2022.

[10] Y. Zhou, A. N. Manea, W. Hua, J. Wu, W. Zhou, J. Yu, and S. Rahman, "Application of distributed ledger technology in distribution networks," *Proceedings of the IEEE*, 2022.

[11] W. Hua, J. Jiang, H. Sun, F. Teng, and G. Strbac, "Consumer-centric decarbonization framework using stackelberg game and blockchain," *Applied Energy*, vol. 309, p. 118384, 2022.

[12] E. D. Zamani and G. M. Giaglis, "With a little help from the miners: distributed ledger technology and market disintermediation," *Industrial Management & Data Systems*, vol. 118, no. 3, pp. 637–652, 2018.

[13] W. Hua and H. Sun, "A blockchain-based peer-to-peer trading scheme coupling energy and carbon markets," in *2019 International Conference On Smart Energy Systems and Technologies (SEST)*. IEEE, 2019, pp. 1–6.

[14] S. Rahmadika, D. R. Ramdania, and M. Harika, "Security analysis on the decentralized energy trading system using blockchain technology," *Jurnal Online Informatika*, vol. 3, no. 1, pp. 44–47, 2018.

[15] F. A. Khan, M. Asif, A. Ahmad, M. Alharbi, and H. Aljuaid, "Blockchain technology, improvement suggestions, security challenges on smart grid and its application in healthcare for sustainable development," *Sustainable Cities and Society*, vol. 55, p. 102018, 2020.

[16] C. Greer, D. A. Wollman, D. Prochaska, P. A. Boynton, J. A. Mazer, C. Nguyen, G. FitzPatrick, T. L. Nelson, G. H. Koepke, A. R. Hefner Jr *et al.*, "Nist framework and roadmap for smart grid interoperability standards, release 3.0," 2014.

[17] S. J. Pee, E. S. Kang, J. G. Song, and J. W. Jang, "Blockchain based smart energy trading platform using smart contract," in *2019 International Conference on Artificial Intelligence in Information and Communication (ICAIIC)*. IEEE, 2019, pp. 322–325.

[18] W. Hua, J. Jiang, H. Sun, and J. Wu, "A blockchain based peer-to-peer trading framework integrating energy and carbon markets," *Applied Energy*, vol. 279, p. 115539, 2020.

[19] L. Thomas, Y. Zhou, C. Long, J. Wu, and N. Jenkins, "A general form of smart contract for decentralized energy systems management," *Nature Energy*, vol. 4, no. 2, pp. 140–149, 2019.

[20] M. L. Di Silvestre, P. Gallo, M. G. Ippolito, E. R. Sanseverino, and G. Zizzo, "A technical approach to the energy blockchain in microgrids," *IEEE Transactions on Industrial Informatics*, vol. 14, no. 11, pp. 4792–4803, 2018.

[21] Y. Li, W. Yang, P. He, C. Chen, and X. Wang, "Design and management of a distributed hybrid energy system through smart contract and blockchain," *Applied Energy*, vol. 248, pp. 390–405, 2019.

[22] M. Mihaylov, S. Jurado, N. Avellana, K. Van Moffaert, I. M. de Abril, and A. Nowe, "NRGcoin: Virtual currency for trading of renewable energy in smart grids," in *11th International conference on the European energy market (EEM14)*. IEEE, 2014, pp. 1–6.

[23] S. Saxena, H. E. Farag, H. Turesson, and H. Kim, "Blockchain based transactive energy systems for voltage regulation in active distribution networks," *IET Smart Grid*, vol. 3, no. 5, pp. 646–656, 2020.

[24] S. Myung and J.-H. Lee, "Ethereum smart contract-based automated power trading algorithm in a microgrid environment," *The Journal of Supercomputing*, vol. 76, no. 7, pp. 4904–4914, 2020.

[25] K. N. Khaqqi, J. J. Sikorski, K. Hadinoto, and M. Kraft, "Incorporating seller/buyer reputation-based system in blockchain-enabled emission trading application," *Applied energy*, vol. 209, pp. 8–19, 2018.

[26] Y. Pan, X. Zhang, Y. Wang, J. Yan, S. Zhou, G. Li, and J. Bao, "Application of blockchain in carbon trading," *Energy Procedia*, vol. 158, pp. 4286–4291, 2019.

[27] A. Richardson and J. Xu, "Carbon trading with blockchain," in *Mathematical Research for Blockchain Economy: 2nd International Conference MARBLE 2020, Vilamoura, Portugal*. Springer, 2020, pp. 105–124.

[28] Q. Tang and L. M. Tang, "Toward a distributed carbon ledger for carbon emissions trading and accounting for corporate carbon management," *Journal of Emerging Technologies in Accounting*, vol. 16, no. 1, pp. 37–46, 2019.

[29] W. Seward, W. Hua, and M. Qadrdan, "Electricity storage in local energy systems," *Microgrids and Local Energy Systems*, vol. 1, p. 127, 2021.

[30] J. K. Boyce, "Carbon pricing: effectiveness and equity," *Ecological Economics*, vol. 150, pp. 52–61, 2018.

[31] M. Swan, *Blockchain: Blueprint for a new economy*. "O'Reilly Media, Inc.", 2015.

[32] V. Buterin *et al.*, "A next-generation smart contract and decentralized application platform," *white paper*, vol. 3, no. 37, 2014.

[33] C. Dannen, *Introducing Ethereum and solidity.* Springer, 2017, vol. 1.

[34] C. Kang, T. Zhou, Q. Chen, J. Wang, Y. Sun, Q. Xia, and H. Yan, "Carbon emission flow from generation to demand: A network-based model," *IEEE Transactions on Smart Grid*, vol. 6, no. 5, pp. 2386–2394, 2015.

[35] J. W. Bialek and P. A. Kattuman, "Proportional sharing assumption in tracing methodology," *IEE Proceedings-Generation, Transmission and Distribution*, vol. 151, no. 4, pp. 526–532, 2004.

[36] S. D. Commission *et al.*, "Lost in transmission?: the role of ofgem in a changing climate," 2007.

[37] J. Aldersey-Williams and T. Rubert, "Levelised cost of energy–a theoretical justification and critical assessment," *Energy policy*, vol. 124, pp. 169–179, 2019.

[38] M. Wohrer and U. Zdun, "Smart contracts: security patterns in the ethereum ecosystem and solidity," in *2018 International Workshop on Blockchain Oriented Software Engineering (IWBOSE)*. IEEE, 2018, pp. 2–8.

[39] D. Vujicic, D. Jagodic, and S. Randic, "Blockchain technology, bitcoin, and ethereum: A brief overview," in *2018 17th international symposium infoteh-jahorina (infoteh)*. IEEE, 2018, pp. 1–6.

[40] D. Li, W.-Y. Chiu, H. Sun, and H. V. Poor, "Multiobjective optimization for demand side management program in smart grid," *IEEE Transactions on Industrial Informatics*, vol. 14, no. 4, pp. 1482–1490, 2017.

[41] A. Gervais, G. O. Karame, K. Wust, V. Glykantzis, H. Ritzdorf, and S. Capkun, "On the security and performance of proof of work blockchains," in *Proceedings of the 2016 ACM SIGSAC conference on computer and communications security*, 2016, pp. 3–16.

[42] N. Z. Aitzhan and D. Svetinovic, "Security and privacy in decentralized energy trading through multi-signatures, blockchain and anonymous messaging streams," *IEEE Transactions on Dependable and Secure Computing*, vol. 15, no. 5, pp. 840–852, 2016.

[43] G. O. Karame, E. Androulaki, and S. Capkun, "Double-spending fast payments in bitcoin," in *Proceedings of the 2012 ACM conference on Computer and communications security*, 2012, pp. 906–917.

[44] X. Yang, W. F. Lau, Q. Ye, M. H. Au, J. K. Liu, and J. Cheng, "Practical escrow protocol for bitcoin," *IEEE Transactions on Information Forensics and Security*, vol. 15, pp. 3023–3034, 2020.

[45] R. Matzutt, J. Hiller, M. Henze, J. H. Ziegeldorf, D. Mullmann, O. Hohlfeld, and K. Wehrle, "A quantitative analysis of the impact of arbitrary blockchain content on bitcoin," in *Financial Cryptography and Data Security: 22nd International Conference, FC 2018, Nieuwpoort, Curacao, February 26–March 2, 2018, Revised Selected Papers 22.* Springer, 2018, pp. 420–438.

[46] S. Zhai, Y. Yang, J. Li, C. Qiu, and J. Zhao, "Research on the application of cryptography on the blockchain," in *Journal of Physics: Conference Series*, vol. 1168, no. 3. IOP Publishing, 2019, p. 032077.

[47] W. Bi, X. Jia, and M. Zheng, "A secure multiple elliptic curves digital signature algorithm for blockchain," *arXiv preprint arXiv:1808.02988*, 2018.

CHAPTER 7

Cyber Physical System Modelling for Energy Internet

This chapter introduces the cyber phyisical system (CPS) modelling method for Energy Internet. A detailed review is performed in Section 7.1, which includes typical CPS modelling methods for the sub-systems involved in Energy Internet. The multi-vector energy system (MVES) is described in Section 7.2, which represents a typical energy internet system. Specifically, the MVES is modelled using the CPS modelling method, where the integration of artificial intelligence (AI) has been investigated.

7.1 REVIEW OF CYBER PHYSICAL SYSTEM MODELLING METHODS

The future energy system is expected to help reduce carbon emissions and improve energy efficiencies. To achieve these goals, the energy system is becoming smarter with the integration of different sub-systems. In the future energy system, both energy flows and information flows will be of similar importance to its successful operations. This has also inspired the new vision of the energy system as the energy internet, which resembles the internet for the computer networks.

On the one hand, the energy system is becoming more versatile in monitoring and controls with the support of smart devices and the application of advanced technologies. On the other hand, the system model is becoming more complicated with the deep integration among the sub-systems. These tight coupling effects are beyond individual sub-systems, where any involved physical devices and the cyber components could be mutually dependent. Existing modelling methods based on single systems are hard to address the new challenges, where a new modelling method called CPS modelling method has shown great potentials. In this part, the CPS modelling will be reviewed from the aspect of the ICT system, the energy system, and the hybrid modelling aspect, which are detailed as follows.

DOI: 10.1201/9781003170440-7

7.1.1 ICT-Based CPS Modelling

In modern energy systems, the ICT system is becoming a critical sub-system to support the real-time data exchange across the system. For a typical ICT system, it consists of both hardware devices and software algorithms. Therefore its integration in energy systems can be regarded as an extension with respect to both hardware and software. In this way, the CPS model can be adapted based on the existing ICT models, while the state-of-the-art research methods and outcomes could be exploited. On the other hand, the ICT system and the CPS are facing some common challenges, such as the cybersecurity challenge.

In this part, the ICT-based CPS modelling will be reviewed from two aspects. The first one is the ICT for CPS, where the CPS model extended based on ICT models will be reviewed. The second one is the cybersecurity for CPS, where the common cybersecurity studies between ICT system and CPS will be reviewed.

7.1.1.1 *ICT for CPS*

It has been noticed that the traditional ICT system is sharing many similarities to the CPS concept. One typical example is the wireless sensor networks (WSNs). The WSNs exploit sensors to interact with the physical environment, where the network configuration and information exchange method can be dynamically optimized via algorithms. With years of development, the sensors in the WSNs are now referring to a wide range of sensing technologies, such as

- environmental sensors, e.g., oxygen gas monitors and carbon monoxide detectors.

- vehicle information sensors, e.g., mobile magnetometers.

- bio-signal sensors, e.g., oximeters.

- location sensors, e.g., the Global Positioning System (GPS) trackers.

- inertial sensors, e.g., accelerometer sensors.

- power system sensors, e.g., Phasor Measurement Units (PMUs).

It is not surprising that these sensors are also key components of many CPS models, and this provides the foundation of the similarity between WSNs and CPSs. Actually if based on the broad definition, the WSN can be also regarded as a special case of the CPS, while the focus is much narrowed down to the ICT system only.

There are several key differences between WSNs and CPSs, which can be summarized as follows:

- Information Flow: in WSN models, the information flow is usually directional from the edge of the network to the information centre. This is because the WSN is mainly used to monitor and acquire knowledge of the environment. In CPS models, the system can be bi-directional, i.e., the information flow could be two-way. This is because the physical devices in the CPS models are not confined to sensors, while there might be also devices for controls.

- Performance Requirement: in WSN models, the most concerned performance indicators are energy savings. This is because a common target application scenario for WSNs are the sensors to be powered by batteries. Usually, the batteries are with constrained energies, and the nodes in the WSNs is with different power consumption performances depending on their tasks. On the contrary, in CPS models, the end performances such as the latency, throughput and reliability of the ICT network are more focused, since these factors have great impacts on the performance of the physical devices.

- Network Topology: in WSN models, the topology of the network is normally homogeneous, where the application scenario is usually dedicated to one geographical area. On the contrary, the CPS models are usually involving the interactions between different systems, where the sensors, actuators, or controllers may not collocate at the same place. The networks might need to be tailored for different sub-scenarios, which would normally lead to a heterogeneous architecture.

Note that the above differences are just in a general sense, where the boundaries could be blurred in some cases. For example the extended notion of Wireless Sensor and Actuator Networks, which further includes the actuators as part of the network. In energy systems, the WSN is also widely used to study some sensing-based application scenarios. For example, in the study of wind energy systems, the WSN can be used to monitor the wind turbines in a real-time manner [1], or help to reach some extreme environment such as the offshore wind farms [2]. A comprehensive review regarding the CPS and WSN is provided in [3].

7.1.1.2 Cyber Security for CPS

One key feature of the CPS is that it is the integration of both cyber components and physical devices. Due to the fact that the cyber components and the physical devices are coupled, the security breach of either the cyber networks might impact the physical systems, and vice versa. To ensure CPS security, it requires efforts from the aspects of both the cyber networks and the physical systems. In the following part, we will review the potential attacking methods in the CPS as well as their corresponding countermeasures.

Based Attacks and Countermeasures   Modern cryptography provides another dimension to improve cybersecurity. By encrypting the data instead of using plain text, the information security can be hard to be deciphered by the adversary. This is usually at the cost of extra complexity to the system, but it can help preserve the privacy of contained information even if the attacker has obtained the encrypted data.

Many cryptographic operations are supported by hardware solutions, where the commonly available methods are listed as follows [4].

- Hash

- Pseudo-random number generators

- Symmetric encryption, e.g. Advanced Encryption Standard (AES)

- Asymmetric encryption, e.g. Rivest-Sharmir-Adleman (RSA), Paillier, EI Gamal

- Elliptic curve cryptography

Based on the hardware/software based cryptography support from individual smart devices, more advanced privacy-preserving schemes have been proposed to support the CPS architecture.

- Homomorphic Encryption and Secret Sharing [5]

- Masking and Brute Forcing [6], Masking and Differential Privacy [7]

- Modified Homomorphic Encryption [8]

- Policy-based Data Sharing [9]

- Local differentially private high-dimensional data publication [10]

User Based Attacks and Countermeasures   The various users are also critical vulnerable parts of the CPS. Although technically the CPS consists of cyber and physical components, the system is designed to serve for people and managed by people. Therefore an adversary can target at the users with access to the system to fulfill the cyber attack to the whole system. The common attacking methods include malware, phishing email, ransomware and cryptojacking [11].

Although cybersecurity is already widely aware, the campaign by National Cyber Security Center showed that 14 out of 1800 malicious emails still successfully triggered the malware installation [12]. Depending on the user's role in the system, the attack can result from mild privacy leakage of individual users to a grid-wide blockage.

The most well-known example is the malware attack on the Ukrainian power grid [13], where the Ukrainian power distribution company was attacked after users opened a malware-rigged attachment in a phishing email. The attackers exploited and committed a series of attacks on the Ukrain power grid, including uploading malicious frameworks at substations, stealing user information and obtaining VPN credentials.

It worth noticing that with the boosting of smart devices, new attacking methods are widely reported targeting at the user's mobile operating system [14]. Especially for the CPS, many smart devices such as smart meters, home gateways and controllers are providing mobile access/control methods to the appliances. The main countermeasures are from two aspects, namely the analysis of the application itself via untrusted application analysis [15] [16] [17] and monitoring the suspicious behaviours via continuous runtime monitoring [18] [19] [20].

Table 7.1 A Communication Protocol layer view of the potential attacks and counter-measures

Layer	Attack Method
Application Layer	Repudiation [24], incomplete information [25], false information [26]
Transport layer	Session hijacking [27], SYN flooding [27]
Network layer	Wormhole [28], blackhole [29], grayhole [30], Byzantine [31], On-off
MAC layer	Traffic monitoring [33], disruption MAC [34]
Physical layer	DoS [35], Spoof-jamming [36], eavesdropping [37]

Communication Network-based Attacks and Countermeasures The advanced communication technologies not only enable the CPS smarter, but also introducing more vulnerable components that can lead to cyber attacks. The communication system based cyber attacks are gaining more attentions with the growing real-world attack incidents. Typical examples are the Distributed Denial of Service (DDoS) in Austrian and German power grid [21], where a mistakenly transmitted test command induced cascaded disaster in the control network and resulted in the power grid nearly knock down.

Due to the layered design of the communication systems, both wired and wireless communication systems are prone to the application layer and network layer attacks. Due to the broadcasting nature of the wireless signal, the wireless communication systems are also prone to the MAC layer and Physical layer attacks. The typical attacks and their corresponding countermeasures on the communication networks can be generally categorized according to its targeted communication protocol layer as illustrated in Table 7.1 [22][23].

Physical Devices Based Security The physical devices such as sensors, controllers, aggregators, servers are subject to both physical security and cybersecurity challenges. The traditional physical security issues are theft, intrusion and vandalism, while it has been observed an increase of new challenges due to emerging technologies. For example, the surveillance incidents by unauthorised drones occupied 16.4% of the total reported physical security issues [38].

Moreover, these physical devices are also key access points to the whole CPS, whose functionally completeness is serving as an important precondition to the dependent cybersecurity. Specifically, five main key aspects are identified as the necessary cybersecurity requirements for the physical devices, which are briefed as follows [39] [40]:

- **Access control**: The access privileges should be differentiated and the smart devices should be able to check them for the access. This ensures the data access request is appropriate to its functional purpose and enables hierarchical data protection schemes.

- **Privacy**: The private information, e.g. user identity and location, should be protected by the smart devices. This includes preventing excessive collection of user information and making sensitive information anonymous.

- **Authentication and identification**: To get access to the CPS, the devices should be able to establish the identity of itself and provide the proof of this identity to the other system participants. In the meantime, the device should also possess the ability to verify the provided identity.

- **Integrity**: When handling (e.g. transmitting, receiving and caching) data, the smart device should be able to check the integrity of the data, i.e. the data are genuine and not modified.

- **Non-repudiation**: The operation history of the smart devices should be able to be audited. The smart device itself should be prevented from denying the truth of its involvement in any activity.

7.1.2   Energy System-Based CPS Modelling

The future energy system is deemed to be an integrated system, which can be predicted based on the trend of fusion with advanced technologies across different research areas. Some existing energy system models have already integrated the other sub-systems, such as the ICT systems, which can be further extended to the more comprehensive CPS models.

Modelling based on Supervisory Control And Data Acquisition (SCADA)   A typical example is the modelling based on Supervisory Control And Data Acquisition (SCADA). The Supervisory Control And Data Acquisition (SCADA) systems describe the system from a control system architecture, which consists of computers for control signal processing, communication networks for data transmissions, and remote terminal units for access of the sensors and actuators. It is widely used in energy system application scenarios, including the electric power transmission domains and distribution domains.

The SCADA system is also evolving with the integration of new technologies in both energy systems and ICT systems. For example, the authors in [41] studied the harmonic impact of a large solar farm in a utility distribution system, whose measurements are fulfilled by SCADA. While in [42], the application of SCADA system for general remote control and observations was studied, where the energy systems with renewable energy resources, such as wind farms, solar farms, and fuel cells, were considered.

From the viewpoint of ICT systems, the SCADA is a framework that many new technologies can be integrated. This is due to the fact that SCADA is applying a high-level description for the ICT systems. For example, the IEC 61850 standard describes the data model and the performance requirements, where it can be implemented with a wide range of communication systems, ranging from TCP/IP via wired networks, to the WiFi networks [43].

The CPS modelling based on SCADA system could be straightforward. The ICT devices and the electric power system devices are the physical part of the CPS, while the data links and the control signal processing systems form the cyber networks. For example, the authors in [44] characterized the SCADA system as an ICT model, and the power system was modelled via DIgSILENT Power Factory. Then the interactions between the physical devices and the cyber components are modelled as a CPS, where a co-simulation testbed was built based on MATLAB Simulink.

ICT impacts on Energy System   Reliability is one of the key performances in the energy system. The traditional analytical framework was for the energy systems, e.g., the transient analysis and steady state analysis for the electric power systems. Even though some models consider the ICT system as comprising components, it usually assumes a perfect ICT system with no errors or faults.

This assumption might be reasonable for some applications, e.g., the wired communication based systems and the traditional control systems based on closed-form results. With the integration of advanced ICT technologies, such as the fifth generation wireless communication technologies, there are new challenges that the ICT systems are also contributing uncertainties to the CPS.

New ICT technologies empower the energy system to have real-time access to almost all possible information, within every corner of the energy system. The benefits of coverage, throughput and flexibility are at the cost of the increased complexity and chances of faults in the ICT systems. On the other hand, the reliability requirements for the ICT systems are also increasing, e.g., new services or applications for real-time monitoring or controls. The time scale change of the monitoring or control applications is also squeezing the tolerance of the latency to acquire the measurements or to deliver the control commands.

To address these new challenges due to the ICT systems, it has stimulated the study of ICT failures' impact on the energy system models. The insufficient situation awareness induced by the ICT failure was considered in [45], while its correlated reliability impact on the power system was studied. In [46], the time dependencies of power demand and injection were studied, whose relation with ICT systems was studied and its impact on the power system reliability was evaluated. In [47], the ICT impacts during the fault or outage situations were considered in a decentralized network automation system. It also considered the failure rates and the time to repair factors, whose impacts on the reliability performance were performed on various medium voltage and low voltage systems. The interdependency between power systems and the ICT systems was considered in [48], where the interdependancy was studied via system operational states and a multi-dimensional state classification approach was proposed. In [49], the ICT system's impact on the power system observability was considered, where an analytical framework was proposed to quantify the coupling effects. The model was then used to optimize the observability sensitivity, probability, and redundancy performances.

7.1.3   Hybrid CPS Modelling

Before the advent of the CPS concept, the physical devices and the cyber components are treated in a relatively separated manner. This is beneficial if the physical parts and the cyber parts are mutually independent, since the study can focus on either of them at one time. The examples are the aforementioned examples, such as the

SCADA. When smart sensors and controllers are widely used in energy systems, the interactions between the physical devices and the cyber components are increasing, which also increases the mutual dependency among the energy system participants.

With the new advanced technologies, such as programmable devices, the boundaries between the functions of the physical devices and the cyber components are also becoming blurred. To consider these new challenges, new concepts such as the CPS has been proposed. Correspondingly, the researchers adapt the existing concepts to cover the new features. Typical examples are the CPS modelling based on the ICT aspect and the energy system aspect, respectively.

Generally speaking, these modelling methods are still based on the core concepts in the existing theories and models. The advantage is apparent, since the new research can be built upon the existing knowledge libraries. But the disadvantage is also clear, since the proposal of new concepts like CPS is usually due to the fact that the new challenges are so prominent, that it requires to call for new efforts beyond the existing concepts.

As for the CPS concept, one of such key challenges is the mutually coupling effect between the physical devices and the cyber components. The end performance results from the synthetic performance of all participating sub-systems, where each sub-system could be adjusted. This requires detailed modelling of not only one part of the system, but also the other parts as well as their correlations.

The corresponding CPS model will hybrid the models from different sub-systems, which characterizes the mutual influences in a quantified manner. For example, in order to improve the power system performance at the transient level, an on-demand communication mechanism was proposed in [50]. The performance was then evaluated with simulations on the fault control scenario, which demonstrated its capability in a more robust performance under system errors. In [51], the Smart Grid system is modelled as a CPS, where the data importance of multiple Smart Grid applications are studied. The data exchange strategy in the ICT system is optimized for the impacts of the energy system, which demonstrated its advantage over traditional ICT-based methods on voltage stability control applications.

In order to study the coupling effects in the CPS, as well as to study the impacts on the performance of the energy systems, it has stimulated the hybrid CPS modelling method via the co-simulation method. Different from the traditional simulation method, the co-simulation method exploits different simulators to model the components for individual sub-systems.

In this way, the sub-system operations can be studied in detail under interdependency. In [52], a co-simulation framework was proposed for the CPS, which exploited the high-level architecture specified by IEEE standard 1516-2010. The framework was demonstrated to be applicable to the study of coupling effects of the ICT system and the power system. In [53], a distributed CPS co-simulation framework was proposed based on the global event-driven mechanism, where the time scale can be adjusted for the wide area measurement and control applications. A distributed CPS was studied in [54], where the ICT system was modelled via the Network Simulator 2 (NS2) and the power system was modelled via the PSCAD. A comprehensive review of the simulation works for the CPS has been provided in [55], where the common challenges and available tools are discussed.

7.2 MULTI-VECTOR ENERGY SYSTEM

Besides meeting the increasing energy demands, modern energy systems are playing a vital role in reducing carbon emissions and tackling climate change challenges [56]. It is common that different energy systems are collocated in the same geographical area to serve the end uses. For example, in a residential area, there are usually electric systems and natural gas systems. This paves the foundation of the integration of different systems, as well as the joint coordination between them.

Moreover, the versatile energy converts can further couple several energy systems, such as the Combined Heat and Power (CHP) generation technologies. Such machines can convert the energies into multiple forms to supply the end-user's demand, which leads to options in optimizing the energy supplies in an integrated energy system. This has also stimulated a multi-vector energy system (MVES) modelling method, which models the energy carriers as multiple energy vectors, and then jointly optimizes them to supply multiple energy demands. In this part, the integrated energy system will be modelled as a MVES. Based on that, the application of AI and blockchain technologies will be studied.

7.2.1 Coordination for Multi-Vector Energy System

The MVES provides a high-level abstract model between the supply and demand in multi-vector energy forms, whose scale could be a district of buildings, a manufacturing site, or a town of residential houses, depending on its corresponding real-world system [59]. With the Internet of Things technologies, the MVES components could be coordinated and managed in an aggregated manner. This includes the versatile energy converters (e.g., the CHP systems), the renewable energy (e.g., the wind farms), the energy storage (e.g., the EVs), and the end-user loads. Especially, these MVES components are interconnected via data links, whose physical locations can be geographically distributed across the system.

The MVES operator acts as a broker between different energy suppliers and the various load demands. It charges the end users by supplying their load demands, and pays the energy markets for the corresponding energy consumption. The integration of renewable energy and energy storage not only further improves the energy efficiency of the MVES [60], but also helps to reduce the operation cost of the MVES and bring profits for their owners (e.g., the MVES and the end users) [61].

7.2.1.1 Multi-Vector Energy System Modelling

The MVES is a general modelling method of the integrated energy system, while there are no firm rules on which type of energy forms or energy converters must be included. In this part, a general system is considered and illustrated in Fig. 7.1, which consists of the common components in the modern energy systems. This includes the energy forms in electricity, heat, and renewable energy.

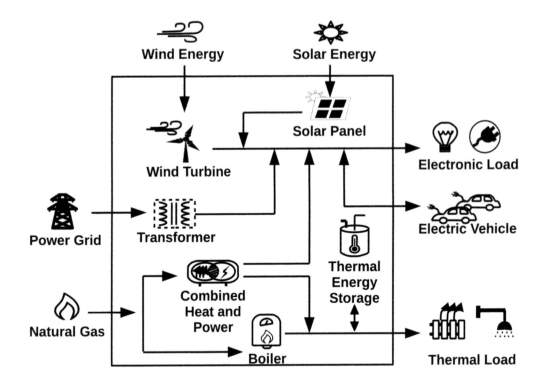

Figure 7.1   A Multi-Vector Energy System (MVES) with typical components, including Transformers, Wind Turbines, Solar Panels, Combined Heat & Power (CHP) systems, Thermal Energy Storages (TESs), Boilers and Electric Vehicles (EVs).

Moreover, typical energy storage, such as Electric Vehicles (EVs) and Thermal Energy Storages (TESs) and EVs are considered. Note that the illustration in Fig. 7.1 is an aggregated view of the modelled integrated energy system, where the wind energy might correspond to a large wind farm, while the EVs can be distributed across a residential area. The named components are also for illustration purposes, while it is to describe the typical functions. For example, in different integrated energy systems, the CHP technologies might be implemented via the Micro Turbines or Fuel Cells, depending on specific cases. In the following part, the MVES model of each component in Fig. 7.1 will be described.

The coordination of the MVES is based on 1 hour time slots, which corresponds to a total of $T = 24$ slots during the 24 hours, namely a whole day. As illustrated in Fig. 7.1, the energy flows in the MVES major consist of three parts, which are the electricity energy flow, the thermal energy flow, and the natural gas flow. The involved energy converters include one transformer, one wind turbine, one solar panel, one CHP system, one boiler, M EVs and N TESs.

Since the energy flow could be two-way in the system, it is assumed that the energy flow is positive when the energy is imported into the energy system. Correspondingly, if the energy is exported out of the energy system, then the energy flow is assumed to be negative.

The main objective of the MVES is to coordinate the three types of energy flows, in order to balance the energy supplies and the energy demands in different forms, which are detailed as follows:

- **The Electricity Load Model:**
  The electricity load is the most typical load form in the integrated energy system. Many residential appliances are powered by electricity, such as fridges, computers and lights. In the considered MVES system, there are multiple sources for the electricity energy, which can be:

  - imported from the power grid,
  - generated by wind turbines,
  - generated by solar panels,
  - generated by CHP systems,
  - discharged from EVs.

  In the meantime, there are multiple electricity energy demands in the MVES, which are consumed to

  - supply electricity loads,
  - charge EVs.

  During each time slot t, it is assumed that the transformer's electricity energy flow, natural gas flow, wind energy flow, PV energy flow and the electricity load demand are denoted as S_E^t, S_G^t, S_W^t, S_{PV}^t and L_E^t, respectively. For the m-th EV among the total M EVs, the energy flow during time slot t is denoted as $S_{EV}^{m,t}$.

  Then the supply and demand of electricity energy in the MVES can be modelled as follows:

  $$\eta_{TF} S_E^t + v_{CHP}^t \eta_{CHP}^E S_G^t + \eta_W S_W^t + \eta_{PV} S_{PV}^t + \sum_{m=1}^{M} S_{EV}^{m,t} = L_E^t, \qquad (7.1)$$

  where v_{CHP}^t denotes the natural gas dispatching factor for the CHP, while the parameters η_{TF}, η_{CHP}^E, η_W and η_{PV} are the energy conversion efficiencies of the transformer, the CHP, the wind turbine and the PV, respectively.

  Since the CHP is able to simultaneously producing electricity energy and thermal energy, their corresponding energy conversion efficiencies are distinguished by the superscripts E and G. Moreover, the natural gas dispatching factors are used to distinguish the energy consumed by the CHP system and the boiler, which are denoted by v_{CHP}^t and v_B^t, respectively. For general cases, the sum of all dispatching factors is 1, and correspondingly in the considered scenario $v_{CHP}^t + v_B^t = 1$.

- **The Thermal Load Model:**
  The thermal load is another common energy form in the residential scenarios, as well as in the industrial scenarios. For example, the hot water supply for domestic heating purposes, or the cooling water for temperature controls in the factory. In the considered MVES examples in Fig. 7.1, the heat load is illustrated.

  The thermal energy can have multiple sources, which can be

    - generated from the CHP,
    - generated from the boiler,
    - discharged from the TESs.

  The supplied thermal energy is consumed by the thermal energy load, which may represent an aggregated thermal loads distributed in the whole energy system.

  Note that it is straightforward to extend the models to include other forms of thermal load if required. For the n-th TES among the total N TESs, the energy flow during time slot t is denoted as $S_{\text{TES}}^{n,t}$. Then the supply and demand of thermal energy in the MVES can be modelled as follows:

  $$v_{\text{CHP}}^t \eta_{\text{CHP}}^{\text{H}} S_{\text{G}}^t + v_{\text{B}}^t \eta_{\text{B}} S_{\text{G}}^t + \sum_{n=1}^{N} S_{\text{TES}}^{n,t} = L_{\text{H}}^t, \qquad (7.2)$$

  where L_{H}^t and η_{B} are the thermal load demand during time slot t and the boiler's energy conversion efficiency, correspondingly.

  It is noticed that the CHP appears in both the thermal energy supply and demand model in (7.2) and the electricity energy supply and demand model in (7.1). This is because the CHP can generate electricity energy by combusting natural gas, while the generated heat during the combustion can be efficiently recovered to supply the thermal loads.

- **The Physical Energy Flow Constraint Models:**
  For each MVES component, it represents a set of energy system devices in an aggregated manner. Therefore the energy flows between them should also satisfy the physical constraints in the real world. For example, for every energy converter, the energy flow will be always operated within the rated range for safety reasons. These physical constraints correspond to the maximum and minimum energy flows, from or to the MVES components.

  In the case that the devices are only supporting one-way energy flow, their specification might only define the maximum energy flows, while the default minimum energy flows are 0 or otherwise described. For example in the considered MVES system as illustrated in Fig. 7.1, the physical energy flow constraint can be modelled as follows:

$$0 \leq S_E^t \leq S_{TF}^{MAX}, \tag{7.3}$$

$$0 \leq S_G^t \leq S_G^{MAX}, \tag{7.4}$$

$$0 \leq S_W^t \leq S_W^{MAX}, \tag{7.5}$$

$$0 \leq S_{PV}^t \leq S_{PV}^{MAX}, \tag{7.6}$$

$$0 \leq v_{CHP}^t S_G^t \leq S_{CHP}^{MAX}, \tag{7.7}$$

$$0 \leq v_B^t S_B^t \leq S_B^{MAX}. \tag{7.8}$$

where the parameters S_{TF}^{MAX}, S_G^{MAX}, S_W^{MAX}, S_{PV}^{MAX}, S_{CHP}^{MAX} and S_B^{MAX} are the maximum energy flow constraints of the transformers, natural gas, wind turbines, solar panels, CHPs and boilers, respectively.

Note that the model has been simplified here for illustration purposes, as the number of each MVES component is considered as one. It is straight-forward to extend the model to a more complicated case, where the MVES component can be an aggregation of more than one element. In such cases, the physical energy flow constraints in $(7.3) - (7.8)$ should be defined for each element. For example, if there are K wind turbines in the energy system, then for any $k = 1, \ldots, K$, its energy flow should satisfy the constraints $0 \leq S_W^{k,t} \leq S_{k,W}^{MAX}$. Note that since the wind turbines might not be in the same models, it is possible to have different physical energy flow constraints for different energy flow terms.

- **The EV Charging and Discharging Model:**
  The EVs are regarded as important solutions in reducing carbon emissions. Currently, there has been a trend in global in encouraging the replacement of fossil-fueled cars by EVs. For some developed countries, there have been clear road maps to completely rule out the fossil-fueled cars. For example, in the UK, it is planned that the sale of fossil-fuel cars are to end by 2030, while there is a grace period for hybrid cars until 2035.

  The EVs are powered by electric batteries, which can be charged and discharged on demand. The capacity of the EVs is typically in the range of 50 kWh to 80 kWh. This means the EV can be a large electricity load when it is in the charging state, while it is also possible to serve as a stable electricity power source when it is in the discharging state. Moreover, when there are multiple EVs that can be coordinated to charge or discharge on demand, the aggregated capacity can reach the megawatts level, which is comparable to some distributed generators.

  Different from battery electric systems that always attached to the grids, there is usually only a range of time periods for an EV to serve as part of the MVES. Here for discussion purpose, it is assumed that the start and end service time for the m-th EV is in a time period from $T_{EV}^{In,m}$ to $T_{EV}^{Out,m}$. Note that here the time range is referring to the period that the EV is providing charging and discharging services to the MVES.

The amount of electricity energy at the service start time is the property of the EV's owner, therefore it will be restored at the end of their service time. In the meantime, any amount of energy charged to or discharged from the EVs during the service time is owned by the MVES. In other words, the MVES is using the EVs' capacity to charge or discharge a certain amount of electricity energy for the system-wise optimizations across their service time. Note that if the EV is connected to the MVES for the purpose of charging its batteries instead of serving the MVES, then this EV should be categorized as an electric load as introduced before.

It is noticed that the direction of the energy flow is defined with regard to the MVES. Therefore the energy flow $S_{EV}^{m,t}$ of the m-th EV (observed at the EV side) can be only in three possible states, where

 – if the EV is discharging, then the energy flow is defined as positive,
 – if the EV is charging, then the energy flow is defined as negative,
 – if the EV is neither charging nor discharging, then the energy flow is 0.

To facilitate the modelling of EVs' charging and discharging operation, the m-th EV's discharging indicator $I_{EV}^{DCh,m,t}$ and charging indicator $I_{EV}^{Ch,m,t}$ are defined as follows:

$$
\begin{aligned}
I_{EV}^{Ch,m,t} &\triangleq \frac{1 - \mathrm{sgn}\{S_{EV}^{m,t}\}}{2}, \\
I_{EV}^{DCh,m,t} &\triangleq \frac{1 + \mathrm{sgn}\{S_{EV}^{m,t}\}}{2},
\end{aligned}
\tag{7.9}
$$

In this way, the charging or discharging state of each EV can be indicated by $I_{EV}^{DCh,m}$, where

 – if the EV is discharging, then $I_{EV}^{DCh,m} = 1$ and $I_{EV}^{Ch,m} = 0$,
 – if the EV is charging, then $I_{EV}^{DCh,m} = 0$ and $I_{EV}^{Ch,m} = 1$,
 – if the EV is neither charging nor discharging, then $I_{EV}^{DCh,m} = 0$ and $I_{EV}^{Ch,m} = 0$.

where the sign operator $\mathrm{sgn}\{\cdot\}$ is defined as follows:

$$
\mathrm{sgn}\{x\} = \begin{cases} 1, & \text{if } x > 0, \\ 0, & \text{if } x = 0, \\ -1, & \text{if } x < 0, \end{cases}
\tag{7.10}
$$

Although the charging and discharging process of the EV is of high efficiency, there are certain energy losses such that there will be a discrepancy between the amount observed at the EV side and the MVES side. To account this energy loss, here we use η_{EV}^{Ch} and η_{EV}^{DCh} to denote the EV's charging and discharging efficiency, respectively.

With the EV's discharging indicator $I_{\text{EV}}^{\text{DCh},m,t}$ and charging indicator $I_{\text{EV}}^{\text{Ch},m,t}$, it can be seen that the energy flow observed at the EV side can be unified as follows:

$$S_{\text{EV}}^{m,t}\left(\frac{I_{\text{EV}}^{\text{Ch},m,t}}{\eta_{\text{EV}}^{\text{Ch}}} + I_{\text{EV}}^{\text{DCh},m,t}\eta_{\text{EV}}^{\text{DCh}}\right),$$

which is valid for all states of the EV, whose equivalent forms can be given as follows:

- if the EV is discharging, then the energy flow observed at the MVES side is $S_{\text{EV}}^{m,t}\eta_{\text{EV}}^{\text{DCh}}$,
- if the EV is charging, then the energy flow observed at the MVES side is $\frac{S_{\text{EV}}^{m,t}}{\eta_{\text{EV}}^{\text{Ch}}}$,
- if the EV is neither charging nor discharging, then the energy flow observed at the MVES side is 0.

From the above formulation, it is seen that the direction of the energy flow at the EV side is following the sign of $S_{\text{EV}}^{m,t}$, which is

- positive, if the EV is discharging to the MVES,
- negative, if the EV is charging from the MVES,
- 0, if the EV is neither charging nor discharging.

It should be noticed that in other works, the direction might be different from the above definitions. The above setup is to make sure the direction of the EV's energy flow is consistent with the whole MVES.

At the service end time $T_{\text{EV}}^{\text{Out},m}$, the EV's energy level should be restored to that of the service start time $T_{\text{EV}}^{\text{In},m}$. This is equivalent to the a total of zero net energy flow during the whole service period $[T_{\text{EV}}^{\text{In},m}, T_{\text{EV}}^{\text{Out},m}]$, which can be modelled as follows:

$$\sum_{t=T_{\text{EV}}^{\text{In},m}}^{T_{\text{EV}}^{\text{Out},m}} S_{\text{EV}}^{m,t} = 0. \tag{7.11}$$

For each EV, the absolute maximum charging energy flow and discharging energy flow during one slot period are denoted as $S_{\text{EV}}^{\text{Ch, MAX}}$ and $S_{\text{EV}}^{\text{DCh, MAX}}$, respectively. Then for the m-th EV, its energy flow constraint can be expressed as follows:

$$S_{\text{EV}}^{\text{Ch, MAX}} \le S_{\text{EV}}^{m,t}\left(\frac{I_{\text{EV}}^{\text{Ch},m,t}}{\eta_{\text{EV}}^{\text{Ch}}} + I_{\text{EV}}^{\text{DCh},m,t}\eta_{\text{EV}}^{\text{DCh}}\right) \le S_{\text{EV}}^{\text{DCh, MAX}}. \tag{7.12}$$

Note that here the EVs are assumed with the same maximum charging and discharging energy flow limit. For cases where there are different types of EVs, there will be different energy flow limits for each EV, which should be applied to constraint each EV's energy flow correspondingly.

The energy capacity of the EV at a given time can be indicated by the parameter called State-of-Charge (SOC). For example, for the m-th EV at time slot t, its energy capacity is given by $\text{SOC}_{\text{EV}}^{m,t}$. Correspondingly, its energy capacity at its start service time $T_{\text{EV}}^{\text{In},m}$ can be given by $\text{SOC}_{\text{EV}}^{m,T_{\text{EV}}^{\text{In},m}}$.

With the energy flow $S_{\text{EV}}^{m,t}$ observed at the EV side, the $\text{SOC}_{\text{EV}}^{m,t}$ of the m-th EV at time slot t can be given as follows:

$$\text{SOC}_{\text{EV}}^{m,t} = \text{SOC}_{\text{EV}}^{m,T_{\text{EV}}^{\text{In},m}} + \sum_{\tau=T_{\text{EV}}^{\text{In},m}}^{t} S_{\text{EV}}^{m,\tau}. \tag{7.13}$$

Since the EV is with fixed electricity energy capacity, the energy capacity $\text{SOC}_{\text{EV}}^{m,t}$ during the service period should be within certain ranges to keep the safety. Without the loss of generality, the EV's maximum and minimum electricity energy capacity are denoted by $\text{SOC}_{\text{EV}}^{\text{MIN}}$ and $\text{SOC}_{\text{EV}}^{\text{MAX}}$, respectively.

Then the physical energy capacity constraint can be modelled as follows:

$$\text{SOC}_{\text{EV}}^{\text{MIN}} \leq \text{SOC}_{\text{EV}}^{m,t} \leq \text{SOC}_{\text{EV}}^{\text{MAX}}. \tag{7.14}$$

Since the core energy storage for the EV is batteries, the above EV model can be extended to characterize a general battery electric system. A key difference between the battery electric system and the EV is that the battery electric system is usually attached to the grid all the time, which corresponds to relaxing the service time to 24 hours each day. In such cases, the above model can be easily adapted by assigning $T_{\text{EV}}^{\text{In},m} = 0$ and $T_{\text{EV}}^{\text{Out},m} = 24$.

- **The TES Charging and Discharing Constraints:**
  TESs have been widely used in the residential and industrial scenario, which have regained the researchers' focus with the new advanced communication and control technologies. The most common form of TESs is the cylinders, which has been already widely used in the UK. The cylinder serves as a thermal storage or thermal buffer, where the hot water leaves the cylinder from the outlet at the top, and the cold water is re-supplied from the inlet at the bottom. With modifications, these cylinders are capable to serve as the distributed thermal energy system and play an active role in the future energy market.

  This is because these TESs are with two beneficial features including,

  - the energy storage medium is water, which can be directly used to supply the user's demand, but also cheap and convenient to obtain,
  - the cylinder system has been already widely used in the UK, where the existing infrastructure can be exploited,
  - the energy capacity for individual users is feasible if aggregated or managed in a large scale.
  - the stored thermal energy can be represented by the temperature of the water, which is also transparent to the users and easy to measure.

In domestic usage scenarios, the water temperature of these TES can also be relaxed to a range, which corresponds to the charging status of the "thermal battery". It has been reported that hot water consumption accounts for 10% to 25% of the total end energy demand in many countries [62] [63]. This flexibility in capacity has been recognized as the potential methods in supporting the power grid operations [64]. In this part, it is assumed the TES is with two-way thermal energy flows, which can be implemented via a large cylinder with hot water supply networks.

TESs are sharing some similarities with the BESs, as well as the EVs. Usually, TESs are attached to the MVES system all the time, but it should be noticed that it does not mean it is always available to serve the MVES. For ease of discussion purpose, the start and end of the service time for the n-th TES are denoted as $T_{\mathrm{TES}}^{\mathrm{In},n}$ and $T_{\mathrm{TES}}^{\mathrm{Out},n}$, respectively. Although the following model can apply to the case where TES can serve 24 hours a day, it is possible that the TES will on personal usage instead of serving the MVES. In these time periods, the TES should be categorized as thermal loads.

At the start service time $T_{\mathrm{TES}}^{\mathrm{In},n}$, the amount of existing thermal energy in the n-th TES is the property of the TES's owner. Therefore this amount of thermal energy should be restored at the end of the service time $T_{\mathrm{TES}}^{\mathrm{Out},n}$. During the service period from $T_{\mathrm{TES}}^{\mathrm{In},n}$ to $T_{\mathrm{TES}}^{\mathrm{Out},n}$, the thermal energy stored in or extracted out the TESs is the property of the MVES. In other words, the MVES is leasing the capacity of the TESs' capacity, which are used as thermal energy buffers to coordinate the system-wise thermal energy supply and demand across their service time.

The state of TES can be reflected by the charging state indicator and discharging state indicator, which are denoted as $I_{\mathrm{TES}}^{\mathrm{Ch},n,t}$ and discharging state indicator $I_{\mathrm{TES}}^{\mathrm{DCh},n,t}$, whose definitions are given as below:

$$
\begin{aligned}
I_{\mathrm{TES}}^{\mathrm{Ch},n,t} &\triangleq \frac{1 - \mathrm{sgn}\{S_{\mathrm{TES}}^{n,t}\}}{2}, \\
I_{\mathrm{TES}}^{\mathrm{DCh},n,t} &\triangleq \frac{1 + \mathrm{sgn}\{S_{\mathrm{TES}}^{n,t}\}}{2}.
\end{aligned}
\tag{7.15}
$$

where $S_{\mathrm{TES}}^{k,t}$ denotes the thermal energy flow of the n-th TES in the slot period t, observed at the TES side. In this way, the charging and discharging state of the TESs can be indicated by the values of $I_{\mathrm{TES}}^{\mathrm{Ch},n,t}$ and $I_{\mathrm{TES}}^{\mathrm{DCh},n,t}$ as follows:

- if the TES is charging, then $I_{\mathrm{TES}}^{\mathrm{Ch},n,t} = 1$ and $I_{\mathrm{TES}}^{\mathrm{DCh},n,t} = 0$,
- if the TES is discharging, then $I_{\mathrm{TES}}^{\mathrm{Ch},n,t} = 0$ and $I_{\mathrm{TES}}^{\mathrm{DCh},n,t} = 1$,
- if the TES is neither charging nor discharging, then $I_{\mathrm{TES}}^{\mathrm{Ch},n,t} = 0$ and $I_{\mathrm{TES}}^{\mathrm{DCh},n,t} = 0$.

There are also thermal energy losses during the charging and discharging process, which are modelled as the TES's charging efficiency and discharging efficiency, denoted as $\eta_{\mathrm{TES}}^{\mathrm{Ch}}$ and $\eta_{\mathrm{TES}}^{\mathrm{DCh}}$, respectively. Therefore the thermal energy flow observed at the MVES side can be represented as follows:

$$S_{\text{TES}}^{n,t} \left(\frac{I_{\text{TES}}^{\text{Ch},n,t}}{\eta_{\text{TES}}^{\text{Ch}}} + I_{\text{TES}}^{\text{DCh},m,t} \eta_{\text{TES}}^{\text{DCh}} \right),$$

where it can be verified that during each time slot t,

- if the TES is charging, the thermal energy flow observed at the MVES side reduces to $\frac{S_{\text{TES}}^{n,t}}{\eta_{\text{TES}}^{\text{Ch}}}$, which is with negative values.
- if the TES is discharging, the thermal energy flow observed at the MVES side reduces to $S_{\text{TES}}^{m,t} \eta_{\text{TES}}^{\text{DCh}}$, which is with positive values.
- if the TES is neither charging nor discharging, the thermal energy flow observed at the MVES side reduces to 0.

Note that the direction of the thermal energy flow $S_{\text{TES}}^{n,t}$ observed at the TES side is also following the same rule as the rest of the MVES, which to make sure the consistency.

The thermal energy flow during any time slot t should also satisfy the maximum charging and discharging constraints, which are denoted as $S_{\text{TES}}^{\text{Ch, MAX}}$ and $S_{\text{TES}}^{\text{DCh, MAX}}$, respectively. Therefore the thermal energy flow constraint for the n-th TES can be modelled as follows:

$$S_{\text{TES}}^{\text{Ch, MAX}} \leq S_{\text{TES}}^{n,t} \left(\frac{I_{\text{TES}}^{\text{Ch},n,t}}{\eta_{\text{TES}}^{\text{Ch}}} + I_{\text{EV}}^{\text{DCh},m,t} \eta_{\text{EV}}^{\text{DCh}} \right) \leq S_{\text{TES}}^{\text{DCh, MAX}}. \tag{7.16}$$

For cases that TESs are with different maximum charging and discharging constraints, e.g., difference TES models, the above constraint can be adapted accordingly. In this part, it is assumed that the maximum charging and discharging constraints are the same for all TESs in the MVES.

The thermal energy capacity of the TES can be also indicated by its SOC. For example, the thermal energy capacity of the n-th TES at time slot t can be given by $T_{\text{TES}}^{t,n}$, while its thermal energy capacity at the service start time $T_{\text{TES}}^{\text{In},n}$ can be given by $\text{SOC}_{\text{TES}}^{n,T_{\text{TES}}^{\text{In},n}}$.

Since at the service end time $T_{\text{TES}}^{\text{Out},n}$, the thermal energy of the n-th TES should be restored to the same level of $\text{SOC}_{\text{TES}}^{n,T_{\text{TES}}^{\text{In},n}}$, it equivalents to a zero net thermal energy flow during the service time $[T_{\text{TES}}^{\text{In},n}, T_{\text{TES}}^{\text{Out},n}]$, which can be modelled as follows:

$$\sum_{t=T_{\text{TES}}^{\text{In},n}}^{T_{\text{TES}}^{\text{Out},n}} S_{\text{TES}}^{n,t} = 0. \tag{7.17}$$

In the meantime, the thermal energy capacity of the n-th TES at time slot t can be derived as follows:

$$\text{SOC}_{\text{TES}}^{n,t} = \text{SOC}_{\text{TES}}^{n,T_{\text{TES}}^{\text{In},n}} + \sum_{\tau=T_{\text{TES}}^{\text{In},n}}^{t} S_{\text{TES}}^{n,\tau}. \tag{7.18}$$

During the service time, the capacity of the TES should be maintained within a range for safety reasons. Without the loss of generality, the maximum and minimum thermal energy capacity of the TESs are denoted as $\text{SOC}_{\text{TES}}^{\text{MIN}}$ and $\text{SOC}_{\text{TES}}^{\text{MAX}}$, respectively.

Then in any time slot t, the TES's SOC should meet the constraints modelled as follows:

$$\text{SOC}_{\text{TES}}^{\text{MIN}} \leq \text{SOC}_{\text{TES}}^{n,t} \leq \text{SOC}_{\text{TES}}^{\text{MAX}}. \tag{7.19}$$

Again here it assumes that the TESs are with the same minimum and maximum thermal energy capacity limits, while it is straight-forward to extend the above model the cases with different limits.

Comparing to the electric load models and thermal load models, it is seen that the EV and TES models are more complex with regard to the physical constraints. This is due to the fact that the EV and TES are with energy storage properties, which can be regarded as devices with "memories".

On the contrary, the electricity loads and thermal loads are modelled as devices without "memories". For the devices with "memories", its energy flow during its service time period would be constrained by several factors including

- the initial state at the service start time,

- the previous energy flows before the current time slot,

- the total capacity of the device,

- other mandatory constraints (e.g., the zero net energy flow constraints in EVs).

Note that the MVES models detailed above are not complete and just for illustration purposes, which shows the general methods to model the common types of devices in MVES. The modelling method can be further extended to characterize the devices with similar features.

7.2.1.2 MVES Coordination Modelling

The MVES models above not only characterize the operations of each MVES device, but also their interconnections from the view of energy vectors. Built upon that, the operation of the MVES can be described, while the coordination of the MVES can be regarded as variables to be optimized in the MVES models.

Real-time coordination modelling In general cases, the performance of the MVES can be defined as costs due to operation results, which are then optimized under the physical constraints in the MVES. Here the cost is a general term, which can be implemented as the carbon emissions measured in tonnes of carbon dioxide equivalent (CO2e), or the real money measured in pounds or dollars.

Without the loss of generality, it is assumed that the cost for the electricity (sourced from power grid), natural gas, wind energy, solar energy, EVs and TESs are denoted as C_E^t, C_G^t, C_W^t, C_{PV}^t, C_{EV}^t and C_{TES}^t, respectively.

There could be different formulations according to the MVES coordination objective. For example, the MVES is to provide a real-time coordination among the devices such that the operation costs are minimized under the physical constraints, then it can be modelled as the following optimization problem below:

$$\mathbb{P}_0: \quad \min. \quad C_E^t S_E^t + C_G^t S_G^t + C_{\text{PV}}^t S_{\text{PV}}^t + C_W^t S_W^t + \sum_{m=1}^{M} C_{\text{EV}}^t S_{\text{EV}}^{m,t} + \sum_{n=1}^{N} C_{\text{TES}}^t S_{\text{TES}}^{n,t}$$

(7.20)

$$\text{s.t. } (7.1) - (7.8), (7.11) - (7.12), (7.14), (7.16) - (7.17) \text{ and } (7.19),$$

It is noticed that in the formulation of $\mathbb{P}_0$, the variables to be optimized are the MVES components' energy flows and the dispatching factors v_{CHP}^t and v_{B}^t at time slot t. Each feasible solution of the problem $\mathbb{P}_0$ corresponds to one coordination method of the MVES. Note that such problems are based on each time slot, therefore it can be regarded as real-time coordination.

However, in the above formulation, it is noticed that the storage type devices are involved, which the objective of the real-time operation depends on not only the studied time slot t, but also each time slot before t as well as each time slot after it. Therefore this formulation is usually applicable to the cases where the operations at each time slot t are known to the storage type devices. In such cases, the energy flow variables $S_{\text{EV}}^{m,t}$ and $S_{\text{TES}}^{n,t}$ are given, so that the problem $\mathbb{P}_0$ reduces to the optimization of the MVES devices other than the EVs and TESs. Some real-world systems can be modelled by the problem $\mathbb{P}_0$. For example, some TESs are pre-programmed with fixed instructions according to the time of the day. Or in other cases, the MVES provides a bid/offer mechanism that a fixed amount of energy flows are to be served by one or multiple EVs.

From the view of the MVES operation across the whole day, the coordination according to the solution of $\mathbb{P}_0$ has been reduced to the MVES system without storage type devices. The benefits of the storage type devices, especially their ability in shifting the energy usage across time slots, are not exploited. However, due to the physical constraints of the storage types devices, the coordination of their energy flow across different time slots is not likely to be optimized in a real-time manner. For example, if only focusing on the time slot t, then some EV's energy flow might give better cost performance than others via solving $\mathbb{P}_0$.

But this choice of the EV's energy flow at the time slot t will influence the choices of other time slots, e.g., the time slot $t + 1$. This leads to a good cost performance at t but bad cost performance at $t + 1$, where the overall cost performance for all time slots is not optimal, comparing to the case that the storage type devices can be jointly coordinated by the MVES together with other devices.

Day-ahead scheduling modelling When the storage type devices are involved in the MVES coordination for all day operations, the overall cost performance can be optimized via well-planned scheduling schemes. In this part, we focus on the daily operation of the MVES, which consists of 24 hours and the operation is hourly based.

Correspondingly, the scheduling decisions are made for the whole 24 hours, which is referred to as day-ahead scheduling.

It is noticed that in the MVES model as introduced above, the renewable energy flows S_W^t and S_{PV}^t are determined by the weather during the given time slot t. In the meantime, the electricity loads and thermal loads L_E^t and L_H^t are determined by the users during the given time slot t. These are all variables determined by the external environments in the future, which are not controlled by the MVES. During the day-ahead scheduling, it is not possible to know these future values, while only forecasts (or sometimes called predictions) can be used.

Without the loss of generality, these forecasts are represented as a forecast vector, where $\mathbf{F}^t = \{L_E^t, L_H^t, S_W^t, S_{PV}^t\}$. For the other variables, they can be scheduled ahead based on the forecasts, including the electricity energy flow S_E^t from the power grid, the natural gas energy flow S_G^t, the m-th EV's energy flow $S_{EV}^{m,t}$, the n-th TES's energy flow $S_{TES}^{n,t}$, and the MVES dispatching factor $\mathbf{v}^t$}. These scheduling variables are denoted as $\mathbf{S}_{Sch}^t = \{S_E^t, S_G^t, S_{EV}^{m,t}, S_{TES}^{n,t}, \mathbf{v}^t\}$.

The scheduling variables are very helpful for the MVES coordination. The storage type devices can be instructed by the scheduled operation based on the m-th EV's energy flow $S_{EV}^{m,t}$ and the n-th TES's energy flow $S_{TES}^{n,t}$, which are optimized on the daily basis for the whole day operation.

As for the electricity energy flow S_E^t from the power grid, the natural gas energy flow S_G^t and the MVES dispatching factor $\mathbf{v}^t$, they are to ensure the demands and supplies are balanced with respect to the forecasts. Beyond that, since the scheduled electricity energy flow S_E^t from the power grid and the scheduled natural gas energy flow S_G^t for each time slot t are known in the day ahead, the MVES can purchase these amount of energy in the day-ahead energy markets, instead of purchasing them in the real-time energy markets. Normally the day-ahead energy markets are with much lower energy prices comparing to the real-time energy markets, which could help further reduce the overall operation costs.

If denote the day-ahead market prices for the electricity and natural gas as $C_E^{0,t}$, $C_G^{0,t}$, then the overall operation cost during time slot t is denoted as C_{All}^t, which can be defined as follows:

$$C_{All}^t = \underbrace{C_E^{+,t}\Delta S_E^t + C_G^{+,t}\Delta S_G^t}_{\text{Real-time Extra Costs}} +$$

$$\underbrace{C_E^{0,t}S_E^t + C_G^{0,t}S_G^t - \sum_{n=1}^{N}C_{TES}|S_{TES}^{n,t}| - \sum_{m=1}^{M}C_{EV}|S_{EV}^{m,t}| - C_{PV}S_{PV}^t - C_W S_W^t,}_{\text{Day-ahead Scheduling Costs}} \quad (7.21)$$

where $C_E^{+,t}$ and $C_G^{+,t}$ denote the real-time energy market prices for the electricity and natural gas, while ΔS_E^t and ΔS_G^t denote the difference between the scheduled energy flow and the actually required energy flow for the electricity and natural gas, respectively.

Since the MVES should balance the supply and demand in real-time, it is expected that any mismatched amount of energy, either more or less than required, will be matched by referring to the real-time energy market. In the next, we will provide a feasible solution as an example to use the real-time energy market to address the mismatched energy amounts.

Considering that it will impact the whole day operations if the energy flows for the storage type devices are adjusted, the EVs and TESs will follow exactly the scheduled operations during each time slot. The scheduled MVES dispatching factor $\mathbf{v}^t$ will be also used for real-time operations, since its changes will also impact multiple devices.

In the meantime, the MVES will adjust the energy flows in the transformer and the boiler for the mismatch of electricity energy and thermal energy, respectively. In this way, the mismatched electricity and natural gas energy flows can be evaluated below:

$$\Delta S_{\mathrm{E}}^t = \frac{1}{\eta_{\mathrm{TF}}} \left(\tilde{L}_{\mathrm{E}}^t - \eta_{\mathrm{TF}} S_{\mathrm{E}}^t - v_{\mathrm{CHP}}^t \eta_{\mathrm{CHP}}^E S_{\mathrm{G}}^t - \tilde{S}_{\mathrm{W}}^t - \tilde{S}_{\mathrm{PV}}^t - \sum_{m=1}^{M} S_{\mathrm{EV}}^{m,t} \right), \quad (7.22)$$

$$\Delta S_{\mathrm{G}}^t = \frac{1}{\eta_{\mathrm{B}}} \left(\tilde{L}_{\mathrm{H}}^t - v_{\mathrm{CHP}}^t \eta_{\mathrm{CHP}}^H S_{\mathrm{G}}^t - v_{\mathrm{B}}^t \eta_{\mathrm{B}} S_{\mathrm{G}}^t - \sum_{n=1}^{N} S_{\mathrm{TES}}^{n,t} \right), \quad (7.23)$$

where the real-time electricity loads and thermal loads are denoted as $\tilde{L}_{\mathrm{E}}^t$ and $\tilde{L}_{\mathrm{H}}^t$, while the real-time wind energy flow and solar energy flow are denoted as $\tilde{S}_{\mathrm{W}}^t$ and $\tilde{S}_{\mathrm{PV}}^t$, respectively.

It is seen that the energy flow mismatches ΔS_{E}^t and ΔS_{G}^t are due to the mismatches between the forecasting vector $\mathbf{F}$ and its corresponding real-time actual values $\tilde{\mathbf{F}}^t = \{\tilde{L}_{\mathrm{E}}^t, \tilde{L}_{\mathrm{H}}^t, \tilde{S}_{\mathrm{W}}^t, \tilde{S}_{\mathrm{PV}}^t\}$. These mismatches are denoted as forecasting errors $\boldsymbol{\delta}^t$. Note that the overall operation cost C_{All}^t during time slot t can be further rewritten as the functions of $\boldsymbol{\delta}^t$, $\mathbf{F}^t$ and $\mathbf{S}_{\mathrm{Sch}}^t$ as follows:

$$C_{\mathrm{All}}^t \triangleq C_{\mathrm{All}}^t \left(\mathbf{F}^t, \boldsymbol{\delta}^t, \mathbf{S}_{\mathrm{Sch}}^t \right).$$

Here a specific MVES scheduling problem is studied. The given inputs are the given day-ahead forecasts $\mathbf{F}^t$, while the objective is to minimize the overall costs across the whole day. Although the real-time actual $\tilde{\mathbf{F}}^t$ cannot be known during the day-ahead scheduling, it is still possible to obtain some knowledge regarding the forecasting errors $\boldsymbol{\delta}^t$.

Here the forecasting errors are defined as follows

$$\boldsymbol{\delta}^t = \frac{\tilde{\mathbf{F}}^t - \mathbf{F}^t}{\mathbf{F}^t}. \quad (7.24)$$

Following this definition, each forecasting errors can be derived accordingly. For example, the electricity load forecasting error $\delta_{\mathrm{E}}^t = \frac{\tilde{L}_{\mathrm{E}}^t - L_{\mathrm{E}}^t}{L_{\mathrm{E}}^t}$.

Note that these historical forecasting errors only provide some statistical knowledge regarding the relationship between the day-ahead forecasts $\mathbf{F}^t$ and the real-time actual $\tilde{\mathbf{F}}^t$, therefore here the objective is set as the minimization of the overall costs in a statistical manner with the knowledge of the forecasting errors $\boldsymbol{\delta}^t$, which can be detailed as follows:

$$\mathbb{P}_1: \quad \min_{\mathbf{S}_{\mathrm{Sch}}^t} \; \mathbb{E}_{\boldsymbol{\delta}^t} \left\{ \sum_{t=1}^{T} C_{\mathrm{All}}^t \left(\mathbf{F}^t, \boldsymbol{\delta}^t, \mathbf{S}_{\mathrm{Sch}}^t \right) \right\} \quad (7.25)$$

$$\text{s.t. } (7.1) - (7.8), (7.11) - (7.12), (7.14), (7.16) - (7.17) \text{ and } (7.19),$$

where $\mathbb{E}_x\{f(x)\} = \int_{-\infty}^{\infty} f(x)\mathrm{PDF}(x)dx$ is the mathematical expectation operation and $\mathrm{PDF}(x)$ is the probability distribution function (PDF) of x.

Comparing the MVES coordination formulations of $\mathbb{P}_0$ and $\mathbb{P}_1$, it is seen that $\mathbb{P}_0$ is focusing on instantaneous operation cost optimization, while $\mathbb{P}_1$ is focusing on long-term whole day operation cost optimization. The two models have their own merits. For example in the former model, the coordination is easier as the MVES is only required to supply the real-time demands, comparing to the later model. While in the latter model, it could achieve lower operation costs with the support of forecasting services and mature energy markets.

7.2.2 Artificial Intelligence Enhancing Multi-Vector Energy System

Based on the MVES models, the coordination can be optimized by solving the mathematical formulations, such as the illustrated examples in $\mathbb{P}_0$ and $\mathbb{P}_1$. In this part, we will focus on the use of the AI algorithms to enhance the MVES performance.

Specifically, the use of AI algorithms in solving $\mathbb{P}_1$ will be given as an example.

7.2.2.1 Addressing Physical Constraints in Artificial Intelligence Algorithms

The energy systems are with strict physical constraints on each of its components, which must be strictly complied with by each provided coordination solution. However, many AI algorithms are model-free solutions, where the AI algorithm is focusing on the data instead of the models behind the screen.

To exploit such data-driven AI algorithms, a common challenge is to make sure that all physical constraints have been enforced, which is a common challenge in the AI enhancing MVES problem. There are no one-for-all solutions to this challenge, but there are indeed some techniques that can be useful in many cases. In this part, we will illustrate several techniques will a data-driven AI architecture as an example.

Without the loss of generality, it is assumed that there exists a Neural Network (NN) $f(\cdot; \boldsymbol{\theta})$ that can solve the problem $\mathbb{P}_1$, namely it can output the day-ahead scheduling decisions $\mathbf{S}_{\mathrm{Sch}}^t$ based on the day-ahead forecasts $\mathbf{F}^t$ as inputs.

This relation can be generally expressed as follows:

$$\mathbf{S}_{\mathrm{Sch}}^t = f(\mathbf{F}^t; \boldsymbol{\theta}). \tag{7.26}$$

Here the NN can be fulfilled by any potential structure, for example, the Convolutional NN (CNN) and Deep NN (DNN). Although NNs have been extensively used to learn the hidden relations in a given problem, the structure of the NN is generally independent from the physical models and meanings, except the input and output layers. Here the input layer and output layer are generally defined, which refers to the general mapping procedure between the values used in the system and the values used in the NN. For example, in the day-ahead scheduling problem defined in $\mathbb{P}_1$, the inputs are forecasts $\mathbf{F}$. The elements of $\mathbf{F}$ are the forecasts for loads and renewable energy flows, where it is common that the values are very different in scales and ranges.

Then the scaling of the inputs to an appropriate range first before using as the inputs of the NN is also regarded as part of the input layer. Note that in other works, this part might be defined as the data pre-processing, but here such signal processing techniques are unified as part of the input layer.

Correspondingly, the outputs of the NN might need some post-processing, before they can be used by the MVES for coordination purposes. Therefore the output layer discussed here is also a general term, which includes any signal processing to map the outputs to the required values by the MVES.

By exploiting signal processing techniques as well as NN architectures, some physical constraints can be addressed by the output layers. But note that not all physical constraints can be addressed by the output layers, for example, the EVs where the variables need to be enforced by multiple constraints at the same time. In the following, we will use the combination of both the NN output layer design and the deep learning design to address the multiple physical constraints as in $\mathbb{P}_1$.

The NN output layer design: In the MVES, some variables are with physical constraints on their value ranges, which can be enforced by some NN functions or layers. Typical examples of such variables are the energy flow constraints for the transformer in (7.3), the natural gas in (7.4), the wind turbine in (7.5), the PV generator in (7.6) and the CHP in (7.7).

The commonly used technique is to firstly enforce the outputs in a known range, then scale them to the required range for the MVES. For example, to enforce the transformer's energy flow via the NN output layer design, it requires the NN output layer to give values in the range of $[0, S_{TF}^{MAX}]$. Then for the parts corresponding to the transformer's energy flow, the NN can first implement a sigmoid layer, which transforms the previous layers output to the range of $[0, 1]$. Then the output of the sigmoid layer is scaled by S_{TF}^{MAX}, which provides exactly the required value range of $[0, S_{TF}^{MAX}]$.

Some outputs, such as the energy flow for EVs in (7.12) and TESs in (7.16), are within ranges with both maximal and minimal bounds. With the m-th EV as an example, its energy flow should be within range $[S_{EV}^{Ch, MAX}, S_{EV}^{DCh, MAX}]$. In such cases, the NN can first implement a tanh layer for the corresponding parts, which transforms the previous output to the range of $[-1, 1]$. Then based on the sign of the output values, they can be scaled to the required ranges.

In other words, if the output value is positive, then the value will be multiplied with $S_{EV}^{DCh, MAX}$. Correspondingly, if the output value is negative, then the value will be multiplied with $S_{EV}^{Ch, MAX}$. In this way, the values from the NN output layer are in the required range of $[S_{EV}^{Ch, MAX}, S_{EV}^{DCh, MAX}]$.

For the storage type devices, their outputs are usually with multiple constraints. The NN output layer design can help address part of the constraints, for example, the EV's energy flows can be confined in the required ranges. But in the EV's models, there are two additional physical constraints associated with the EV's energy flows, other than the value range constraints in (7.12). One of the two physical constraints is the zero net energy flows constraints across their service time as defined in (7.11).

This kind of constraints describes some statistical features of multiple outputs, such as the sum or the average of them. To address their physical constraints, it can apply post-signal processing based on the multiple outputs. In the EV's case with zero net energy flows constraints in (7.11), the NN can first calculate the average of the EV's energy flows from the previous layers, and then subtract this average from each of the EV's energy flows. This manipulation will ensure that the EV's energy flows are always with zero average. The same technique can be applied to the TES, where the zero net energy flow constraints in (7.17) can be enforced.

Note that since there might be more than one technique applied to the same output values, it should be checked that the techniques are not conflicting with each other, or further manipulations will be required. In the case of the EV's energy flows applying both the value range scaling and average removing, it can be checked that the scaling and removing average might have potential conflicts. For example, when removing the average from the values, its value range will be changed.

These two techniques can be jointly considered as follows:

1. the EV's energy flows are firstly mapped to the range of $[-1, 1]$ by the tanh layers,

2. the average is then removed from the EV's energy flows,

3. the maximum absolute value of the EV's energy flows is calculated and denoted as $\max |S_{\mathrm{EV}^{m,t}}|$,

4. the absolute value of the scaling factor is calculated as $\frac{\min\{|S_{\mathrm{EV}}^{\mathrm{Ch, MAX}}|, |S_{\mathrm{EV}}^{\mathrm{DCh, MAX}}|\}}{\max |S_{\mathrm{EV}}^{m,t}|}$,

5. the EV's energy flows are scaled as follows: if the value is positive, then it is multiplied by $\frac{\min\{|S_{\mathrm{EV}}^{\mathrm{Ch, MAX}}|, |S_{\mathrm{EV}}^{\mathrm{DCh, MAX}}|\}}{\max |S_{\mathrm{EV}}^{m,t}|}$; else the value is multiplied by $-\frac{\min\{|S_{\mathrm{EV}}^{\mathrm{Ch, MAX}}|, |S_{\mathrm{EV}}^{\mathrm{DCh, MAX}}|\}}{\max |S_{\mathrm{EV}}^{m,t}|}$.

It can be verified that with the above manipulations, both physical constraints can be enforced at the same time. Note that the above technique is just one feasible solution, where other variations can be used as long as the constraints are enforced.

The loss function design: Besides using the NN layers to implement the physical constraints, another commonly used technique is to exploit the learning feature of the NNs, which embeds the constraints as part of the learning goals. In other words, the NN is required to learn not only how to optimize the problem's original objectives, but also how to satisfy the constraints.

One practical method of such techniques is to append the physical constraints with penalty factors at the end of the original loss functions, where any constraint violations will incur a positive penalty increase in the loss during the training. The philosophy behind this method is to take advantage of the NN training procedure, where the training is mathematically fulfilled by reducing the losses (defined by the loss function). Since the constraint violation will increase the losses, it is expected that a well-trained NN is able to learn not only the original problem's objective, but also a "strategy" to meet the physical constraints.

An example of such techniques can be given based on the MVES day-ahead scheduling problem defined in $\mathbb{P}_1$. Note that with the relation $\mathbf{S}_{\mathrm{Sch}}^t = f(\mathbf{F}^t; \boldsymbol{\theta})$ in (7.26), the overall MVES operating cost during time slot t can be further rewritten as follows:

$$C_{\mathrm{All}}^t \triangleq C_{\mathrm{All}}^t \left(\mathbf{F}^t, \boldsymbol{\delta}^t, f(\mathbf{F}^t; \boldsymbol{\theta}) \right). \tag{7.27}$$

Without the consideration of the physical constraints, the loss function can be defined as the overall operating costs as follows:

$$\mathcal{L}(\boldsymbol{\theta}) = \mathbb{E}_{\boldsymbol{\delta}^t} \left\{ \sum_{t=1}^{T} C_{\mathrm{All}}^t \left(\mathbf{F}^t, \boldsymbol{\delta}^t, f(\mathbf{F}^t; \boldsymbol{\theta}) \right) \right\},$$

but the above formulation is with two deficiencies that need to be addressed: a) the trained NN will depend on $\mathbf{F}^t$, i.e., for each $\mathbf{F}^t$, it might require to train a totally different NN, and b) the physical constraints are not considered.

The first problem can be addressed by training the NN to be applicable for all possible $\mathbf{F}^t$ with one NN structure, this can be achieved by modifying the loss function as follows:

$$\mathcal{L}(\boldsymbol{\theta}) = \mathbb{E}_{\boldsymbol{\delta}^t, \mathbf{F}^t} \left\{ \sum_{t=1}^{T} C_{\mathrm{All}}^t \left(\mathbf{F}^t, \boldsymbol{\delta}^t, f(\mathbf{F}^t; \boldsymbol{\theta}) \right) \right\},$$

where the mathematical expectation is taken not only with regard to $\boldsymbol{\delta}^t$, but also $\mathbf{F}^t$. In this way, the outputs of the trained NN are expected to optimize the overall MVES operating costs with different forecasting values and forecasting errors, using the same trained NN parameters.

Next, in order to consider the physical constraints as part of the training procedure, we can define their corresponding penalty terms. For example, for the EVs' SOC constraints defined in (7.14) , its penalty term can be defined as follows:

$$C_{\mathrm{EV}}^{\mathrm{P},t}(\mathbf{F}^t, f(\mathbf{F}^t; \boldsymbol{\theta})) = \max\{\mathrm{SOC}_{\mathrm{EV}}^{\mathrm{MIN}} - \mathrm{SOC}_{\mathrm{EV}}^{m,t}, \mathrm{SOC}_{\mathrm{EV}}^{m,t} - \mathrm{SOC}_{\mathrm{EV}}^{\mathrm{MAX}}, 0\}, \tag{7.28}$$

which will always return a non-negative value. Specifically, it can verify that

$$C_{\mathrm{EV}}^{\mathrm{P},t}(\mathbf{F}^t, f(\mathbf{F}^t; \boldsymbol{\theta})) = \begin{cases} \mathrm{SOC}_{\mathrm{EV}}^{\mathrm{MIN}} - \mathrm{SOC}_{\mathrm{EV}}^{m,t}, & \text{if } \mathrm{SOC}_{\mathrm{EV}}^{m,t} < \mathrm{SOC}_{\mathrm{EV}}^{\mathrm{MIN}}, \\ \mathrm{SOC}_{\mathrm{EV}}^{m,t} - \mathrm{SOC}_{\mathrm{EV}}^{\mathrm{MAX}}, & \text{if } \mathrm{SOC}_{\mathrm{EV}}^{m,t} > \mathrm{SOC}_{\mathrm{EV}}^{\mathrm{MAX}}, \\ 0, & \text{otherwise.} \end{cases} \tag{7.29}$$

where it is clear that $C_{\mathrm{EV}}^{\mathrm{P},t}$ will be zero if the EVs' SOC constraints have been met, or its values will be a positive value indicating the severity of the constraint violation. Note that the definition of $C_{\mathrm{EV}}^{\mathrm{P},t}$ may have different variations (by means of the absolute difference between $\mathrm{SOC}_{\mathrm{EV}}^{m,t}$ and the violated constraint bound).

Similarly, the corresponding penalty term for the TESs' energy flow constraints in (7.19) can be defined as follows:

$$C_{\mathrm{TES}}^{\mathrm{P},t}(\mathbf{F}^t, f(\mathbf{F}^t; \boldsymbol{\theta})) = \max\{\mathrm{SOC}_{\mathrm{TES}}^{\mathrm{MIN}} - \mathrm{SOC}_{\mathrm{TES}}^{n,t}, \mathrm{SOC}_{\mathrm{TES}}^{n,t} - \mathrm{SOC}_{\mathrm{TES}}^{\mathrm{MAX}}, 0\}, \tag{7.30}$$

Then by defining a positive penalty parameter λ, the final loss function can be formulated as follows:

$$\begin{aligned}
\mathcal{L}(\boldsymbol{\theta}) =& \mathbb{E}_{\mathbf{F}^t, \boldsymbol{\delta}^t} \Bigg\{ \sum_{t=1}^{T} C_{\text{All}}^t \left(\mathbf{F}^t, \boldsymbol{\delta}^t, f(\mathbf{F}^t; \boldsymbol{\theta}) \right) \\
&+ \lambda \sum_{t=1}^{T} \left(C_{\text{EV}}^{\text{P},t}(\mathbf{F}^t, f(\mathbf{F}^t; \boldsymbol{\theta})) + C_{\text{TES}}^{\text{P},t}(\mathbf{F}^t, f(\mathbf{F}^t; \boldsymbol{\theta})) \right) \Bigg\}.
\end{aligned} \tag{7.31}$$

Remarks: When using NNs to solve problems for the MVES, it is a common challenge to address the physical constraints before the NN can be trained. The above formulations give an example of how to exploit some techniques to enforce the physical constraints via NN structure design and loss function design.

Note that this formulation is not unique, where other formulations are also feasible. For example, the EV's zero net energy flow constraint in (7.11) can be also addressed via the loss function design, instead of the NN structure design. In such cases, the corresponding penalty term can be defined as follows:

$$C_{\text{EV}}^{\text{P}_{\text{Net}},t} = \Bigg| \sum_{t=T_{\text{EV}}^{\text{In},m}}^{T_{\text{EV}}^{\text{Out},m}} S_{\text{EV}}^{m,t} \Bigg|, \tag{7.32}$$

which can be then appended to the loss function as an additional penalty term.

In general cases, the physical constraints can be usually addressed via the loss function design. But the NN structure design technique depends on situations, where sometimes it might not be applicable. For example for the three physical constraints related to the EV's model, the previous examples have used the NN architecture design to address two of them, but it is unlike to address all of them.

The expected performance from these two techniques is also different. For the loss function designs, the constraints are enforced in a "soft" manner, where the NN is punished via the penalty term during the learning. It is referred to as the "soft" manner because it is still possible that the NN will give outputs violating the constraints, since there is no firm enforcement to ensure the constraints are satisfied.

To mitigate this challenge, it can assign a large penalty parameter λ. During the training procedure, the NN is expected to learn that any constraint violation will be very "expensive" in the sense of the overall loss. The main drawback is that the parameter λ may be hard to choose. Because if λ is not large enough, the constraint violations might be prominent. Or if λ is too large, it might compromise the main learning objective, because the NN might learn to give outputs in an over-conservative manner.

For the NN structure designs, the constraints are enforced in a "hard" manner, which is guaranteed in a firm manner. This will be preferred in the MVES or other physical constraint related problems, because in practice these physical constraints are not supposed to be violated in any situations.

Besides the challenges in finding an appropriate design for specific constraints, the NN structure designs will also have great impact on the training and learning performances. This is because a change of the NN structure will lead to different calculations during the training procedure, which would lead to totally different NN parameters and the corresponding final performances.

7.2.2.2 Deep Learning Enhanced Multi-Vector Energy System

By addressing the multiple physical constraints via the NN structure design and loss function design, the original MVES day-ahead scheduling problem defined in $\mathbb{P}_1$ is transformed to an unconstrained deep learning problem defined as follows:

$$\min_{\boldsymbol{\theta}} \mathcal{L}(\boldsymbol{\theta}), \tag{7.33}$$

where during the training procedure, the NN is expected to learn to meet the EVs' constraints in (7.14) and TESs' SOC constraints in (7.19), while the other constraints identified in $\mathbb{P}_1$ are enforced via the NN output layer designs.

It is noticed that (7.33) is a general form for deep learning problems. Therefore in order to make the learning problem much clearer, (7.33) can be further rewritten in its equivalent form as follows:

$$
\begin{aligned}
\min_{\boldsymbol{\theta}} \mathbb{E}_{\mathbf{F}^t, \boldsymbol{\delta}^t} \Bigg\{ & \sum_{t=1}^{T} C_{\text{All}}^t \left(\mathbf{F}^t, \boldsymbol{\delta}^t, f(\mathbf{F}^t; \boldsymbol{\theta}) \right) \\
& + \lambda \sum_{t=1}^{T} \left(C_{\text{EV}}^{\text{P},t}(\mathbf{F}^t, f(\mathbf{F}^t; \boldsymbol{\theta})) + C_{\text{TES}}^{\text{P},t}(\mathbf{F}^t, f(\mathbf{F}^t; \boldsymbol{\theta})) \right) \Bigg\}.
\end{aligned}
\tag{7.34}
$$

which is obtained by substituting the detailed representation of the loss function of (7.31) into (7.33).

With the representation in (7.34), the deep learning based MVES day-ahead scheduling can be summarized as follows:

- The NN $f(\cdot; \boldsymbol{\theta})$ is built to learn to solve the problem in $\mathbb{P}_1$, whose inputs are the day-ahead forecasting vectors $\mathbf{F}^t$ and outputs are the day-ahead scheduling vectors $\mathbf{S}_{\text{Sch}^t}$.

- The NN output layers are following the design detailed in Section 7.2.2.1 to address some of the physical constraints, while the other physical constraints are addressed via the loss function design as specified in (7.31).

- The training procedure is following an unsupervised learning manner, where the training performance at each step is evaluated by the loss function (7.31), with historical or simulated forecasting errors $\boldsymbol{\delta}^t$.

- The NN parameters $\boldsymbol{\theta}$ are updated via the Gradient Decent algorithm [65].

To evaluate how the described deep learning method can help reduce the MVES operating costs, in the next we will use a dummy example with 4 EVs and 2 TESs. Specifically, the NN is fulfilled by a 5 layer DNN with the shape of $96 \times 768 \times 576 \times 384 \times 216$, where ReLU function is used as the activation layer at each hidden

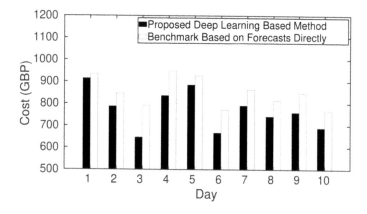

Figure 7.2 The daily MVES operating costs simulated with the data from September 1st, 2017 to September 10th, 2017, comparing against the benchmark method based on forecasts directly.

layer. Some typical values are assumed for the energy converter, with $\eta_{TF} = 0.98$, $\eta_{CHP}^{E} = 0.404$, $\eta_{CHP}^{H} = 0.566$, $\eta_B = \eta_{EV}^{Ch} = \eta_{EV}^{DCh} = \eta_{TES}^{Ch} = \eta_{TES}^{DCh} = 0.9$.

The maximal energy flow bounds are assumed as follows, where $S_{TF}^{MAX} = S_{G}^{MAX} = S_{B}^{MAX} = 1200$ kWh, $S_{W}^{MAX} = S_{PV}^{MAX} = 200$ kWh, $S_{CHP}^{MAX} = 300$ kWh. For the EVs, the physical constraints on the energy flows are assumed as $\{S_{EV}^{Ch,\ MAX}, S_{EV}^{DCh,\ MAX}, SOC_{EV}^{MIN}, SOC_{EV}^{MAX}\} = \{80$ kWh, 80 kWh, 40 kWh, 80 kWh$\}$. For the TESs, he physical constraints on the energy flows are assumed as $\{S_{TES}^{Ch,\ MAX}, S_{TES}^{DCh,\ MAX}, SOC_{TES}^{MIN}, SOC_{TES}^{MAX}\} = \{50$ kWh, 50 kWh, 40 kWh, 200 kWh$\}$.

Fixed prices are assumed for the day-ahead energy market, where $C_{E}^{0} = 0.062$ GBP/kWh, $C_{G}^{0} = 0.026$ GBP/kWh. For the real-time energy market, the prices are differentiated according to situations, where $C_{E}^{+}=0.054$ GBP/kWh if the MVES needs to buy more electricity, or $C_{E}^{+}=0.012$ GBP/kWh if the MVES needs to refund for the unconsumed electricity. Similarly, for the natural gas, $C_{G}^{+}=0.018$ GBP/kWh if the MVES needs to import more natural gas, or $C_{G}^{+}=0.004$ GBP/kWh if the MVES needs to refund for the unconsumed natural gas.

By training with the forecasting and actual data in 2019 according to the U.K. electricity and natural gas dataset [66] [68], the model is then applied to the data in 2017 to test its performance.

The trained NN is first applied to test the daily MVES operating costs, whose simulation results are illustrated in Fig. 7.2. The data for the simulation are from September 1st, 2017 to September 10th, 2017. Since the data are from the year 2017 and the trained data are from 2019, it is clear that these data are not involved in the training process.

It is seen that for the sequential 10 days, the proposed deep learning based method is capable of reducing the daily MVES operating costs for each day, comparing to the benchmark method which directly uses forecasts for day-ahead scheduling. The cost saving performance is varying with days, which is expected since the forecasting errors and forecasting vectors are both random.

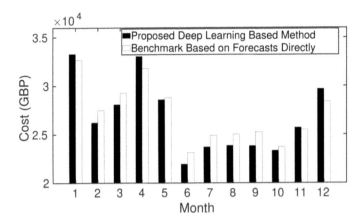

Figure 7.3 The monthly MVES operating costs simulated with the data from January 2017 to December 2017, comparing against the benchmark method based on forecasts directly.

A monthly view of the MVES operating cost is provided in Fig. 7.3. It is seen that for most months, except for January, April, November, and December, the proposed deep learning based method outperforms the benchmark method based on forecasts directly.

In other words, comparing to the benchmark method, the proposed method is better for the majority of cases, instead of all cases. This is because the formulation in $\mathbb{P}_1$ is expecting statistically good performance on the MVES operating cost, where the long-term performance should be good but there might be fluctuations for individual cases.

Besides the MVES operating costs, the physical constraints are also of great interest here. In Fig. 7.4, it illustrates the SOC of one EV operated following the day-ahead scheduling decisions made by the NN. During the whole day operation, this EV's SOC has been successfully constrained within the range of $[\text{SOC}_{\text{EV}}^{\text{MIN}}, \text{SOC}_{\text{EV}}^{\text{MAX}}]$, which corresponds to the range of $[40, 80]$ kWh in Fig. 7.4.

In addition, the SOC of one of the TESs has been illustrated in Fig. 7.5, which also demonstrates that this TES's SCO has been successfully constrained within its physical constraint range. Note that the outputs of the NN are the day-ahead scheduled energy flows for the EVs and TESs, and the meeting of SOC constraint is learned by the NN instead of NN structure design. This illustrates that the exploited techniques in the previous formulations have successfully addressed the physical constraints.

Remarks: From the examples above, it demonstrates the great potentials of using AI algorithms for the enhancement of MVES performances. If without the AI algorithms, the traditional solution in such cases might be using the forecasts directly (i.e., the benchmark algorithm). This choice is the heuristic solution in common sense when forecasts are available, which relies on the accuracy of the forecasts.

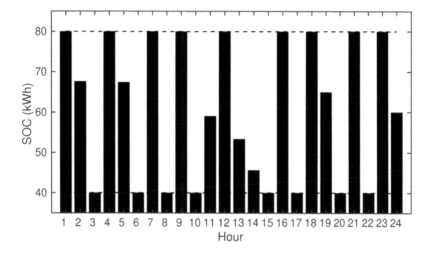

Figure 7.4 The hourly SOC performance of one EV simulated with the data from September 1st, 2017.

The accuracy of the forecasts is normally referring to the forecast itself, while the impacts due to the forecasting errors are implicit in most cases. By using the AI algorithms, it helps the MVES to learn from the forecasting errors, so that the day-ahead scheduling can be statistically optimized for the operating costs.

Besides, the above example has also exploited another important feature of the AI algorithm, which is end-to-end learning. Instead of worrying about any intermediate variables, the NN in the example directly learns how to improve the end performance (i.e., the MVES operating costs) with given forecasts. This could be important to the enhancement of the MVES operations, since from the models it can tell the MVES is essentially a cross-discipline system. In such cases involving complicated coordination

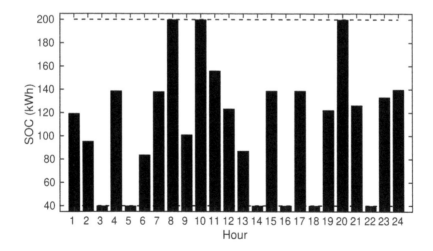

Figure 7.5 The hourly SOC performance of one TES simulated with the data from September 1st, 2017.

among multiple systems, the AI algorithm might even be the most promising solution, where traditional analytical methods might be short-handed.

Additionally, it might be also beneficial to separate the forecasting task from the MVES coordination. For research purposes, it might show some advantages via the joint optimization of the forecasting performance and the MVES end performance (e.g., the operating costs). But it should be minded that the forecasting task is a very complicated and dynamic task. It may require huge volumes of historical data, the access of multiple environmental factors (e.g., weather), and more importantly, the frequent maintenance or upgrade of the forecasting models to adapt to new changes.

Comparing to the subscription of forecasting services from third parties, it might incur more costs to include the forecasting as part of the MVES coordination. The above example demonstrates another option when forecasting is involved, where it admits the forecasting errors instead of trying to reduce them. It leaves the problem of forecasting accuracy improvement to professional parties, while focusing on the problem of how to improve the MVES performance with given forecasting accuracy performance.

7.2.3 Remarks of Challenges

The CPS modelling for Energy Internet provides a new research direction to an integrated model of the different participants in the Energy Internet, especially for its emphasis on the interaction between cyber models and their physical counterparts. However, the CPS modelling method is also subject to several key challenges.

The first challenge is on the time scale of the involved models, which is due to the essence of the integration of different energy system components. Each component could have its own unique temporal features, where their corresponding models can describe their behaviours in different time scales. For example, the appliances (boilers for instance) usually respond to controls at minutes level, while the ICT systems (WiFi transceivers for instance) are working in the sub-second level. The integration of the models with different time scales would require a joint consideration of their own temporal features, the coordination against other models, and the interaction between their physical counterparts. This demands a reconsideration of the traditional modelling method for each individual Energy Internet components. Their CPS model may require a multiple time scale model, instead of their current single time scale models.

The second challenge is on the gap between cyber models against their physical counterparts. The cyber models are essentially mathematical descriptions of their physical processes, where there are assumptions or simplifications leading to a gap between models and the exact processes. This would lead to challenges to the CPS models, because the CPS models should characterize not only the behaviours for each components, but also the cooperation between them. The modelling of cooperation usually demands a joint consideration of the features from different systems, while it will involve non trivial efforts in inter disciplinary research and development.

The final but not the last challenge is on the flexibility of CPS models for scale-up studies. CPS modelling for Energy Internet is a novel research direction, where most existing studies are still focusing on small scale systems as proof of concepts. When the system scales up to involve significant more devices, the interaction between cyber models and their physical counterparts will become more complex. This will demand the CPS system to be flexible in characterizing systems at different scales, where further modelling and evaluation efforts are required to complement the models.

Bibliography

[1] Xingzhen Bai, Xiangzhong Meng, Zhaowen Du, Maofa Gong, and Zhiguo Hu, "Design of wireless sensor network in scada system for wind power plant," in *2008 IEEE International Conference on Automation and Logistics*, 2008, pp. 3023–3027.

[2] F. Zhixin and Y. Yue, "Condition health monitoring of offshore wind turbine based on wireless sensor network," in *2012 10th International Power & Energy Conference (IPEC)*. IEEE, 2012, pp. 649–654.

[3] F.-J. Wu, Y.-F. Kao, and Y.-C. Tseng, "From wireless sensor networks towards cyber physical systems," *Pervasive and Mobile Computing*, vol. 7, no. 4, pp. 397–413, 2011.

[4] Z. Erkin, J. R. Troncoso-Pastoriza, R. L. Lagendijk, and F. Pérez-González, "Privacy-preserving data aggregation in smart metering systems: An overview," *IEEE Signal Processing Magazine*, vol. 30, no. 2, pp. 75–86, 2013.

[5] R. L. Lagendijk, Z. Erkin, and M. Barni, "Encrypted signal processing for privacy protection: Conveying the utility of homomorphic encryption and multiparty computation," *IEEE Signal Processing Magazine*, vol. 30, no. 1, pp. 82–105, 2012.

[6] K. Kursawe, G. Danezis, and M. Kohlweiss, "Privacy-friendly aggregation for the smart-grid," in *International Symposium on Privacy Enhancing Technologies Symposium*. Springer, 2011, pp. 175–191.

[7] G. Ács and C. Castelluccia, "I have a dream!(differentially private smart metering)," in *International Workshop on Information Hiding*. Springer, 2011, pp. 118–132.

[8] Z. Erkin and G. Tsudik, "Private computation of spatial and temporal power consumption with smart meters," in *International Conference on Applied Cryptography and Network Security*. Springer, 2012, pp. 561–577.

[9] R. Bobba, H. Khurana, M. AlTurki, and F. Ashraf, "Pbes: a policy based encryption system with application to data sharing in the power grid," in *Proceedings of the 4th International Symposium on Information, Computer, and Communications Security*. ACM, 2009, pp. 262–275.

[10] X. Ren, C.-M. Yu, W. Yu, S. Yang, X. Yang, J. A. McCann, and S. Y. Philip, "Lopub: High-dimensional crowdsourced data publication with local differential privacy," *IEEE Transactions on Information Forensics and Security*, vol. 13, no. 9, pp. 2151–2166, 2018.

[11] S. F. Aboelfotoh and N. A. Hikal, "A review of cyber-security measuring and assessment methods for modern enterprises," *JOIV: International Journal on Informatics Visualization*, vol. 3, no. 2, pp. 157–176, 2019.

[12] National Cyber Security Centre, "Phishing attacks: defending your organisation," 2019. [Online]. Available: https://www.ncsc.gov.uk/guidance/phishing\ #downloads

[13] D. U. Case, "Analysis of the cyber attack on the Ukrainian power grid," *Electricity Information Sharing and Analysis Center (E-ISAC)*, 2016.

[14] D. J. Tan, T.-W. Chua, V. L. Thing *et al.*, "Securing android: a survey, taxonomy, and challenges," *ACM Computing Surveys (CSUR)*, vol. 47, no. 4, p. 58, 2015.

[15] S. Arzt, S. Rasthofer, C. Fritz, E. Bodden, A. Bartel, J. Klein, Y. Le Traon, D. Octeau, and P. McDaniel, "Flowdroid: Precise context, flow, field, object-sensitive and lifecycle-aware taint analysis for android apps," in *Acm Sigplan Notices*, vol. 49, no. 6. ACM, 2014, pp. 259–269.

[16] Z. B. Celik, E. Fernandes, E. Pauley, G. Tan, and P. McDaniel, "Program analysis of commodity iot applications for security and privacy: Challenges and opportunities," *ACM Computing Surveys (CSUR)*, vol. 52, no. 4, p. 74, 2019.

[17] X. Sun, J. Dai, P. Liu, A. Singhal, and J. Yen, "Using bayesian networks for probabilistic identification of zero-day attack paths," *IEEE Transactions on Information Forensics and Security*, vol. 13, no. 10, pp. 2506–2521, 2018.

[18] W. Enck, P. Gilbert, S. Han, V. Tendulkar, B.-G. Chun, L. P. Cox, J. Jung, P. McDaniel, and A. N. Sheth, "Taintdroid: an information-flow tracking system for realtime privacy monitoring on smartphones," *ACM Transactions on Computer Systems (TOCS)*, vol. 32, no. 2, p. 5, 2014.

[19] Y. He, X. Yang, B. Hu, and W. Wang, "Dynamic privacy leakage analysis of android third-party libraries," *Journal of Information Security and Applications*, vol. 46, pp. 259–270, 2019.

[20] Q. Lin, J. Mao, F. Shi, S. Zhu, and Z. Liang, "Detecting android side channel probing attacks based on system states," in *International Conference on Wireless Algorithms, Systems, and Applications*. Springer, 2019, pp. 201–212.

[21] C. Wueest, "Targeted attacks against the energy sector," *Symantec Security Response, Mountain View*, CA, 2014.

[22] B. Wu, J. Chen, J. Wu, and M. Cardei, "A survey of attacks and countermeasures in mobile ad hoc networks," in *Wireless Network Security*. Springer, 2007, pp. 103–135.

[23] J.-H. Cho, A. Swami, and R. Chen, "A survey on trust management for mobile ad hoc networks," *IEEE Communications Surveys & Tutorials*, vol. 13, no. 4, pp. 562–583, 2010.

[24] S. Buchegger and J.-Y. Le Boudec, "Nodes bearing grudges: Towards routing security, fairness, and robustness in mobile ad hoc networks," in *Proceedings of 10th Euromicro Workshop on Parallel, Distributed and Network-based Processing*. IEEE, 2002, pp. 403–410.

[25] L. Capra, "Towards a human trust model for mobile ad-hoc networks," 2004.

[26] W. J. Adams, G. C. Hadjichristofi, and N. Davis, "Calculating a node's reputation in a mobile ad hoc network," in *PCCC 2005. 24th IEEE International Performance, Computing, and Communications Conference, 2005*. IEEE, 2005, pp. 303–307.

[27] I. Stojmenovic, *Handbook of Wireless Networks and Mobile Computing*. Wiley Online Library, 2002.

[28] R. Maheshwari, J. Gao, and S. R. Das, "Detecting wormhole attacks in wireless networks using connectivity information," in *IEEE INFOCOM 2007–26th IEEE International Conference on Computer Communications*. IEEE, 2007, pp. 107–115.

[29] H. Yih-Chun and A. Perrig, "A survey of secure wireless ad hoc routing," *IEEE Security & Privacy*, vol. 2, no. 3, pp. 28–39, 2004.

[30] C. Karlof and D. Wagner, "Secure routing in wireless sensor networks: Attacks and countermeasures," in *Proceedings of the First IEEE International Workshop on Sensor Network Protocols and Applications, 2003*. IEEE, 2003, pp. 113–127.

[31] L. Zhang, G. Ding, Q. Wu, Y. Zou, Z. Han, and J. Wang, "Byzantine attack and defense in cognitive radio networks: A survey," *IEEE Communications Surveys & Tutorials*, vol. 17, no. 3, pp. 1342–1363, 2015.

[32] N. Bhalaji and A. Shanmugam, "Reliable routing against selective packet drop attack in DSR based manet." *Journal of Software*, vol. 4, no. 6, pp. 536–543, 2009.

[33] C. Kolias, G. Kambourakis, A. Stavrou, and S. Gritzalis, "Intrusion detection in 802.11 networks: empirical evaluation of threats and a public dataset," *IEEE Communications Surveys & Tutorials*, vol. 18, no. 1, pp. 184–208, 2015.

[34] H. Yang, H. Luo, F. Ye, S. Lu, and L. Zhang, "Security in mobile ad hoc networks: Challenges and solutions," IEEE Wireless Communications, vol. 11, no. 1, pp. 38–47, 2004.

[35] P. L. Campbell, "The denial-of-service dance," *IEEE Security & Privacy*, vol. 3, no. 6, pp. 34–40, 2005.

[36] K. Gai, M. Qiu, Z. Ming, H. Zhao, and L. Qiu, "Spoofing-jamming attack strategy using optimal power distributions in wireless smart grid networks," *IEEE Transactions on Smart Grid*, vol. 8, no. 5, pp. 2431–2439, 2017.

[37] C. Valli, A. Woodward, C. Carpene, P. Hannay, M. Brand, R. Karvinen, and C. Holme, "Eavesdropping on the smart grid," in *The Proceedings of the 10th Australian Digital Forensics Conference*, Perth, Western Australia, pp., 54–60 2012.

[38] North American Electric Reliability Corporation, "State of reliability," 2019.

[39] W. Han and Y. Xiao, "Privacy preservation for V2G networks in smart grid: A survey," *Computer Communications*, vol. 91, pp. 17–28, 2016.

[40] S. Tan, D. De, W.-Z. Song, J. Yang, and S. K. Das, "Survey of security advances in smart grid: A data driven approach," *IEEE Communications Surveys & Tutorials*, vol. 19, no. 1, pp. 397–422, 2017.

[41] R. K. Varma, S. A. Rahman, T. Vanderheide, and M. D. Dang, "Harmonic impact of a 20-MW PV solar farm on a utility distribution network," *IEEE Power and Energy Technology Systems Journal*, vol. 3, no. 3, pp. 89–98, 2016.

[42] K. Sayed and H. A. Gabbar, "Scada and smart energy grid control automation," in *Smart Energy Grid Engineering*. Elsevier, 2017, pp. 481–514.

[43] P. M. M. A. Silva, C. M. Serodio, and J. L. Monteiro, "Ubiquitous scada systems on agricultural applications," in *2006 IEEE International Symposium on Industrial Electronics*, vol. 4. IEEE, 2006, pp. 2978–2983.

[44] M. D. Ilić, L. Xie, U. A. Khan, and J. M. Moura, "modelling of future cyber–physical energy systems for distributed sensing and control," *IEEE Transactions on Systems, Man, and Cybernetics-Part A: Systems and Humans*, vol. 40, no. 4, pp. 825–838, 2010.

[45] M. Panteli, P. A. Crossley, D. S. Kirschen, and D. J. Sobajic, "Assessing the impact of insufficient situation awareness on power system operation," *IEEE Transactions on Power Systems*, vol. 28, no. 3, pp. 2967–2977, 2013.

[46] D. Schacht, D. Lehmann, H. Vennegeerts, S. Krahl, and A. Moser, "Modelling of interactions between power system and communication systems for the evaluation of reliability," in *2016 Power Systems Computation Conference (PSCC)*. IEEE, 2016, pp. 1–7.

[47] K. Kamps, F. Möhrke, M. Zdrallek, P. Awater, and M. Schwan, "modelling of smart grid technologies for reliability calculations of distribution grids," in *2018 Power Systems Computation Conference (PSCC)*. IEEE, 2018, pp. 1–7.

[48] A. Narayan, M. Klaes, D. Babazadeh, S. Lehnhoff, and C. Rehtanz, "First approach for a multi-dimensional state classification for ICT-reliant energy systems," in *International ETG-Congress 2019; ETG Symposium.* VDE, 2019, pp. 1–6.

[49] M. You, J. Jiang, A. M. Tonello, T. Doukoglou, and H. Sun, "On statistical power grid observability under communication constraints," *IET Smart Grid*, vol. 1, no. 2, pp. 40–47, 2018.

[50] E. Moradi-Pari, N. Nasiriani, Y. P. Fallah, P. Famouri, S. Bossart, and K. Dodrill, "Design, modelling, and simulation of on-demand communication mechanisms for cyber-physical energy systems," *IEEE Transactions on Industrial Informatics*, vol. 10, no. 4, pp. 2330–2339, 2014.

[51] M. You, Q. Liu, and H. Sun, "New communication strategy for spectrum sharing enabled smart grid cyber-physical system," *IET Cyber-Physical Systems: Theory & Applications*, vol. 2, no. 3, pp. 136–142, 2017.

[52] H. Georg, S. C. Müller, C. Rehtanz, and C. Wietfeld, "Analyzing cyber-physical energy systems: The inspire cosimulation of power and ICT systems using HLA," *IEEE Transactions on Industrial Informatics*, vol. 10, no. 4, pp. 2364–2373, 2014.

[53] H. Lin, S. S. Veda, S. S. Shukla, L. Mili, and J. Thorp, "Geco: Global event-driven co-simulation framework for interconnected power system and communication network," *IEEE Transactions on Smart Grid*, vol. 3, no. 3, pp. 1444–1456, 2012.

[54] K. Hopkinson, X. Wang, R. Giovanini, J. Thorp, K. Birman, and D. Coury, "Epochs: A platform for agent-based electric power and communication simulation built from commercial off-the-shelf components," *IEEE Transactions on Power Systems*, vol. 21, no. 2, pp. 548–558, 2006.

[55] P. Palensky, E. Widl, and A. Elsheikh, "Simulating cyber-physical energy systems: Challenges, tools and methods," *IEEE Transactions on Systems, Man, and Cybernetics: Systems*, vol. 44, no. 3, pp. 318–326, 2013.

[56] S. Skarvelis-Kazakos, P. Papadopoulos, I. G. Unda, T. Gorman, A. Belaidi, and S. Zigan, "Multiple energy carrier optimisation with intelligent agents," *Applied Energy*, vol. 167, pp. 323–335, Apr. 2016.

[57] H. Wang, C. Gu, X. Zhang, and F. Li, "Optimal CHP planning in integrated energy systems considering network charges," *IEEE Systems Journal*, pp. 1–10, Jun. 2019.

[58] M. Moeini-Aghtaie, P. Dehghanian, M. Fotuhi-Firuzabad, and A. Abbaspour, "Multiagent genetic algorithm: an online probabilistic view on economic dispatch of energy hubs constrained by wind availability," *IEEE Transactions on Sustainable Energy*, vol. 5, no. 2, pp. 699–708, Apr. 2014.

[59] C. Tang, C. Gu, J. Li, and S. Dong, "Optimal operation of multi-vector energy storage systems with fuel cell cars for cost reduction," *IET Smart Grid*, vol. 3, no. 6, pp. 794–800, Dec. 2020.

[60] National Grid ESO. [Online]. Available: http://fes.nationalgrid.com/fes-document/

[61] Y. Xiang, H. Cai, C. Gu, and X. Shen, "Cost-benefit analysis of integrated energy system planning considering demand response," *Energy*, vol. 192, p. 116632, Feb. 2020.

[62] P. Bertoldi and B. Atanasiu, "Electricity consumption and efficiency trends in the enlarged European Union," *IES–JRC. European Union*, 2007.

[63] A. Allouhi, Y. El Fouih, T. Kousksou, A. Jamil, Y. Zeraouli, and Y. Mourad, "Energy consumption and efficiency in buildings: current status and future trends," *Journal of Cleaner Production*, vol. 109, pp. 118–130, 2015.

[64] A. Balint and H. Kazmi, "Determinants of energy flexibility in residential hot water systems," *Energy and Buildings*, vol. 188, pp. 286–296, 2019.

[65] I. Sutskever, J. Martens, G. Dahl, and G. Hinton, "On the importance of initialization and momentum in deep learning," in *Proceedings on International Conference on Machine Learning*, Atlanta, USA, 2013, pp. 1139–1147.

[66] Balancing Mechanism Reporting Service. [Online]. Available: *https://www.bmreports.com/*

[67] L. Gelažanskas and K. A. Gamage, "Forecasting hot water consumption in residential houses," *Energies*, vol. 8, no. 11, pp. 12 702–12 717, Nov. 2015.

[68] National Grid Gas. [Online]. Available: *https://www.nationalgridgas.com/balancing*

[69] D. P. Kingma and J. Ba, "Adam: A method for stochastic optimization," in *3rd International Conference on Learning Representations, San Diego, CA, USA*, 2015, pp. 1–15.

III

Testbeds for Smart Energy Systems

CHAPTER 8

Developing Testbeds for Smart Energy Systems

I n energy systems, the researchers from both academic and industry are interested in developing all kinds of testbeds which are used as proof of concepts, especially when new technologies are involved. The key point of a testbed is to provide a safe and practical environment for experiments, where the data regarding new concepts or technologies can be collected and analysed, while the performance can be evaluated and verified against theories.

The future smart energy system is a fusion of advanced technologies from disciplines beyond traditional energy systems, including information communication and technologies (ICTs) and artificial intelligence (AI). The design of a testbed for such smart energy systems demands new perceptions and new methods, where the fusion is not only regarding the individual technologies, but also the cooperation between them. In this chapter, the development of a testbed for smart energy system will be addressed, with focuses on both the implementation of individual technologies and their cooperation in the whole system.

8.1 REVIEW OF ENERGY SYSTEMS TESTBEDS

The concept of energy systems has evolved from the traditional electric power system, where the energy refers not only to electric power, but also to other energy forms, such as heat and renewable energy. The role of an energy system is also changed from simply supplying the end-user's demand, to greater commissions involving the improvement of energy efficiency, reduction of carbon emission, and boost of economics.

These changes are relying on the fusion of new technologies from various disciplines, and in turn, this fusion demands the evaluation of smart energy systems from a cross-disciplinary aspect. Correspondingly, the smart energy system testbeds are required to be capable of such experiments. It should be noted that the smart energy system testbeds vary in forms and scales, but in general, can be categorized into the hardware-based testbed, the simulator-based testbed, and the hybrid testbed. In the following subsection, state-of-the-art testbeds for the smart energy systems are reviewed.

DOI: 10.1201/9781003170440-8

8.1.1 Hardware-Based Designs

Hardware-based testbeds provide a scaled energy system that runs in the real world, which contains a range of energy system components, such as the generators, the loads, and the circuits. The scale of a hardware-based testbed depends on the purpose of experiments, which can be small-scale hosted in a comprehensive laboratory, while some large-scale testbeds can be fulfilled by a real-world industrial site. This is because in hardware-based testbed, all components are physically implemented, which provides practical experimental data with real-world measurements.

The large-scale hardware-based energy system testbeds can even be full-scale energy system capable to provide an isolated real-world energy system for experiments. The most typical example is the Jeju Island built in South Korea [1]. The whole island forms the basis of the testbed, where the devices participate in the real-world energy system operations. The testbed is capable to support five major research areas, including the smart power grid, smart place, smart transportation, smart renewable, and smart electricity service studies [1].

The small-scale hardware-based testbed could also be versatile, especially with the advent of the concepts such as the microgrid. One key feature of the microgrid is that it is capable to operate in an island mode when disconnected from the main grid, which naturally fits the small-scale hardware-based testbeds. Typical examples are the microgrid testbed hosted at Zhejiang University [2], the testbed at Illinois Institute of Technology in Chicago [3], and Smart Energy Integration Lab [4]. These testbeds are capable to support research on a range of topics in microgrid and beyond, including fault controls, distributed energy resources, and isolation mechanisms.

In hardware-based designs, the design complexity and corresponding cost grow significantly as the scale increases. This is also due to the fact that very large-scale systems involve more supporting sub-systems, safety measures, and infrastructure investments. Therefore the aim of the hardware-based testbed is usually dedicated to a pre determined set of experiments, while the further extension or scale-up of the testbed is not the main focus.

8.1.2 Software-Based Designs

With the increase of computing capability and decrease of corresponding cost, a wide range of simulation software has been invented to simulate the complex procedures or processes, whose data are traditionally obtained via hardware-based experiments. By characterising the real-world components via mathematical models and solving numerically with the computing devices, the software-based simulations provide a fully controlled experiment environment, which is friendly to the researchers and developers with regard to not only data acquisitions, but also repeatable experiments.

Traditionally, software simulators are designed for users in specific disciplines. Typical examples in the power system include stand-alone solutions such as Opal-RT and Real-time Digital Simulator (RTDS) for real-time simulations, and software toolboxes such as Simscape Electrical based on MATLAB. As Smart Energy System involves a wide range of topics, the testbed based on software-based designs usually focuses on specific areas.

For example, to study the cybersecurity of the Smart Energy System, various testbeds have been developed, including the Virtual Control System Environment (VCSE) [5], Virtual Power System Testbed (VPST) [6], intrusion and defense testbed [7] , Industrial Internet of Things testbed [8], and testbed for plug-in hybrid electric vehicles (PHEVs) and plug-in electric vehicles (PEVs) [9].

With the extension from traditional power systems to the more compound notion of energy systems, the integration of different sub-systems from other disciplines has largely enabled the energy systems. This trend also reflects in the testbed designs, where the co-simulation method has been introduced to study the coupled effects regarding the cooperation between different sub-systems.

Smart energy systems can be generally viewed as the compound of energy flow and data flow. This underpins the key idea of the co-simulation methods, where the energy flows are implemented via the energy simulators, while the data flows are fulfilled by simulators in ICT. Examples are the Testbed for Analyzing the security of SCADA Control Systems (TASSCS) [10], SCADASim [11], and the Mosaid-based testbeds [12] [13] [14].

Software-based designs are versatile because these convert most system setups into more flexible parameter configuration procedures. The designs based on software-based testbeds are also friendly to further expansions, where new devices or experiments can be supported with an upgrade of the dependent simulators. It also worth noticing that software-based simulators are not simply the digital replica of their hardware counterparts. The numerical models can provide some critical analysis that is hard or impossible for hardware-based designs, for example, the transience analysis and fault-related studies.

8.1.3 Hybrid Designs

The hybrid design method is a new solution other than the hardware-based and software-based testbed designs. The key idea of the hybrid testbed design is a mixed structure, where some system parts are real-world hardware devices, while others are implemented by software simulators. For a given testbed design task, this hybrid of software and hardware provides a trade-off between the system complexity, practicality, and cost.

With the flexibility in software simulators, the hybrid design method usually implements the focused sub-systems in hardware for practicality studies, while the scalable and cost-intensive sub-systems are implemented via the software simulators. According to the research the hybrid design varies in forms, where typical examples are ScorePlus testbed [15], Network Intrusion Detection System (NIDS) [16], University of South Florida (USF) Smart Grid Power System Lab (SPS) testbed [17], and GreEn-ER1 Industrial Control Systems Sandbox (G-ICS) testbed [18].

Generally speaking, the energy system forms the basic infrastructure of the final testbed, which usually involves cost-intensive investment in research areas including energy transmissions and renewable energies. The advent of hybrid design could help mitigate this cost challenge, where the system infrastructure can be simulated without the need for physical construction. In such systems, it can still support the study focus on the hardware part, e.g., to study the behaviours of the hardware devices via the hardware-in-the-loop (HIL) techniques.

Typical HIL techniques exploit the standard interface to connect the hardware and software parts. Typical real-time power system simulators with such features are the Opal-RT and RTDS. Some typical examples of this real-time power system simulator-based hybrid design include the co-simulator testbed [19], Cyber Physical testbed at Iowa State University [20], Exo-GENI testbed [21], PowerCyber testbed [22], and the testbed developed by Texas A&M University [23].

The hybrid design is not a simple process of co-locating software simulators and hardware devices at the same place. Instead, it provides a new observation of the energy system, which benefits from the consistent efforts in standardisation on the system models and interfacing models. This involves a wide range of parties, such as the standards organizations International Electrotechnical Commission (IEC) and National Institute of Standards and Technology (NIST). Due to the standard characterisation of system components, the hybrid design is capable to treat its sub-system in a modular manner, which results in the possibility of replacement between hardware and software components.

8.1.4 Remarks of Challenges

The categorisation of testbed design method into hardware-based, software-based, and hybrid design is from a broad and general view, while each design method has its own advantages and disadvantages. It should be noticed that every testbed is the compromise between the research goals and the design costs, where the selection of the design method should be based on individual cases.

To facilitate the comparison between the three design methods, a general comparison is made on the main features regarding the cost, flexibility, and practicality as follows:

- *Cost:*
  The cost to build a hardware-based testbed could vary largely depending on its architecture, which is mainly subject to its infrastructure construction and device purchases. On the contrary, a simulator-based testbed is relatively cheap, where the cost is usually spent on the simulators and their associated software license. The cost may be further reduced via exploiting open-source software and general computing platforms. The hybrid testbed is usually in the middle, where only focused components are implemented while some costly components are replaced by software counterparts.

- *Flexibility:*
  The desired testbed is expected to be expandable and modifiable to support different experimental considerations. The simulator-based testbeds are generally very flexible, where the configurations can be tuned by software models or parameters. However, the hardware-based testbed usually requires the installation of new devices for new features. For a large-scale hardware-based testbed such as Jeju Island, the infrastructure is less likely to be changed frequently.

The hybrid testbeds take the advantages of the high flexibility of software simulators, while the main constraints are due to the interfacing method between hardware and software components [22].

- *Practicality:*
  Testbeds are designed to conduct experiments, which should provide the evaluations as close to the real world as possible. The hardware-based testbed supports the real operations on its components, which provides the most practical experiment environment. The simulator-based testbed relies on the accuracy of the models, which normally provide approximated or simplified performance. The testbeds combining hardware and software components could provide real operations on the hardware implemented parts, but the simulator parts still depend on the practicality of models.

As discussed above, no testbed architecture can outperform the other with all three aspects of the cost, flexibility, and practicality, instead it is usually a trade-off among these aspects according to the testbed dedicated purposes. With the development of interfacing techniques, the boundaries between these three kinds of testbeds are blurring. Especially with the high-speed analogue/digital converters, more simulators are supporting HIL tests, while the hardware devices are providing more versatile simulator interfaces.

8.2 TESTBED DESIGN AND IMPLEMENTATION FOR ENERGY SYSTEMS

The energy system is a broad concept in modern societies, which consists of not only the supply and demand-side of the energy, but also the related supporting services and devices. This section is focusing on the testbed design and implementation for energy systems, where the focus is placed on the key sub-systems involved in the energy flow and data flow. Specifically, we will introduce the testbed design and implementation from three major sub-systems, namely the ICT systems, power systems, and advanced signal processing systems.

As illustrated in Figure 8.1, the three focused sub-systems represent the three key aspects to form the whole energy system.

- The power system characterises the power flow of the energy system regarding the power generation, storage, transformation, delivery, and end-consumption. The infrastructure of the power system spans the whole energy system, which also defines the physical basis of the services in the energy systems. This also includes all kinds of sensors and controllers for system monitoring and controls.

- The ICT system characterises the data flow of the energy system regarding the information generation, aggregation, and transmission. The topology of the ICT system aligns with the infrastructure of the power system, which supports the necessary data flows between the physical components within the energy system.

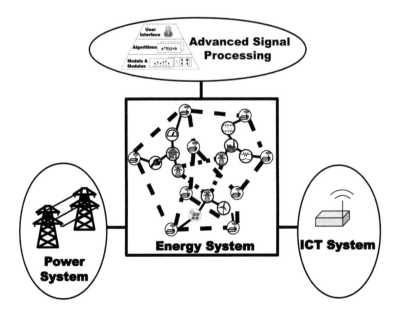

Figure 8.1 An illustration of energy system from a testbed design aspect, which consists of ICT systems, power systems, and the advanced signal processing systems.

- The advanced signal processing system characterises the models, algorithms, and methods for the energy system operations. The inputs are the measurements generated by the power system sensors, while the outputs are the controls for the system operations and high-level system status information reported to system operators.

In the following subsections, we will describe the design and implementation of each sub-system in a hybrid testbed design aspect.

8.2.1 ICT Implementation

The ICT system is the key sub-system in the energy system to enable the data flow between different components. It is noticed that traditionally the ICT system and the power system are generally two different disciplines, where the research focuses, terminologies, and theories are largely different from each other. In this part, several techniques with regard to the integration of ICT systems are discussed.

8.2.1.1 Integration via Layered Architecture

Thanks to the standardization efforts in both areas, the integration of ICT systems and power systems has been much facilitated. A typical example is the IEC 61850 standard for electric power systems, which defines the communication protocols for intelligent electronic devices in an abstract manner at a high level. In other words, the IEC 61850 defines the data models and requirements for the communication protocols, while it allows any applicable communication protocols to be implemented, as long as the requirements are met.

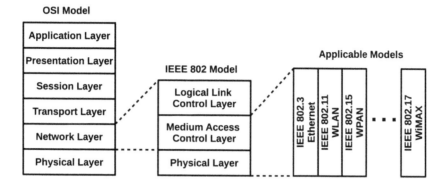

Figure 8.2 An illustration of OSI and IEEE 802 reference models.

From the view of the communication systems, the communication model described in IEC 61850 can be categorized in the layers above network layer in the Open Systems Interconnection (OSI) model as illustrated in Figure 8.2, which is compatible to different lower layer protocols in the data link layer or physical layer. For example, the IEEE 802 model is compatible with the OSI models with regard to the network layer and physical layer, where there is a wide range of protocol choices from the IEEE 802 protocol family. This includes the widely used protocols, such as IEEE 802.3 (Ethernet) and IEEE 802.11 (e.g., WiFi, Bluetooth, and ZigBee).

By applying the OSI model in describing the data flows in the energy system, the testbed design can regard the whole energy system as a stack of protocols, where the network layer and physical layer are specified by the ICT sub-system, while the power sub-system and advanced signal processing sub-system forms the other layers.

In this way, the ICT sub-system can be decoupled from the whole system with regard to the layered design in the OSI model. Correspondingly, traditional design methods in the ICT discipline can be applied, where each layer is transparent to the other layers and each layer is only responsible for its only layer as well as the interfaces between other layers.

However, it should be noticed that the above layered design is only the necessary conditions for an ICT sub-system can be integrated into the whole energy system, while the sufficient condition is whether the required communication performance can be met. It is pointed out by the IEEE Task Force on Interfacing Techniques for Simulation Tools in [24] that, it is necessary to jointly consider ICT systems and power systems to describe the entire energy system.

This is because in the energy system, the data flow and energy flow are coupled, where correspondingly the ICT systems and power systems are coupled. On the one hand, the performance of the ICT system directly affects that of the power system, because any delay or failure due to the ICT systems will affect the successive processing of the data in the power systems. On the other hand, the power system is the data source and data sink of the ICT systems, which characterizes the data size, pattern, importance and urgency that determines the choice and configuration of the ICT systems.

Therefore although the ICT system can be decoupled from the entire energy system with regard to the layers in the OSI model, the design and implementation should still be considered jointly with the power system. Especially the mutual influence between the ICT system and the power system is of great interest in the viewpoint of testbed design, which could lead to not only a better selection and optimization of the ICT systems for real-world deployment, but also a further improvement of the energy systems with regard to the services' end performances and overall system reliability.

8.2.1.2 *Software-Defined Radio Implementations*

In the following, a Software-Defined Radio (SDR) based wireless communications system is introduced as an illustration of the design and implementation of the communication sub-system in the energy system. In SDR based wireless communication systems, most functions are defined by software modules and fulfilled by general purpose computing devices, while its counterpart concept, namely the hardware defined radio, exploits dedicated hardware chips or modules for corresponding functional purposes.

Besides the general purpose computing devices and antennas, the typical SDR solution consists of a Radio Frontend (RF) device, which converts the baseband signals from the general purpose computing devices to the broadcasting waveforms for the antennas. Note that different from pure simulators, the SDR based solutions are capable to work as real-world devices, e.g., to be exploited as the full set of ICT sub-systems in the energy system testbed.

The most important feature of the SDR is its modular architecture, where each function can be capsuled into a functional module, and to be loaded as required. Since the modules are described by software blocks, it is possible to experiment with different combinations of the modules with the same hardware setup by software configurations.

With this concept, a protocol pool can be built, which is illustrated in Fig. 8.2. In the protocol pool, each communication sub-system implementation can be defined as a combination of protocols for each layer in the IEEE 802 model or the OSI model, while each protocol can be described by a software module. During the testbed evaluation phase, the implemented communication can be re-configured by module combinations within the protocol pool.

A counterpart concept in the hardware defined radio is the gateway method, where each of the protocols is fulfilled by corresponding hardware chips and devices, while these chips and devices are physically co-located together to realize a similar function as of the protocol pool.

Note that from the testbed design aspect, the difference between the protocol pool based on SDR and the gateway method is very large. The difference is not in their functions because their performance is largely determined by the protocols, and these protocols can be exactly the same for either solution. The key point is in the protocol pool, all involved processes can be investigated by the researchers, which are basically black boxes in the gateway method.

Beyond the basic feature such as the implementation of existing communication protocols, the testbed design based on SDR is also powerful in the integration and evaluation of advanced communication technologies. With the advent of the fifth generation wireless communication technologies, wireless communication technologies are expected to be compatible with the wired communication technologies, with regard to throughput, latency and reliability.

There are also service optimized protocols dedicated for the energy systems, which, however, are mostly verified and tested via simulations only. This largely confines the application of these advanced technologies in the energy systems, because reliability is the top criteria in communication technology selection, while the new technologies usually lack practical test supports. This is worsened in the hardware defined communication systems, where it is very hard, and usually impossible, to modify an existing hardware chip or device to implement the new technologies.

With the SDR based testbed design, the experiments with new and advanced technologies, especially in the communication systems, are much facilitated. In the development of hardware chips and devices, the low-level hardware oriented languages are used, such as Field Programmable Gate Arrays (FPGA) and Application Specific Integrated Circuit (ASIC).

Different from that, the state-of-the-art SDR environments could support most existing programming languages and developing methods. For example, the GNU Radio, which is an SDR toolkit and part of the well-known open source GNU family, could support programming methods including Python and C++, as well as the popular languages such as MATLAB via interfacing tools.

8.2.1.3 Protocol Pool Method

To implement these advanced technologies and evaluate jointly with the whole energy system, the modified protocols are implemented in software modules and integrated into the protocol pool. The key barrier of the experiment with new technologies is then changed from hardware compatibility challenges to software compatibility issues, where the latter is much easier to be resolved in the layered models, and the fact that software modules are friendly to researchers and developers.

Next, an example is given on how to implement advanced spectrum sharing communication technologies via the SDR based testbed design. To support the communications among widely deployed smart devices and all kinds of real-time services, there is an increasing demand for the performance of the communications in the Smart Grid. The available spectrum resource is a key bottleneck in performance improvements of the communication systems.

Traditionally there are two options, either to obtain exclusive spectrum licenses or to use the Industrial, Scientific, and Medical (ISM) radio bands. The former usually involves high cost, while the latter can be very noisy and crowded for good performance. As a promising solution, the cognitive radio technology can exploit the unused spectrum resources for opportunistic data transmissions. By sharing the spare spectrum resources in the dimensions of space, frequency, time or code, the cognitive

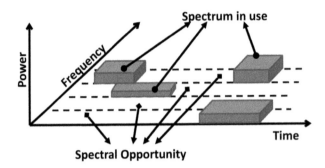

Figure 8.3 An illustration of spectrum sharing mechanism, where the unused spectrum is potential to be opportunistically exploited by the unlicensed users.

radio enabled testbed provides wider choices for data transmissions and enhances the overall communications performance.

An illustration of the spectrum sharing concept is illustrated in Figure 8.3. To this end, a cognitive radio system is implemented as the new communication protocol and integrated into the protocol pool.

The new spectrum sharing specifies the medium access control layer and physical layer as illustrated in Figure 8.3, which describes the same layers as the other protocols including the WiFi and ZigBee.

In the new spectrum sharing protocol, each round of data transmission is with a fixed period of time, which is illustrated in Figure 8.4. During each round of data transmission, it consists of three phases, which are detailed as follows:

- In the spectrum sensing phase, the communication system observes the surrounding spectrum environment and estimates the spectrum usage status. The spectrum sensing relies on signal processing techniques based on the measurements with regard to the spectrum, where methods like energy detection can be used.

- The signal processing phase is between the spectrum sensing phase and the data transmission and reception phase, which corresponds to the time period of the guardian interval in Figure 8.4. During this phase, the communication system estimates the surrounding spectrum status, and makes decisions on its spectrum usage for the data transmission and reception phase.

- During the data transmission and reception phase, the communication systems exchange the energy system data with the selected spectrum opportunities. These data are sourcing from (or delivering to) other sub-systems in the energy system.

Note that the key point of the spectrum sharing is to determine how the communication system is to access the spectrum resources, where the data transmission and reception can re-use any existing protocols. In this way, by specifying the aforementioned phases via sub-modules and integrating them as part of the protocol pool,

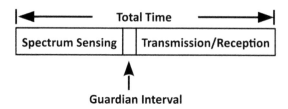

Figure 8.4 An illustration of the spectrum sharing protocol, where a dedicated time period is allocated for the spectrum sensing and the data transmission and receipt are then using the identified opportunistic spectrum.

the new spectrum sharing protocol can be implemented as the communication sub-system, and compatible with the rest of the energy system testbed.

8.2.2 Power System Implementation

In the energy system, the energy may have different energy carriers, such as electricity, natural gas and renewable energies. In this part we focus on the power system as a typical example of the energy system, while the testbed design method can be further extended to include other energy forms.

8.2.2.1 Simulator-Based Implementation

The power system is a broad concept, which includes all aspects with regard to the generation, transmission, distribution, and consumption of electricity. In a hardware-based testbed, the implementation of the power system has to address all these aspects, which involves a very complex design and implementation procedure.

Instead, here the software-based method and hybrid design method are interested, where the infrastructure of the power system is implemented via software modules or simulators, and only the control sub-systems or end users are implemented in hardware. Specifically, the RTDS is used as an example to illustrate how to implement the power system via the software-based method or the hybrid design method.

The RTDS is a state-of-the-art power system simulator, which simulates the real-time operations of the power system operation and supports electromagnetic transient simulation with time steps as small as $1 - 50$ μs. The key feature of the RTDS is its modular structure, where the computing resources and interfaces are implemented in individual modules which are referred to as the functional cards. This architecture is sharing some similarities with the SDR in the communication systems, where the hardware devices are generalized and can be largely re-configured by the software.

A standard implementation procedure with the RTDS follows similar steps as in most simulators. The system is constructed with the provided software suite RSCAD, which consists of necessary libraries including common components modelling from basic resistors and capacitors to the wind turbines and solar panels. With the graphical user interface (GUI), it allows the users to draw the desired power system in

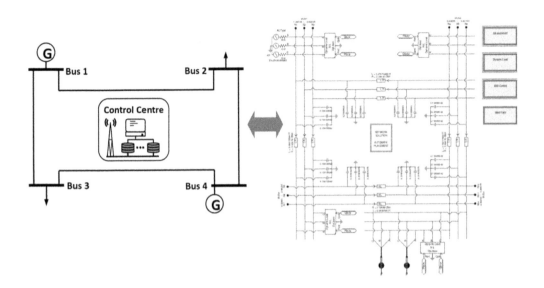

Figure 8.5 An illustration of the IEEE 4 bus power system implementation in the RTDS, where the power system is expanded with wind energies.

different hierarchies, which are then compiled as numerical simulation tasks to be executed on the hardware devices.

An example of IEEE 4 bus power system is illustrated on the left-hand side of Figure 8.5, whose corresponding implementations via the RTDS are presented on the right-hand side, where the generator is replaced with wind energies to study the renewable energies. It is seen that the implementation is quite straightforward with the GUI inputs. For example, the power transmission lines correspond to the graphical lines connecting each bus, whose parameters can be explicitly represented by the resistors and capacitors attached to the lines.

Another example of the PV generator implementation is illustrated in Figure 8.6. In this example, the PV module is reusing the PV model in the renewable energy library in the RSCAD. Besides the common inputs and outputs such as the insolation and temperature, the model can be tailored to simulate common PV products in the markets. Note that during the integration of such detailed models with the rest of

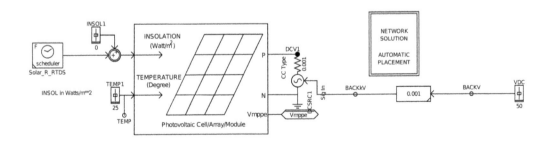

Figure 8.6 An example of the PV implementation in the RTDS.

the power system models, it usually requires additional circuits such as converts and transmitters.

Since the RTDS is also a model-based simulator, it also supports the simulation of power system devices at different levels. For example, to implement the wind turbine, it can be implemented via a complex model with the considerations of turbine design, converter design and fault controls, which can be reconfigured based on a prototype model from the RSCAD library.

The detailed implementation of the wind turbine is essential for the studies of wind energy related studies, where features such as electromagnetic transient analysis are very helpful for the data acquisition and analysis. On the other hand, if the study is focusing on the integration of wind energy as a distributed generator, while the focus is on the power flow studies in the whole system, then the wind turbine can be regarded as a power converter from the wind power to the electrical power.

In such cases, a lot of details within the wind turbine can be overlooked, while its end performances such as power efficiency, power factor, and active and reactive power generations are more interested in the power flow studies. Therefore the implementation of the wind turbine in power flow studies can be abstracted as a power injector, where the wind energy generation can be scheduled with regard to the time of the day.

Note that the key point in the power system implementation is to select the appropriate level of model to characterize the key features. It is indeed possible to exploit the detailed model for transient analysis for general studies such as the power flow analysis, which can be fulfilled by the interfacing component during the implementation.

But it should be noticed that this is not beneficial in most cases, because simulations are to capture and reflect the impact from the key factors, while the many trivial details are not helpful for this purpose. Actually, the implementation of a model with excessive details could be even troublesome in practice.

A common issue is that the model details are exhaustive in the computing resources during the implementation, for example, the 1 ms time resolution could be sufficient for the power flow analysis, while it requires 50 μs time resolution if the transient analysis is involved. This means if the transient models are implemented in the power flow analysis, the whole testbed has to spend at least 20 times more calculations than the actual requirement.

8.2.2.2 Real-Time Simulations

Due to the high simulation resolution, the RTDS is able to simulate the events in the power system in the same time scale as in the real world. This is the key point that the RTDS is claiming real-time simulations as its main feature. Actually, it is the versatile functional cards that make the RTDS versatile for power system simulation, especially with regard to the hardware-in-the-loop experiments.

The optional cards include both digital and analogue inputs and outputs, communication protocol selections and high voltage and low voltage interfaces, which can be re-configured to interact with other devices for the purpose of data input or

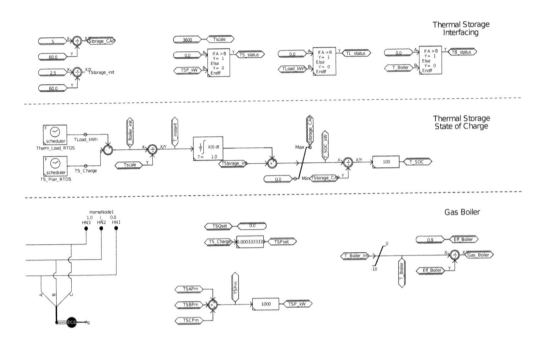

Figure 8.7 An example of the thermal energy system implementation in the RTDS.

output. A typical example is the simulation via the RSCAD. In fact, the RSCAD splits the models and the simulation scripts, which need to be separately defined.

The implementation as illustrated in Figure 8.6 is regarded as building models in the RSCAD, where a simulation script can be defined via the GUI and almost all values in the models can be configured to be observable during the simulation. The demanded simulation data on the RTDS hardware are transmitted to the hosts running RSCAD, which is fulfilled by the Ethernet connections and communication card.

A wide range of power system implementations can be made with the RTDS, where the build-in library in the RSCAD provides many prototype models for references. This also includes some models or protocols for the new technologies, such as the wind turbine models, Photovoltaics (PV) models, battery models, Phasor Measurement Unit (PMU) models and IEC 61850 protocol. It also supports self-defined models that are not available from the libraries, which can be defined in C++ and integrated into the implementations.

For most cases, the new systems or models can be built without the low-level modelling method based on C++ programming. The RSCAD provides an extensive library that includes the most commonly used electrical components, control devices, and mathematical operators. Therefore with a given model for the new models, it is usually possible to fulfill the implementation task by the combination of these existing models. For example, the thermal energy system is an important counterpart of the electricity system, which is widely existing in real-world energy systems.

Although RTDS is targeted for the electric power system, the thermal energy system can be modelled, where an example is illustrated in Figure 8.7. In this example, the purpose of the experiment is to study the supply and demand with regard to both electricity and thermal energy, where the thermal energy devices can be abstractly modelled and implemented.

For example, in the study of thermal storage operation for the demand-side management, the operations can be modelled by the charging and discharging parameters to reflect the interaction between the thermal energy storage with other system components, as well as the stage of charge to characterize the operating status of the thermal energy storage. As illustrated in Figure 8.7, this model can be implemented with the control circuits, and integrated with the electricity power systems.

For advanced modelling of thermal energy system devices, it is also possible to implement a more complicated mathematical model following a similar method. This is actually transforming the models to a set of numerical calculations, which is supported by the RTDS since this is in line with its electrical modelling method.

From the view of the whole energy system, the support of the modelling of other energy forms is of great importance, because it would save the researchers from referring to additional simulators for these tasks. Note that from the aspect of the testbed design and implementation, it is not trivial to enforcing similar modelling methods for the same sub-systems, which helps to reduce the complexity in both the model development and implementation.

8.2.3   Artificial Intelligence Integration

The cost for computing resources has been reducing every year. On the one hand, it has stimulated all kinds of software simulators, which is because we are now able to simulate the physical process at high accuracy with much lower cost than hardware implementations.

On the other hand, it has also boosted the technologies such as artificial intelligence (AI) to a new high level. Actually, AI is not a new term, which can trace back to 1956 when it was coined by John McCarthy at a workshop. The current research topics and research methods in the AI field have been much different since the last decade, which are more data intensive thanks to the powerful computers.

With regard to energy systems, the advances of AI technologies have also indicated potential solutions to the many long-standing challenges. For example with the power of learning, AI algorithms have shown advantages in addressing the uncertainties in the energy system, such as energy forecasting and load forecasting. More importantly, the learning ability of some AI algorithms shows potentials in further improving the system stability, where the system with AI could evolve with changes.

In this part, the focus is on the integration of AI algorithms with the testbed design and implementation. It will first introduce an example of the integration of AI algorithm with the energy system testbed, and then proceed to more general discussions on the integration techniques.

8.2.3.1   An Integration of Reinforcement Learning with the Testbed

In Section 8.2.1, it has been introduced the integration of the ICT system with the testbed. Specifically, the potential use of spectrum sharing techniques to improve the ICT system performance has been discussed.

In Fig. figure ch8 spectrum sharing, it is seen that there are three consisting parts for a system exploiting spectrum sharing protocol, namely the spectrum sensing, guardian interval and transmission/reception. In simple implementations, for example, when there is only one spectral opportunity, the testbed can easily decide its access or not, based on traditional algorithms such as the spectrum energy detection.

Now let us consider a more advanced scenario, where there are multiple wireless channels (i.e., the spectral opportunities) to this testbed. On these wireless channels, there might be other users or devices. The testbed can transmit its data if there are no others using it. If there are any other users using one of the channels, the testbed should avoid this specific channel, or both the testbed and the other users are interfered with a performance degrade.

This system model can be implemented into a wide range of practical scenarios, for example, the opportunistic spectrum is in the 2.4GHz WiFi band and there might be others occasionally using WiFi. In such cases, the decision on which channel to use can be more active than the passive energy detection method. The other users might present some patterns in usage. For example, it might show a periodic pattern when streaming a video, or a sporadic pattern when browsing the web pages.

With an unknown usage pattern by other users, it could be a challenge for the testbed to access the channels for data transmission. With the energy detection method, it can only know the current channel usage status, and therefore pick the best known channel for data transmissions.

Strictly speaking, any known channel usage via the energy detection method is the knowledge in the past, because there is always a time lag between the time when things happen and the time when it is detected. Therefore without cooperation, the other users might interrupt at any time even it is not detected to be presented.

Another traditional method with more active solutions to avoid conflict with other users is the random access method. In the random access method, the testbed will first detect the current channel usage via the energy detection method, and then randomly pick one of the available channels for data transmissions.

This might not guarantee the testbed to access the best channel, but it is potential to mitigate collisions via randomness. Since the decision is based on randomness, the chance of the collision is also random, and might be good for some time while bad for others.

Therefore there are two specific challenges in accessing the channels, the first one is the channel quality, and the second is the chance of collision with other users. Comparing to the energy detection method and the random access method, a more active method can be achieved via the AI based method.

Here the reinforcement learning algorithm Multi-Armed Bandit (MAB) is considered. The MAB algorithm is an online learning algorithm, which is detailed in Figure

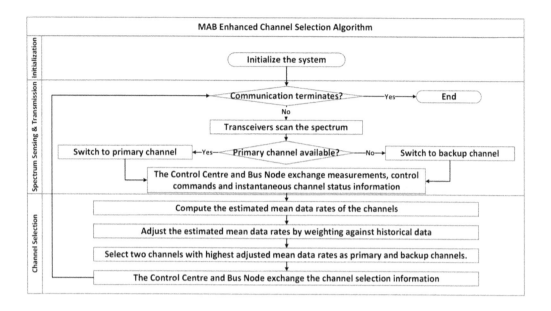

Figure 8.8 The Multi-Armed Bandit algorithm, which serves as an example of reinforcement learning algorithm for the testbed implementation.

8.8. In general, the MAB algorithm empirically predicts the next best available channels based on historical observations.

The testbed considers a scenario where a Bus Node is to communicate with a Control Centre, where the data to be exchanged are the measurements from the Bus Node to the Control Centre, and the control commands from the Control Centre to the Bus Node. The testbed exploits the spectrum sharing method as detailed in Section 8.2.1, where the MAB algorithm is used for the spectrum access decisions. Its calculation is completed during the guardian interval as indicated in Figure 8.8.

As can be seen from Figure 8.8, the presented MAB algorithm has been integrated with the functionalities in the testbed, and also helps to coordinate the ICT system operation. To avoid the potential collisions with other users, the MAB algorithm dynamically selects two channels, namely the primary channel for the next round of data transmission, and the backup channel in case the primary channel is interrupted by other users.

Along with the data transmissions, the good selections are positively rewarded and encouraged by the MAB algorithm, while the bad selections are negatively rewarded and discouraged. In this way, the channel selections are reinforced with good channel conditions and no collisions.

To integrate the MAB algorithm, it can be implemented as a signal processing module in the advanced signal processing systems in Figure 8.1. It is then interfaced with the spectrum sensing module and the data transmission module in the ICT system.

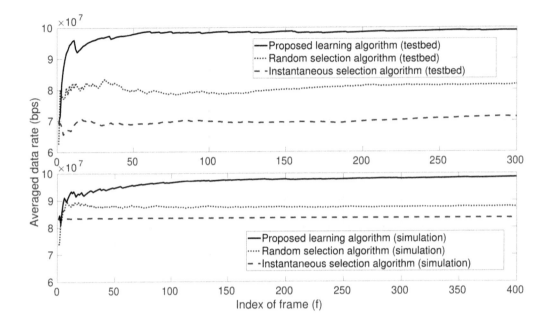

Figure 8.9 The experiment results of the Multi-Armed Bandit algorithm on the testbed, comparing against simulations. The benchmark algorithms are the energy detection algorithm and random channel selection algorithm.

In Figure 8.9, the channel selection performance is evaluated via the averaged data rate in bit per second (bps). The MAB algorithm results are referred to as the proposed learning algorithm, and two benchmark algorithms are used for comparison purposes, which are the energy detection algorithm and the random selection algorithm as described above.

The offline simulation results are presented at the bottom of Figure 8.9, while the real-world testbed implementation results are presented at the top. Since the implemented MAB algorithm is an online learning algorithm, it requires a training period to converge to a stable performance.

This corresponds to the first 100 rounds as illustrated in Figure 8.9. Actually, after the first 50 rounds, it can be already spotted that the real-world implementation already stabilized to about 10×10^7 bps, which is the best performance comparing to the two benchmark algorithms.

For the MAB algorithm, it is also seen that the simulation results are generally agreeing with the real-world testbed implementations. In the meantime, the two benchmark algorithms degrade more than the MAB algorithm. This is due to the fact that simulations are under certain assumptions, which can approximate the real-world scenarios but the differences still exist.

This also indicates the importance of the testbed implementation, since the testbed experiments are more close to the end-performance as products, while the simulations are not sufficient to evaluate the algorithm nor the testbed completely.

8.2.3.2 Remarks on the Integration of Artificial Intelligence

The first remark is with regard to the environment to implement the AI algorithm as a module. In the previous example, the testbed integrates the reinforcement learning algorithm, namely the MAB algorithm, as a module in the advanced signal processing system.

Apparently, the relation between the modules in the advanced signal processing system is in a logical manner, instead of physically grouped together. In fact, according to Fig. 8.9, the reinforcement learning algorithm is closely interacting with the ICT system modules instead of other advanced signal processing modules.

Similarly, if the integrated AI algorithm is for the forecasting in the power system, then it is highly likely that the AI algorithm will closely interact with the power system modules. Therefore from the testbed design and implementation aspect, the AI algorithm modules are preferable to be implemented in the same environment as its application scenario, other than treating them as independent data-oriented environments. For example, let us consider a dummy scenario that the AI algorithm is for power system applications developed in MATLAB.

To integrate the AI algorithm with the energy system testbed, there are two general options. The first is to treat the power system as the data source and data sink of the AI algorithm, which decouples the necessity of implementing the AI algorithm in the MATLAB environment. Under this condition, the AI algorithm module can be implemented in any available environment beyond MATLAB, such as Pytorch and Tensorflow.

Then the options to implement the AI algorithm are much flexible, where the development of the AI algorithm can be even independent of the whole testbed. This is similar to the implementation of ICT system and power system, where mature techniques or existing solutions can be borrowed directly, or with light efforts on the adaptions. But the disadvantage is also clear, since there might be interfacing requirements between the AI module with the other modules.

The second option is to implement the AI module exactly the same as the application environment. For example in the previous dummy example, the AI module can be implemented in MATLAB similar to the power system modules. The advantage of this option is that there will be little interfacing overheads, as the AI modules and its interacting modules are in the save environment.

This is quite important for some time-sensitive applications. For example, the spectrum sharing example in the previous sub-section demands the decision to be made within the time of the guardian interval, as the inputs (spectrum sensing results) are not available before the sensing finishes, and the outputs must be ready before the data transmissions. The disadvantage of this integration option is that it might also restrict the development of the AI modules.

Similar to the power system and ICT system, the researchers and developers are normally sticking to the environment they are most familiar with. Note that when considering the cost to design and implement a testbed, there are also hidden costs such as the time to learn unfamiliar things. In some cases, the learning curve for a new development environment could be very long.

The second remark is regarding the numerical aspects of the AI algorithm. Testbeds are special application environments for AI algorithms, where the involved numerical calculations should be carefully treated. All AI algorithms require to learn first before use, and not all testbeds are supporting both the learn and inference (i.e., the use of the trained AI algorithm or models) phases.

For some online-learning algorithms such as the MAB algorithm, the train and inference are integrated processes. In such cases, there will be a negligible performance difference between train and inference due to numerical calculations. For many data-intensive applications, such as load forecasting in power systems, it requires an offline training phase to address the big data challenge.

Although the training process can be very computing-intensive, the application of the trained AI algorithm or model in the testbed is, on the contrary, much relaxed on the computing resources. This is because during the inference process, the trained AI models are with fixed structures and parameters, and the involved calculations are quite straightforward.

However, this is the point where special cares should be taken, as the numerical calculation environment could be different from its training one. For example, the default numerical precision for MATLAB is double-precision floating point (following IEEE Standard 754), while the ASIC or FPGA modules are normally based on fixed point numerical calculations.

The support of floating point might be default for PCs, but might be very luxurious for some controllers, or even impossible. The gap between the floating point and the fixed point is huge with regard to the AI algorithms, which could lead to unexpected performances if without treatments.

This difference could lead to a completely different AI algorithm design for the same problem, therefore should be considered from the start of the testbed design. For example, many AI algorithm involves the operation to normalize the output in the range from 0 to 1, where the frequently used non-linear activation layer is the "Sigmoid" layer, defined as $\text{Sigmoid}(x) = \frac{1}{1+e^{(-x)}}$.

The involved calculation is not a problem at all in a floating-point system, but it could be a challenge in the fixed-point system. This is because the accuracy of the calculation will change with different inputs due to the fixed-point precision, thus it needs to be intentionally calibrated, e.g., with approximation solutions by sections according to the range of the input values.

It should be minded that if the floating point data type is supported by two AI environments, it is better to double-check that the precision is acceptable with regard to the performance, in order to avoid any potential issues. To have a better idea about the numerical precision issues, numerical computing with MATLAB is used as examples in the following discussions.

The default precision for the floating-point type numbers is double-precision in MATLAB, while it also supports single-precision numbers if otherwise specified. Since all calculations in the CPUs are in the binary format, the floating-point numbers correspond to the method to represent binary bits.

In MATLAB, the single-float numbers are represented with 32 bits, while the double-float numbers are represented with 64 bits. Due to the limit of the finite length of bits in representation, the calculations using single-float and double-float will have differences in accuracy. For example, what is the difference between the value 0.1 in single-float and in double-float? This corresponds to the following line in MATLAB as below:

```
single(0.1) - double(0.1)
```

If you are not familiar with the numerical precision, it will surprise you as the result is not 0, but 1.4901×10^{-9} in MATLAB. Here is what happens behind the screen. The constants are perceived by MATLAB as double-float by default, so *single(0.1)* converts the value 0.1 from double-float to single-float, with a round-up to the nearest single-float representation.

This conversion has caused a difference 1.4901×10^{-9} as observed in the above example. Someone may wonder that the difference is so small, how could it affect the calculations? Let us see another example as follows:

```
(0.1 + 0.2) == 0.3
```

In the above example, it is to compare if the values $0.1+0.2$ and 0.3 in double-float are equal. In common sense, the result is apparently true, but in MATLAB it gives a *FALSE* as the result. Again, this is due to the fact that all numbers are represented in binary bits, but not all numbers can be represented in finite number of bits.

Considering that even the most simple summation operation like $0.1 + 0.2$ will lead to some errors due to precision issues, what about the complicated AI models with a combination of other operations? And what if there are precision conversions required if the testbed is using a different precision system than the AI training environment?

There are no one-for-all solutions to the numerical precision induced problems, but there are indeed some partial mitigation solutions. For example, the numerical precision challenge can be considered as the quantization errors, where a faking quantization environment can be exploited during the training in some AI frameworks such as PyTorch.

8.2.4 Interfacing Techniques

The operation of the smart energy system is with the support of multiple sub-systems. This determines that the testbed design is deemed to face challenges in the coordination of different systems, such as the ICT sub-systems and electrical power sub-systems. It should be noticed that each sub-system could represent a well-established research area, where the coordination not only requires each of the functions correctly, but at the same pace in a timely manner against the others.

To get a much clear idea about the coordination between the sub-systems, it can be imagined that the whole energy system is an orchestra, and each sub-system is representing one specific instrument, which could be the violin, the horn, or the piano. The system operation is like the ensemble of the orchestra, where each instrument needs to be conducted to meet the right tempo.

More importantly, a failure of one instrument will ruin the whole ensemble, and this is also true in the case of modern energy systems. The key point of the above metaphor is that in the viewpoint of the whole energy system, each sub-system is not stand-alone but closely coupled with other sub-systems. This feature determines that the coordination between different sub-systems is a key research problem to the final success of the testbed design as well as its implementation.

From the viewpoint of testbed design and implementations, the sub-systems are usually designed and implemented as a module, where the coordination between modules is usually with standard interfaces. Note that this is usually true during the design and implementation of parts within a given sub-system, where the communication between different sub-modules is achieved via interfaces, instead of a direct talk between their consisting components. This is due to the trend of the standardization and the modular design, whose typical examples are IEEE 802 models for the wireless communication systems as detailed in the previous subsections.

The advantage of modular design is that it enables certain independence between modules. On the one hand, it decouples the total sub-system design and implementation tasks into smaller ones, which can be tested and debugged independently. On the other hand, the individual modules (or sub-modules) are only responsible for their own functions and transparent to the other modules (or sub-modules), which makes it challenging to jointly optimise their design and implementation with very limited interfaces between them.

In the case of the testbed design for the energy system, since the most common purpose of the testbed is for the proof-of-concept, it is usually an essential requirement that the testbed is versatile in providing data, especially those for status monitoring and post-processing purposes.

Therefore in most software-based or hybrid testbed designs, each sub-system is likely to be fulfilled by the specified simulator designed by and for its own discipline. For example, one of the most popular communication system simulators is the Network Simulator 3 (NS-3), which supports the study of the communication network topology and protocol with a simulation of a vast number of transmitters and receivers. In many energy system designs when detailed communication systems are necessary, the NS-3 has been widely used [25], where some advanced wireless communication protocols such as the Long-Term Evolution (LTE) standards.

8.2.4.1 Software-Based Interfacing Techniques

Generally speaking the interfacing techniques for the energy system testbed design and implementation can be generally categorized as software-based interfacing techniques, and hardware-based techniques, which will be detailed as follows.

In this book, the *software-based interfacing techniques* are referred to as the application programming interface (API) based method in interfacing two modules in the testbed. The concept of API refers to a software interface of functional modules, which provides abstract calls or requests to their contents or functionalities. The current form of API definition is much broader than its initial idea of the interface for application programs, which is now widely used in various disciplines, such as web development and communication protocols.

For software-based testbeds, the API is usually playing the key role, no matter it is implemented within one simulator or with conjunction of multiple simulators. One of the most popular software-based testbed designs is to transform the operations of energy systems to the numerical simulations of the underlying events, which are usually referred to as event-based simulators. By abstracting and formulating the required simulations into numerical calculations, it is possible to implement the testbed within one simulator, or more generally one simulation environment.

For example, MATLAB is a general purpose numerical computing environment, which encourages matrix-based manipulations during calculations. It has its own programming language which is called MATLAB programming, but it shares many similarities with the commonly used C/C++ programming language. The advantages of MATLAB as the implementation of the energy system testbed due to the merit of its versatile embedded toolboxes and their associated APIs. These toolboxes cover the frequently involved sub-systems in the energy systems, such as the control system toolbox, communications toolbox, and signal processing toolbox.

For example, in the case of the energy system is exploiting ZigBee as the low-cost solution for smart meter applications, it needs to implement the physical layer of the ZigBee protocol, which transforms the message bits to the waveform to broadcast in the air. The transmitter can be achieved by simply one line code in MATLAB with the API of ZigBee toolbox as follows:

```
ZigBeeWave = lrwpan.PHYGeneratorOQPSK(Message, 4, '2450 MHz')
```

Correspondingly, the receiver that converts the waveform to message bits can be achieved by one line code as follows:

```
Message = lrwpan.PHYDecoderOQPSKNoSync(ZigBeeWave, 4, '2450 MHz')
```

Although there are only two lines given here, actually they implement a series of protocol-specified manipulations in the ZigBee physical layers, such as the spread spectrum, modulation, demodulation, and pulse shaping. With the other high-level APIs for counterparts of ZigBee, the communication sub-systems can be implemented as well-known protocols such as WiFi, LTE, and even 5G.

These high-level APIs are also available in other sub-systems. Another example given here is the electric power system implementation via the MATPOWER, which is a third-party toolbox for MATLAB. The power flow analysis is commonly used in power systems, which is the basis of many advanced controls and services, such as system reliability analysis, power loss reductions, and voltage magnitude controls.

The implementation of a standard IEEE 30 bus system and its power flow analysis can be fulfilled by the one-line code in MATLAB via MATPOWER as follows:

```
PowerFlowResults = runpf('case30')
```

The operation status of the power system can be readily obtained by accessing the corresponding sub-fields in the results. For example, whether the power flow analysis converges to a stable solution with a given power system configuration can be indicated by the sub-field called 'success' in the PowerFlowResults, which can be obtained as follows:

```
PowerFlowAnalysisStatus = PowerFlowResults.success
```

while the voltage magnitude of each bus can be obtained as follows:

```
BusStatus = PowerFlowResults.bus(:,8)
```

Now let us consider a simple energy system consisting of the above IEEE 30 bus power system and ZigBee communication system, where the monitoring and measurements are assumed to be at the individual buses, whose data should be sent via communication systems to a control center for power flow analysis.

Then the implementation of the power system and communication system can be fulfilled by the APIs above, and it only requires interfacing between these two sub-systems to complete this specified simulation task. The interfacing might involve signal processing based on the measurements, or the system specifications mapped from the scenarios.

For a simple scenario, if the voltage magnitude 'BusStatus' of each bus is to be transmitted via the ZigBee communication system, then the interfacing between them can be fulfilled by signal processing APIs. One feasible method is to convert each measurement as a stream of bits, then sends the stream of bits via the ZigBee system.

The conversion of measurements to a stream of bits can be fulfilled by a series of signal processing via APIs in MATLAB as follows:

```
Message = uint32(double(string(vec(dec2bin(num2str(BusResults))))))
```

This formed 'Message' is ready to be transmitted in the implemented ZigBee communication system.

It is seen that in the above examples based on MATLAB, the interfacing between the electric power system and the communication system is simple and easy. This is due to the fact that all systems are implemented within one software environment and programmed under the same MATLAB programming language, where there are no barriers for the two sub-systems to exchange data.

Note that this is not always the case if they are implemented in their hardware counterparts, as the mismatch between the inputs and outputs of data can be a challenge to address.

With one simulator or software environment, it is also possible to exploit different modelling methods to implement the testbed. MATLAB provides a model-based design method, which is called Simulink.

The aforementioned implementation of ZigBee systems and IEEE 4 Bus system can be also built via the embedded Simulink Simscape models, which converts the previous script-based implementation in MATLAB to the graphic model based implementation in Simulink.

It should be noticed that since the numerical calculations are with the same engines, these two testbed implementation methods can achieve the same performance.

Actually, these two modelling methods can be easily combined, where the script can be transformed to a model to be connected in Simulink, or the Simulink models can be converted to codes to run in scripts.

As another example of using different modelling methods in one software environment, here a web-based data sub-system is considered. Note that this sort of data system is widely used in energy systems with distributed devices. The central system may broadcast the operation-related information in a periodical manner, or can serve as a data centre and provide the information on demand. One typical example of such systems is the data services provided by the National Grid, where the regional and global data regarding load and generation information can be requested via web API interfaces.

To request the actual load demands for the 24 hour period on December 31, 2018, the core implementation of such systems in MATLAB can be given as follows:

```
url = 'https://api.bmreports.com/BMRS/B0610/V1?APIKey=DUMMYKEY...
       &SettlementDate=2018-12-31&Period=*&ServiceType=csv'
LoadDemands = webread(url)
```

The above example calls the RESTful API 'webread' for the access of the web-based dataset provided by the National Grid.

Note that in the above example of web-based data sub-system, the RESTful API is for web services, which is a different language comparing to the communication systems, and the electric power systems. But thanks to the API integrated into the MATLAB, this web-based data sub-system can be implemented similarly at a high level as the communication systems and electric power systems.

For the testbed implemented via multiple simulators or software environments, the interfacing between the different software is usually not as simple as the single simulator or software environment. For simulators with the same software environment, the barrier will be much reduced.

This is because, for different software environments, the descriptions and definitions of the data and processes could be essentially different, while in the same software environment it has already been well addressed. The typical example is MATLAB and Simulink, although with different modelling methods, the mutual interfacing has been supported with embedded APIs.

Each research discipline would prefer some routine software environment for simulations, which results in a common situation that different sub-systems are developed independently. Although it is apparent that the single software environment is simple in interfacing different sub-systems, it could be hard, or even impossible, to have a full translation of all sub-systems into one software environment.

This is because in most cases, the sub-system is described by a set of APIs specified by the software environment. These APIs depend on the low-level dependent libraries, which are depending on other libraries therein. If comparing the implemented sub-system as a tree, the APIs are the trunks and leaves, while the dependent libraries in the software environment are the intertwined roots.

Migrating the sub-system from one software environment is similar to moving the tree parts above the ground without the roots, while the success of the transplant depends on whether it can live with the 'new' roots in a different software environment. Note that this is usually hard but not impossible.

An example can be given based on the aforementioned dummy energy system, where the sub-systems can be transplanted from the MATLAB environment to the Python environment. For example, the implementation of the IEEE 4 bus system in Python can be fulfilled by PYPOWER, which is a Python counterpart of MATPOWER for the power flow analysis.

Then the power flow analysis in MATLAB can have an equivalent python implementation, certainly with APIs in the context of Python and PYPOWER. Note that this is not a special case by chance, but instead, it is the results enabled by the various researchers and developers.

For some dependent libraries and toolboxes, they are developed to be compatible or mimic the classical features of their counterparts in a different software environment, where the PYPOWER and MATPOWER serves as one typical example. Some later toolboxes for electrical power system studies in Python also followed this compatible solution, such as the pandapower toolbox, which is compatible with both PYPOWER and MATPOWER.

For the cases where no easy substitution of dependent libraries is available, the cases will be more complicated and usually relies on the dedicated intermediary interfacing solutions. Although the detailed realization of such fundamental low-level interfacing solutions could be very complex, it is, however, usually with much simplified operations thanks to the high-level APIs.

The first example is the usage of Python functions in MATLAB. With a proper configuration in MATLAB, the Python functions and scripts can be directly called inline with the MATLAB programs. For example, if a dummy control algorithm named 'funcPythonCtrl()' is implemented in a Python file named 'MyPythonScript.py', and it is to be called in the MATLAB-based programs, then it can be fulfilled by the following line:

```
Result = py.MyPythonScript.funcPythonCtrl()
```

Similarly, the reveres case, i.e., to call a dummy control algorithm named 'funcMatlabCtrl()' implemented in a MATLAB file 'funcMatlabCtrl.m' in a Python based program, then it can be fulfilled by the following lines:

```
import matlab.engine
eng = matlab.engine.start_matlab()
Result = eng.funcMatlabCtrl()
```

After a long time of evolution, MATLAB has now supported a range of different programming languages to run inline with its MATLAB language, and the run of MATLAB in those different programing languages. These APIs are embedded in the MATLAB cores, which include commonly used programming languages such as C, C++, Java and .NET.

8.2.4.2 *Hardware-Based Interfacing Techniques*

Unlike the software scenarios which might be changed in a flexible way, the interfaces for a hardware-based device are normally fixed and cannot be changed. Therefore it should take extra care when there are interfacing requirements in the hardware-based testbed implementations. For the most common cases, this issue of interfacing between hardware-based devices within the testbed is addressed during the design phase, where compatibility is the decisive factor that if one specific device is to be used or not.

For example, let us consider an energy system that all components are interconnected via the wired networks based on TCP/IP network. From the testbed implementation aspect, this wired network can be fulfilled by Ethernet switches and Ethernet cables. The selection of Ethernet switches and Ethernet cables is required to meet the demands of the energy system, otherwise, it could cause interfacing challenges during the implementation.

The key parameter that characterizes the Ethernet switches and Ethernet cables is the throughput, and the devices could differ largely from each other for different throughput parameters. Generally speaking, the usual choice of cables for throughput below 1000 Mbps is the twisted pairs of copper wire with RJ45 connectors. These cables are commonly used in the home or office scenarios, for example, the cable wires connecting the 'Router' to your own computer are usually of this kind.

If the throughput demands for the energy system are 1000 Mbps, then a feasible communication network structure can be implemented similar to the connection between your 'Router' and the multiple devices. However, this might be that simple as it seems. In the case of a 1000 Mbps throughput is required, then the 'Router', which is usually fulfilled by an Ethernet switch, is required to have at least 1000 Mbps throughput.

Therefore among the various types of switches, it is applicable to use switches with throughput featuring 1000 Mbps/100 Mbps, and the testbed will not work with the 100 Mbps/10 Mbps models. Correspondingly, the cable needs to support at least 1000 Mbps throughput, which means the cables should at least meet the Category 6 in the ISO/IEC 11801:2002 standards, while the testbed will not work with the Cat5 cables (e.g., the default cables provided along with the 'Router' from most internet service providers).

The bad choices of either the Ethernet switches or the Ethernet cables might only cause some confusion in normal life, e.g., 'why my network is so slow than what I've paid for', but it could be critical to determine whether the testbed can work or not.

The appropriate match between the demands and device choices is the key in interfacing devices. It is clear that any incompatible device in the whole testbed could lead to an implementation failure, but actually, it is not always beneficial to exploit a device with 'excessive' capabilities than required.

The above example can be extended to illustrate this point as follows. For Ethernet switches and cables, it is common that they are backward compatible with lower throughput applications. For example, Cat6 cables are generally with no problem to be used as Cat5 cables, while it is also true that a 1000 Mbps switch can work if the peak throughput requirement is only 100 Mbps.

Then the question is, is it also alright to implement a 10 Gbps or even a 100 Gbps compatible network for the need of 1000 Mbps throughput? The answer is probably not, with both concerns of both costs and interfacing challenges. For the throughput demand above 1 Gbps, the common choices are the fiber optics-based produces, instead of the twisted copper wire-based ones.

Although the fiber optic switches and cables can support larger throughput, they are usually more expensive than twisted copper wire products. Besides the costs, the two kinds of products are usually not likely to be directly mixed-used without extra adapters. This could be a serious interfacing issue during the implementation. For example, if a 10 Gbps throughput network is to be implemented for the energy system with 100 Mbps demands, then the physical network interface for the energy system devices is usually with a Registered Jack 45(RJ45) connector.

Meantime, the 10 Gbps cables are usually with the physical network interface Small Form-factor Pluggable+ (SFP+) interface module. These two interfaces are not compatible, which will require extra adapters to convert SFP+ to RJ45, and the extra adapters would be required for each networked device in the energy system. Although from the aspect of network throughput, the 10 Gbps throughput implementation might provide better networking performances, such as the low latency, it is not the optimal choice in costs and would result in interfacing issues.

One important feature of the testbed, comparing to the final products, is that it could provide more detailed operating information for research and development purposes. For hardware-based components, these pieces of information are usually output from the interfaces on the devices.

Therefore the form and standard of the physical interfaces on the devices are critical to the interfacing between hardware devices in the whole testbed, as well as to the final success of the experiments on the testbed. Besides the aforementioned network interfaces, some commonly used interfaces are illustrated in Table 8.1.

If the hardware devices are with the same interface types in Table 8.1, then the interfacing between them could be quite straightforward by following corresponding interfacing protocols. There are two specific points that need to be kept in mind when considering these interfaces in the testbed.

The first point is whether the interfaces are following master/slave architecture. Some interfaces, the USB for instance, are following a master/slave architecture. When two devices are connected via USB, it requires one of them to play the master role, while the other to play the slave role.

Note that usually the role of the device is fixed when they are produced and cannot be changed on demand, unless in some cases they are specially designed. This means, it will fail to exploit USB as the connections between two USB slave devices, nor two USB master devices.

A simple example is that it is not possible to use direct USB connections without external adapters between two computers, because they are both USB masters. In the meantime, if two sensors are providing USB interfaces for external measurements,

TABLE 8.1 Commonly used interfaces in the testbed design

Interface Name	Bandwidth	Comments
USB	1.5 Mbps – 20 Gbps	USB (Universal Serial Bus) connector form might not compatible between versions.
COM	100 bps – 10 Mbps	Commonly used COM (COMmunication port) interfaces are RS-232/422/485.
I^2C	100 kbps – 5 Mbps	The I^2C (Inter-Integrated Circuit) is typically used for Low-speed peripheral Integrated Circuits (ICs) to the micro-controllers.
SPI	upto 50 Mbps	The SPI (Serial Peripheral Interface) is typically used for data transmission between peripheral ICs and micro-controllers.
CAN	125 kbps – 5 Mbps	The CAN (Controller Area Network) is a robust vehicle bus standard.
Analogue I/O	—	The conversion between voltage levels and digital values as inputs/outputs.
Digital I/O	—	Binary voltage levels corresponding to digital bits as inputs/outputs.

then they are likely to be both USB slave devices, and it would need other interfaces other than the USB to exchange information between them.

The second point is the match between different protocol or standard versions and connector types for a specific interfacing method. For example, currently the USB has 3 major versions (generations) in use, and there are a total of 10 connector types with totally different physical forms.

Although the solutions might be as easy as an adapter, it should remind the researchers and developers that the shape corresponds to their versions, and versions correspond to their performances. If a sensor is providing a USB Type-C interface, then it means the device is at least with version USB 3.1 or above, and correspondingly its throughput might be as high as 10 Gbps.

When tackling the issue of connecting a host device (e.g., a computer) with a USB Type-A interface, then the key issue is not only on the conversion from Type-C to Type-A, but also whether the host device is with compatible USB versions, and more importantly, the processing capacity to handle the measurements from this sensor.

Besides the direct connections via the provided interfaces on the devices, a flexible aggregator method is introduced in the next part. In the aggregator method, it exploits an external device for interfacing functionalities in the testbed. The use of a specified aggregator for interfacing purposes is beneficial in many ways, for example when there are multiple kinds of sensors with different interfaces, or there are gaps between the interfaces of some devices.

The former case is easy to understand, as it is also common in our normal lives that each device is with its own interfaces. Even for the same kind of device, e.g., the monitor, its interface might be VGA, HDMI, DisplayPort, or USB Type-C. On the other hand, it will be a pity to abandon some devices from the testbed implementation due to their interfaces, instead of their performances.

With further consideration of the potential upgrades of the testbed implementation, new devices may be integrated for extended evaluations and experiments. In such conditions, the interfacing problems are potential issues, which can be addressed by a versatile interfacing aggregator.

In the following, an interfacing aggregator, named as Data Acquisition and Actuation (DAA) module, is detailed as an example of this type of interfacing technique. The key motivation to implement the DAA module is to exploit a general-purpose micro-controller as a re-configurable intermediary for the different interfacing requirements.

There are various choices for such general-purpose micro-controllers on the market, which are at the cost of around a few tens of sterling pounds. This price range is attractive as interfacing modules in the testbed, which are comparable to the dedicated adapters but cannot be re-configured. The examples are Raspberry PI, MBED, and Arduino. These micro-controllers are sharing some similarities, such as the support of most listed interfaces in Table 8.1 and user-friendly development toolkits and environments.

Let's now consider a testbed interfacing requirement, where it requires to interface a laptop to the RTDS via the DAA module. Specifically, the laptop is to commit some control algorithm, while the control inputs are some continuous measurements from the power system, and the control outputs are some continuous control values to the power system devices.

The power system is simulated via the RTDS, and the measurement point is fulfilled by the Analogue output interface on the RTDS. In the meantime, the control input is fulfilled by the Analogue input interface on the RTDS. In this case, the interfacing requirement is to bridge the laptop and the Analogue inputs and outputs on the RTDS.

For normal cases, it will require an external data acquisition device for such purposes, as laptops cannot read analogue voltage levels directly with any default interfaces. However, this can be easily done with DAA modules. Using the MBED NXP LPC1768 model as an example, it is based on a 32-bit ARM Cortex-M3 core running at 96MHz. The 26 re-configurable pins can be programmed to support interfaces including Ethernet, USB, CAN, SPI, I2C, ADC, DAC, PWM, and Digital I/O.

In this example case, the analogue outputs from the RTDS can be measured via the ADC pins, while the analogue inputs to the RTDS can be fulfilled via the DAC pins. The MBED can be then programmed to read the ADC pins and write the DAC pins on demand. The implementation of the ADC part of the function is illustrated in Fig. 8.10.

During the implementation, it is highly recommended to be extra careful about the details. Firstly, before making any physical implementations such as connections with wires, it needs to be double-checked for safety reasons. Here the safety is not only referring to the safety of the researchers, but also the safety of the devices, because one minor wrong operation may damage all devices at once.

For example in the implementation of the above example, it is always recommended to make some mandatory checks before turning on the powers, including the

Figure 8.10 An example of hardware-based interfacing technique with the micro-controller MBED .

power supply to the MBED, and the grounding status of the whole circuit. Secondly, there might require some supporting circuits for certain connections, which should be checked during the design and implementation phases. For most interfacing methods such as USB and Ethernet, they are using constant voltages, which allows a direct connection using wires.

For some interfacing methods with analogue outputs, such as the analogue inputs and outputs, the outputs are the varying voltage, which needs supporting circuits in use. For example, in the implementation of the above example, the analogue output is a varying voltage with the reference to the ground of the RTDS.

Therefore in order to enable the MBED to read the correct voltages via the ADC pins, it needs to common the grounds between the MBED and the RTDS. In addition, some auxiliary circuits might improve this interfacing implementation, such as a rheostat if there is voltage range mismatch, or some capacitors to filter the high frequency noises from the RTDS.

To enable the laptop to read the values on the MBED ADC pins in an on-demand manner, the MBED is implemented to work in a Remote Procedure Call (RPC) mode over the USB connections. This can be fulfilled by registering the ADC pins with RPC functions, where a feasible implementation can be given as follows:

```
#include "mbed.h"
#include "mbed_rpc.h"

# register the analogue input and output pins
RpcAnalogIn p20(p20,"p20");

Serial pc(USBTX, USBRX);
```

```
int main() {
    RPC::add_rpc_class<RpcAnalogIn>();
    // receive PC commands and response
    char buf[256], outbuf[256];
    while(1) {
        pc.gets(buf, 256);
        RPC::call(buf, outbuf);
        pc.printf("%s\n", outbuf);
    }
}
```

In the above example codes, it registers the ADC pin 'p20' that can be 'called' via the RPC methods. After compiling the above script with both MBED and RPC libraries with the MBED online compiler, it generates a binary execution file. By copying this binary file to the MBED and reset the MBED to take effect, the configuration of the MBED part is completed.

For the laptop side, it is now ready to read the voltages on the MBED ADC pins via the RPC method. Note that the RPC is a general interfacing method, which provides supports to commonly used software environments, including MATLAB, Python, .NET and Java. The implementation of such reading functions is quite straight-forward, where the implementation in Python is illustrated as follows:

```
from mbedrpc import *
import time

# initialize the mbed as a serial
serdev = "/dev/ttyACM0"
mbed=SerialRPC(serdev, 9600)
interface = mbed_interface(mbed, "interface")

# create the ADC to use
interface.new("AnalogIn","adc_input","p20")
ADC_reading=AnalogIn(mbed,"adc_input")

read_count=0
while(read_count<100):
    read_count+=1
    time.sleep(0.1)
    ADC_results=ADC_reading.read()
    print ADC_results
```

In the above example codes, the laptop is configured to read the voltages on the MBED ADC pin "p20" every 0.1 seconds for 100 times. The above dummy example can be easily adapted for general data interfacing purposes between a PC and the RTDS. The interfacing requirements such as PWM for precise controls and SPI to integrate other peripheral ICs, can be implemented following a similar procedure.

8.2.5 Remarks of Challenges

The testbed design and implementation for energy systems is a systematical task, it depends on a sophisticated consideration of all details in the testbed. From the examples along with the discussions, it can be seen that the design process and the implementation process are sometimes deeply coupled that they must be jointly considered.

The general purpose of the testbed is for the proof-of-concept and performance verification, but the design and implementation of the testbed could involve many more theories and concepts beyond the one to prove or verify.

The general solutions for testbed design and implementation: The most significant challenge is still the lack of general solutions in the design and implementation of the testbed. This is firmly rooted in the fact that testbeds are generally case-specified. The research problems, even in the energy system alone, can be so different that the experience of one testbed design and implementation could be totally non-useful in other cases.

In the meantime, thanks to the technology explosion nowadays, there could be numerous testbed solutions for one specific design task. These facts together make it hard to get a complete manual to each testbed design, whether the testbed is in the context of software-based, hardware-based, or hybrid architectures.

But there are indeed some commonly used techniques or general methods that can facilitate the design and implementation procedure, which can be summarized as general purpose hardware and modular design detailed as follows:

- *General purpose hardware:* The key point of using general purpose hardware in the testbed design is the transform of the research problems into numerical calculation problems. It is not to diminish the meaning of dedicated hardware devices in testbed design, but actually, it is to emphasise the general trend as witnessed in all kinds of simulators.

  The state-of-the-art simulators can be regarded as examples, either software-based simulators such as MATLAB and PowerFactory, or hybrid simulators such as RTDS and OpalRT. Such solutions are separating hardware and software functionalities, where general hardware devices are providing powerful computing capacities, while the software functions define the studied cases.

- *Modular design:* The key point of modular design is to standardise the consisting components in the testbed. This is not to say it is a compulsory requirement on each part of the testbed, but a general direction to make efforts. This is in line with the standardisation progress since the industrial revolution, which has brought us numerous benefits such as a reduction in cost. By modular design, the testbed modules are expected to be independent, scalable and reusable, whose interfaces are standard and well-defined. This not only helps to clarify the required functionalities in the testbed, but also improves the maintenance of the whole testbed.

The integration of multiple sub-systems: Since energy systems are an integrated concept based on multiple research disciplines, the design and implementation of the testbed for an energy system cannot avoid the integration of different sub-systems. These sub-systems, such as the discussed ICT systems, power systems, and AI systems, are with essentially different features, ranging from methodologies to terminologies.

Especially with the new technologies such as recent advanced wireless communication technologies, there might be new challenges to integrate them which might be also challenges in the wireless communication subject itself. For such new concepts, it could be hard to find existing solutions to implement them in a testbed design. Actually many testbeds are destined to such challenges, as they are to serve as prototypes, namely the very first real-world implementation of the concepts.

For testbed design and implementation to address such integration of multiple sub-systems, it also requires efforts to reconcile the cooperation between them, beyond the success of the implementation of each sub-system. To address such challenges, it would require an overall planning as part of the testbed design. It might fill some gaps between different sub-systems if more common grounds can be placed between them, e.g., the same software environment, or compatible interfaces.

It is a challenging task to design and implement a testbed for the energy system, but the benefits could still be worth the efforts, or even more than that. Given the evolving speed in each involved discipline in the energy system, it is not expected that there could be a complete solution that fits all testbed designs and implementations. However, the discussed examples, solutions, and techniques in this chapter are expected to serve as some basic tools, or to enlighten the readers regarding the ways to consider these challenges.

Bibliography

[1] South korea: Jeju island smart grid test-bed. [Online]. Available: *https://www.gsma.com/iot/wp-content/uploads/2012/09/cl_jeju_09_121.pdf*

[2] B. Zhao, X. Zhang, and J. Chen, "Integrated microgrid laboratory system," *IEEE Power Energy Society General Meeting*, 2013, pp. 1–1.

[3] M. Shahidehpour and M. Khodayar, "Cutting campus energy costs with hierarchical control: The economical and reliable operation of a microgrid," *IEEE Electrification Magazine*, vol. 1, no. 1, pp. 40–56, Oct. 2013.

[4] M. Prodanovic, A. Rodríguez-Cabero, M. Jiménez-Carrizosa, and J. Roldán-Pérez, "A rapid prototyping environment for DC and AC microgrids: Smart energy integration lab (SEIL)," in *IEEE 2nd International Conference on DC Microgrids (ICDCM)*, 2017, pp. 421–427.

[5] M. J. McDonald, G. Conrad, R. Cassidy *et al.*, "Cyber effects analysis using VCSE," *Sandia National Laboratories*, 2008.

[6] D. C. Bergman, D. K. Jin, D. M. Nicol, and T. Yardley, "The virtual power system testbed and inter-testbed integration," in *USENIX Conference on Cyber Security Experimentation and Test*, Montreal, Canada, pp. 1–6, Aug. 2009.

[7] J. Hong, S.-S. Wu, A. Stefanov, A. Fshosha, C.-C. Liu, P. Gladyshev, and M. Govindarasu, "An intrusion and defense testbed in a cyber-power system environment," in *2011 Power and Energy Society General Meeting*, 2011, pp. 1–5.

[8] J. Wan, S. Tang, Z. Shu, D. Li, S. Wang, M. Imran, and A. V. Vasilakos, "Software-defined industrial internet of things in the context of industry 4.0," *IEEE Sensors Journal*, vol. 16, no. 20, pp. 7373–7380, May 2016.

[9] W. Su, W. Zeng, and M.-Y. Chow, "A digital testbed for a PHEV/PEV enabled parking lot in a smart grid environment," in *2012 IEEE PES Innovative Smart Grid Technologies (ISGT)*, 2012, pp. 1–7.

[10] M. Mallouhi, Y. Al-Nashif, D. Cox, T. Chadaga, and S. Hariri, "A testbed for analyzing security of SCADA control systems (TASSCS)," in *2011 IEEE PES Innovative Smart Grid Technologies (ISGT)*, 2011, pp. 1–7.

[11] C. Queiroz, A. Mahmood, and Z. Tari, Scadasim: a framework for building SCADA simulations," *IEEE Transactions on Smart Grid*, vol. 2, no. 4, pp. 589–597, Sep. 2011.

[12] A. M. Kosek, O. Lunsdorf, S. Scherfke, O. Gehrke, and S. Rohjans, "Evaluation of smart grid control strategies in co-simulation: integration of IPSYS and mosaik," in *2014 Power Systems Computation Conference*, 2014, pp. 1–7.

[13] M. Buscher, A. Claassen, M. Kube, S. Lehnhoff, K. Piech, S. Rohjans, S. Scherfke, C. Steinbrink, J. Velasquez, F. Tempez, and Y. Bouzid, "Integrated smart grid simulations for generic automation architectures with RT-LAB and mosaik," in *2014 IEEE International Conference on Smart Grid Communications (SmartGridComm)*, 2014, pp. 194–199.

[14] S. Lehnhoff, O. Nannen, S. Rohjans, F. Schlogl, S. Dalhues, L. Robitzky, U. Hager, and C. Rehtanz, "Exchangeability of power flow simulators in smart grid co-simulations with mosaik," in *2015 Workshop on modelling and Simulation of Cyber-Physical Energy Systems (MSCPES)*, 2015, pp. 1–6.

[15] S. Tan, W. Z. Song, S. Yothment, J. Yang, and L. Tong, "Scoreplus: An integrated scalable cyber-physical experiment environment for smart grid," in *2015 12th Annual IEEE International Conference on Sensing, Communication and Networking (SECON)*, 2015, pp. 381–389.

[16] G. Koutsandria, R. Gentz, M. Jamei, A. Scaglione, S. Peisert, and C. McParland, "A real-time testbed environment for cyber-physical security on the power grid," in *Proceedings of the First ACM Workshop on Cyber-Physical Systems-Security and/or Privacy*, 2015, pp. 67–78.

[17] H. G. Aghamolki, Z. Miao, and L. Fan, "A hardware-in-the-loop SCADA testbed," in *2015 North American Power Symposium (NAPS)*, 2015, pp. 1–6.

[18] M. Kabir-Querrec, S. Mocanu, J. Thiriet, and E. Savary, "A test bed dedicated to the study of vulnerabilities in IEC 61850 power utility automation networks," in *IEEE 21st International Conference on Emerging Technologies and Factory Automation (ETFA)*, 2016, pp. 1–4.

[19] P. Palensky, A. van der Meer, C. Lopez, A. Joseph, and K. Pan, "Applied cosimulation of intelligent power systems: Implementing hybrid simulators for complex power systems," *IEEE Industrial Electronics Magazine*, vol. 11, no. 2, pp. 6–21, Jun. 2017.

[20] A. R. R. Matavalam and V. Ajjarapu, "Implementation of user defined models in a real-time cyber physical test-bed," in *2016 National Power Systems Conference (NPSC)*, 2016, pp. 1–6.

[21] A. Chakrabortty and Y. Xin, "Hardware-in-the-loop simulations and verifications of smart power systems over an Exo-GENI testbed," in *2013 Second GENI Research and Educational Experiment Workshop (GREE)*, 2013, pp. 16–19.

[22] V. Venkataramanan, P. Wang, A. Srivastava, A. Hahn, and M. Govindarasu, "Interfacing techniques in testbed for cyber-physical security analysis of the electric power grid," in *2017 Workshop on modelling and Simulation of Cyber-Physical Energy Systems*, 2017, pp. 1–6.

[23] O. Bassey, B. Chen, K. L. Butler-Purry, and A. Goulart, "Implementation of wide area control in a real-time cyber-physical power system test bed," in *2017 North American Power Symposium (NAPS)*, 2017, pp. 1–6.

[24] I. T. F. on Interfacing Techniques for Simulation Tools, S. C. Müller, H. Georg, J. J. Nutaro, E. Widl, Y. Deng, P. Palensky, M. U. Awais, M. Chenine, M. Küch, M. Stifter, H. Lin, S. K. Shukla, C. Wietfeld, C. Rehtanz, C. Dufour, X. Wang, V. Dinavahi, M. O. Faruque, W. Meng, S. Liu, A. Monti, M. Ni, A. Davoudi, and A. Mehrizi-Sani, "Interfacing power system and ICT simulators: Challenges, state-of-the-art, and case studies," *IEEE Transactions on Smart Grid*, vol. 9, no. 1, pp. 14–24, Jan. 2018.

[25] M. U. Tariq, B. P. Swenson, A. P. Narasimhan, S. Grijalva, G. F. Riley, and M. Wolf, "Cyber-physical co-simulation of smart grid applications using NS-3," in *Proceedings of the 2014 Workshop on ns-3*, 2014, pp. 1–8.

Index

Note: Locators in *italics* represent figures and **bold** indicate tables in the text.